Unsupervised and Semi-Supervised Learning

Series Editor
M. Emre Celebi, Computer Science Department, Conway, AR, USA

Springer's Unsupervised and Semi-Supervised Learning book series covers the latest theoretical and practical developments in unsupervised and semi-supervised learning. Titles – including monographs, contributed works, professional books, and textbooks – tackle various issues surrounding the proliferation of massive amounts of unlabeled data in many application domains and how unsupervised learning algorithms can automatically discover interesting and useful patterns in such data. The books discuss how these algorithms have found numerous applications including pattern recognition, market basket analysis, web mining, social network analysis, information retrieval, recommender systems, market research, intrusion detection, and fraud detection. Books also discuss semi-supervised algorithms, which can make use of both labeled and unlabeled data and can be useful in application domains where unlabeled data is abundant, yet it is possible to obtain a small amount of labeled data.

Topics of interest in include:

- Unsupervised/Semi-Supervised Discretization
- Unsupervised/Semi-Supervised Feature Extraction
- Unsupervised/Semi-Supervised Feature Selection
- Association Rule Learning
- Semi-Supervised Classification
- Semi-Supervised Regression
- Unsupervised/Semi-Supervised Clustering
- Unsupervised/Semi-Supervised Anomaly/Novelty/Outlier Detection
- Evaluation of Unsupervised/Semi-Supervised Learning Algorithms
- Applications of Unsupervised/Semi-Supervised Learning

While the series focuses on unsupervised and semi-supervised learning, outstanding contributions in the field of supervised learning will also be considered. The intended audience includes students, researchers, and practitioners.

More information about this series at http://www.springer.com/series/15892

Adil M. Bagirov • Napsu Karmitsa • Sona Taheri

Partitional Clustering via Nonsmooth Optimization

Clustering via Optimization

Springer

Adil M. Bagirov
School of Science, Engineering
& Information Technology
Federation University Australia
Ballarat, VIC, Australia

Napsu Karmitsa
Department of Mathematics and Statistics
University of Turku
Turku, Finland

Sona Taheri
School of Science, Engineering
& Information Technology
Federation University Australia
Ballarat, VIC, Australia

ISSN 2522-848X ISSN 2522-8498 (electronic)
Unsupervised and Semi-Supervised Learning
ISBN 978-3-030-37828-8 ISBN 978-3-030-37826-4 (eBook)
https://doi.org/10.1007/978-3-030-37826-4

This Springer imprint is published by the registered company Springer Nature Switzerland AG.
The registered company address is: Gewerbestrasse 11, 6330 Cham, Switzerland

To our children:

Zarifa, Nushaba, Uma, and Selin

Preface

Cluster analysis deals with the problem of organizing objects in a data set into clusters based on their similarities. It is among the most important tasks in data mining. Clustering is also known under different names such as unsupervised data classification, numerical taxonomy, and automatic data classification.

Nowadays clustering is applied in many areas of science, business, and engineering. Such areas include biology, sociology, chemometrics, psychometrics, economics, ecology, medical sciences, finance, text mining, cybersecurity, and artificial intelligence.

In cluster analysis no prior information about data is required and, in general, unlabelled data is considered. It is expected that the objects in a cluster are similar to each other and dissimilar to the objects in other clusters.

All clustering algorithms involve some process for measuring the similarity of objects in a data set. In data sets with only numeric attributes, different norms can be used to define the *similarity measure*—in general, any Minkowski norm can be utilized. Clustering problems with such similarity measures can be formulated as *global optimization problems*. However, not all Minkowski norms lead to efficiently tractable clustering problems. Throughout this book, we use the *squared Euclidean norm* as well as the L_1- and L_∞-*norms* to define the similarity measures.

The first clustering algorithm, the k-means algorithm, was independently discovered in various scientific areas: by Steinhaus in 1956, by Lloyd in 1957, by Ball and Hall in 1965, and by MacQueen in 1967. Since then cluster analysis has become an important research direction in machine learning. There are different types of clustering problems, and many techniques based on completely different approaches have been developed to solve them. In this book we concentrate on optimization—in particular, nonsmooth optimization—approaches and algorithms for solving clustering problems. We omit other approaches including those based on statistical techniques.

This book has three parts. Part I consists of introduction to clustering, theoretical results from nonsmooth analysis used to model the clustering problems, and methods of nonsmooth optimization applied to design algorithms for solving clustering problems. In this part, we do not provide proofs of theoretical results as they can be found elsewhere. We present these methods with their flowcharts. Part II contains optimization formulations of the clustering problems as well as traditional clustering and optimization-based clustering algorithms. Here, we first provide descriptions of heuristic algorithms such as the k-means, the global k-means, the k-medians, and the k-medoids algorithms. Then we describe some metaheuristics and evolutionary clustering algorithms like tabu search, simulated annealing, genetic, and artificial bee colony clustering algorithms. However, our main focus is on the clustering algorithms, which are based on nonsmooth optimization approaches. All these algorithms are presented using their flowcharts and step-by-step descriptions. Finally, Part III is devoted to implementation and evaluation of clustering algorithms using real-world data sets.

The book is ideal for everyone who is studying and learning cluster analysis. In particular, it is intended for researchers and practitioners interested in optimization-based approaches for solving clustering problems. Furthermore, it can be used as a reference text by anyone including experts dealing with clustering.

Acknowledgments

The authors acknowledge Professor Alexander Rubinov with whom one of the authors, Adil M. Bagirov, had a privilege to work with. He started to work on the subject of this book with Professor Rubinov almost 20 years ago.

We are grateful to the following colleagues and collaborators, all of whom have influenced on the contents of the book: Tero Aittokallio, Kaisa Joki, Marko M. Mäkelä, Karim Mardaneh, Ehsan Mohebi, Burak Ordin, Gurkan Ozturk, Sattar Seifollahi, Julien Ugon, Dean Webb, Adilson Elias Xavier, and John Yearwood. The authors are also grateful to Springer's team (Brian Halm, Mary James, Zoe Kennedy, Abhishek Ravi Shankar, and Nandhakumar Sundar) for their support and patience during the preparation of this book.

The work was financially supported by the Australian Government through the Australian Research Council's Discovery Projects funding scheme (Project No. DP190100580), the Academy of Finland (Project No. 289500, 319274), Federation University Australia, and University of Turku, Finland.

Ballarat, VIC, Australia — Adil M. Bagirov
Turku, Finland — Napsu Karmitsa
Ballarat, VIC, Australia — Sona Taheri
October 2019

Introduction

Data clustering is one of the most important tasks in data mining. It deals with the problem of partitioning a given data into clusters (groups) based on similarity of objects, that is, all objects within the cluster are similar while objects in different clusters are dissimilar. Different approaches and algorithms have been developed to solve various clustering problems [174, 183, 216, 272, 279]. This book is specifically devoted to optimization-based approaches for solving clustering problems. More precisely, nonsmooth optimization techniques are used to develop partitional clustering algorithms.

The book consists of three parts. In the first part, the basic concepts of cluster analysis are defined and the necessary information on the theory and methods of nonsmooth optimization is provided. This part contains three chapters. In Chap. 1, we introduce notations used throughout the book, describe some concepts of cluster analysis, discuss different types of clustering problems, and outline important application areas of clustering. Chapter 2 contains necessary theoretical results from nonsmooth analysis used to model and solve clustering problems. In this chapter, we do not provide any proofs since they can be found in provided references. Numerical methods of nonsmooth optimization, which are applied to design clustering algorithms, and the basic ideas of their convergence analysis are described in Chap. 3. The flowcharts of these methods are also given.

Part II consists of various optimization formulations of the clustering problems as well as descriptions of heuristic, metaheuristic, and optimization-based clustering algorithms. This part contains six chapters. In Chap. 4, different models of partitional clustering problems are discussed. In particular, we give the nonsmooth formulation of the clustering and the auxiliary clustering problems, their difference of convex (DC) representations, and study the optimality conditions for these problems.

Heuristic clustering algorithms such as the k-means, k-medians, k-medoids algorithms as well as clustering algorithms based on mixture models and self-organizing map are described in Chap. 5. The tabu search, simulated annealing, genetic, artificial bee colony, particle swarm, and ant colony optimization algorithms for clustering are presented in Chap. 6.

Our main interest is on clustering algorithms which are based on (nonsmooth) optimization models and techniques. These algorithms are designed based on the combination of the local search optimization algorithms with the so-called incremental approach. In Chap. 7, we describe the basic idea of the incremental approach as well as three clustering algorithms based on the combination of this approach and the heuristic clustering algorithms. Then in Chap. 8, we discuss five nonsmooth optimization-based clustering algorithms. These algorithms combine the incremental approach with the nonsmooth optimization techniques described in Chap. 3. Finally, in Chap. 9, we present three clustering algorithms where we utilize the DC structure of the clustering problems and combine the incremental approach with the DC optimization methods. All algorithms in Chaps. 7–9 are presented using their flowcharts and step-by-step descriptions.

Part III of this book is devoted to implementation and evaluation of clustering algorithms using some real-world data sets. This part consists of three chapters. Different evaluation measures for clustering, including cluster validity indices, are discussed in Chap. 10. In Chap. 11 we discuss the implementation of clustering algorithms considered in this book and give the description of data sets used to evaluate these algorithms. The results of numerical experiments are reported in Chap. 12 and finally, some concluding remarks are given in Chap. 13.

Contents

Acronyms and Symbols

Symbols and Notations

$\mathbb{R}^n$	n-dimensional Euclidean space
$\mathbb{N}$	Set of natural numbers
$\bar{\mathbb{R}}_+$	Nonnegative numbers, $\bar{\mathbb{R}}_+ = \{r \in \mathbb{R} : r \geq 0\}$
$a, b, c, \alpha, \varepsilon, \lambda$	Scalars
$\boldsymbol{x}, \boldsymbol{y}, \boldsymbol{z}$	Vectors
$\boldsymbol{x}^T$	Transposed vector
$\boldsymbol{x}^T\boldsymbol{y}$	Inner product of $\boldsymbol{x}$ and $\boldsymbol{y}$
$\\|\boldsymbol{x}\\|$	Euclidean norm of $\boldsymbol{x}$ in $\mathbb{R}^n$, $\\|\boldsymbol{x}\\| = \left(\boldsymbol{x}^T\boldsymbol{x}\right)^{\frac{1}{2}}$
x_i	i-th component of vector $\boldsymbol{x}$
$\{\boldsymbol{x}_h\}$	Sequence of vectors
$\mathbf{0},\ \mathbf{0}_n$	Zero vector in $\mathbb{R}^n$
$A = \{\boldsymbol{a}_1, \ldots, \boldsymbol{a}_m\}$	Data set $\boldsymbol{a}_i \in \mathbb{R}^n,\ i = 1, \ldots, m$
$\boldsymbol{a},\ \boldsymbol{a}_1, \ldots, \boldsymbol{a}_m$	Data points
$A^1, \ldots, A^k$	Clusters
$\bar{A} = \{A^1, \ldots, A^k\}$	Set of clusters, partition of A
m_j	Number of data points in the cluster A^j
$d_p(\boldsymbol{x}, \boldsymbol{y})$	Distance function, $p > 0$
$B,\ G$	Matrices
$(B)_{ij},\ b_{ij}$	Element in the row i, the column j of the matrix B
B^T	Transposed matrix
B^{-1}	Inverse of the matrix B
$\\|B\\|_F$	Frobenius norm $\\|B\\|_F = \left(\sum_{i=1}^m \sum_{j=1}^n \|b_{ij}\|^2\right)^{\frac{1}{2}}$
I	Identity matrix
$\boldsymbol{e}_i$	i-th column of the identity matrix
$B(\boldsymbol{x}; r)$	Open ball with the radius r and the central point $\boldsymbol{x}$
$\bar{B}(\boldsymbol{x}; r)$	Closed ball with the radius r and the central point $\boldsymbol{x}$
S_1	Sphere of the unit ball

(a, b)	Open interval
$[a, b]$	Closed interval
$[a, b), (a, b]$	Half-open intervals
$H(\boldsymbol{p}, \alpha)$	Hyperplane
$H^+(\boldsymbol{p}, \alpha), H^-(\boldsymbol{p}, \alpha)$	Halfspaces
$S,\ U$	Sets
$\mathrm{cl}\, S$	Closure of the set S
$\mathrm{int}\, S$	Interior of the set S
$\mathrm{bd}\, S$	Boundary of the set S
$\mathrm{conv}\, S$	Convex hull of the set S
$\cup_{i=1}^m S_i$	Union of sets $S_i,\ i = 1, \ldots, m$
$\cap_{i=1}^m S_i$	Intersection of sets $S_i,\ i = 1, \ldots, m$
$\mathrm{lev}_\alpha\, f$	Level set of f with the parameter α
$S \times U$	Cartesian product of sets S and U
$\mathcal{I}, \mathcal{J}, \mathcal{K}$	Sets of indices
$\lvert\mathcal{I}\rvert$	Number of elements in the set $\mathcal{I}$
$f(\boldsymbol{x})$	Value of the objective function f at $\boldsymbol{x}$
$\nabla f(\boldsymbol{x})$	Gradient of the function f at $\boldsymbol{x}$
$\frac{\partial}{\partial x_i} f(\boldsymbol{x})$	Partial derivative of the function f with respect to x_i
$\nabla^2 f(\boldsymbol{x})$	Hessian matrix of the function f at $\boldsymbol{x}$
$\frac{\partial^2}{\partial x_i \partial x_j} f(\boldsymbol{x})$	Second order partial derivative of the function f with respect to x_i and x_j
$C^m(\mathbb{R}^n)$	Space of the function $f : \mathbb{R}^n \to \mathbb{R}$ with continuous partial derivatives up to order m
D_h	(Generalized) variable metric approximation of the inverse of the Hessian matrix
$f'(\boldsymbol{x}; \boldsymbol{d})$	Directional derivative of the function f at $\boldsymbol{x}$ in the direction $\boldsymbol{d}$
$f^\circ(\boldsymbol{x}; \boldsymbol{d})$	Generalized directional derivative of the function f at $\boldsymbol{x}$ in the direction $\boldsymbol{d}$
$\partial_c f(\boldsymbol{x})$	Subdifferential of the convex function f at $\boldsymbol{x}$
$\partial f(\boldsymbol{x})$	Clarke subdifferential of the function f at $\boldsymbol{x}$
$\partial_\varepsilon f(\boldsymbol{x})$	ε-subdifferential of the convex function f at $\boldsymbol{x}$
$\partial_\varepsilon^G f(\boldsymbol{x})$	Goldstein ε-subdifferential of the function f at $\boldsymbol{x}$
$\boldsymbol{\xi} \in \partial f(\boldsymbol{x})$	Subgradient of the function f at $\boldsymbol{x}$
$\mathcal{D} f(\boldsymbol{x})$	Quasidifferential of the function f at $\boldsymbol{x}$, $\mathcal{D} f(\boldsymbol{x}) = \left[\underline{\partial} f(\boldsymbol{x}), \overline{\partial} f(\boldsymbol{x})\right]$
$\underline{\partial} f(\boldsymbol{x})$	Subdifferential of the quasidifferentiable function f at $\boldsymbol{x}$
$\overline{\partial} f(\boldsymbol{x})$	Superdifferential of the quasidifferentiable function f at $\boldsymbol{x}$
$\boldsymbol{\Gamma}(\boldsymbol{x}, \boldsymbol{g}, \boldsymbol{w}, \lambda, \alpha)$	Discrete gradient of the function f at $\boldsymbol{x}$
Ω_f	Set in $\mathbb{R}^n$ where the function f is not differentiable
$\hat{f}_k(\boldsymbol{x})$	Piecewise linear cutting-plane model of the function f at $\boldsymbol{x}$
$t \downarrow 0$	$t \to 0_+$
$\mathrm{argmin}\ f(\boldsymbol{x})$	Point where the function f attains its minimum value

Argmin $f(\boldsymbol{x})$	Set of points where the function f attains its minimum value
argmax $f(\boldsymbol{x})$	Point where the function f attains its maximum value
Argmax $f(\boldsymbol{x})$	Set of points where the function f attains its maximum value
$\lceil \cdot \rceil$	Ceiling of a number
$\mathcal{C}(n, k) = \binom{n}{k}$	Binomial coefficient $\frac{n!}{k!(n-k)!}$
$O(\cdot)$	Time complexity or space requirement of an algorithm
$U[0, 1]$	Uniform distribution with the domain [0, 1]

Abbreviations

ABC	Artificial bee colony
ACO	Ant colony optimization
ANN	Artificial neural networks
ARn index	Adjusted Rand index
BH index	Ball and Hall index
BIC	Bayesian information criterion
BMU	Best matching unit
CCCP	Concave–convex procedure
CH index	Calinski–Harabasz index
CLR	Clusterwise linear regression
DC	Difference of convex
DCA	DC algorithm
DCD-BUNDLE	DC diagonal bundle method
DCDB-CLUST	DC diagonal bundle clustering algorithm
DB index	Davies–Bouldin index
Dn index	Dunn index
DG-CLUST	Discrete gradient clustering algorithm
DGM	Discrete gradient method
EM	Expectation maximization
FGKM	Fast global k-means algorithm
FMGKM	Fast modified global k-means algorithm
GA	Genetic algorithm
GKM	Global k-means algorithm
H index	Hartigan index
HSM	Hyperbolic smoothing method
IDCA-CLUST	Incremental DCA for clustering
IKMED	Incremental k-medians algorithm
INC-CLUST	Incremental clustering algorithm
IS-CLUST	Smooth incremental clustering algorithm
KDD	Knowledge discovery in databases
KL index	Krzanowski–Lai index
LLC	Locally Lipschitz continuous

LMB-CLUST	Limited memory bundle method for clustering
LMBM	Limited memory bundle method
MGKM	Modified global k-means algorithm
ML	Maximum likelihood
MSAC	Minimum sum-of-absolutes clustering
MSINC-CLUST	Multi-start incremental clustering algorithm
MSSC	Minimum sum-of-squares clustering
NDC-CLUST	Incremental nonsmooth DC clustering algorithm
NDCM	Nonsmooth DC method
NMI	Normalized mutual information
NSO	Nonsmooth optimization
PAM	Partitioning around medoids
PBM	Proximal bundle method
PSO	Particle swarm optimization
Rn index	Rand index
SA	Simulated annealing
Sm index	Sym index
SOM	self-organizing map
TS	Tabu search
XB index	Xie–Beni index

Part I
Preliminaries

Chapter 1
Introduction to Clustering

1.1 Introduction

Over the last two decades there has been a significant growth in the amount of data generated. This trend can be observed in various sources such as social media, online transactions, network sensors, satellite and astronomical information. The exponential increase in the volume of data poses significant challenges in the decision making process. Analyzing data allows us to discover meaningful patterns in data and to make better decisions. Therefore, it is imperative to develop new approaches and computational tools in data analysis.

We start by introducing the commonly used tasks and terminologies in data analysis. Then we define the clustering problem and describe the similarity measures. Finally, we give short surveys on different types of clustering algorithms and various applications of clustering.

Knowledge discovery in databases (KDD) is a comprehensive process of discovering and extracting useful knowledge (information) in data. The definition of "useful information" depends on the end users' requirements. The KDD applications cover almost all areas of human activities including marketing, finance, information security, telecommunication, and engineering [103, 136]. The main steps in the KDD process are as follows:

- data collection and selection;
- data pre-processing (e.g., cleansing and preparation);
- data transformation;
- data mining;
- incorporating prior knowledge on data; and
- data evaluation and knowledge presentation.

A. M. Bagirov et al., *Partitional Clustering via Nonsmooth Optimization*,
Unsupervised and Semi-Supervised Learning,
https://doi.org/10.1007/978-3-030-37826-4_1

Data mining is a field of computer science—a predominantly computational process—that aims to discover and extract previously unknown information (knowledge) from data. It is among the most important steps in the KDD process. Indeed, sometimes the term "data mining" is used as a synonym to KDD. However, in the KDD process data mining is considered as a method (algorithm) to extract unknown knowledge from data. Data mining has wide range of applications including information retrieval [17, 232], image analysis [58, 63], bioinformatics [55], signal processing, and internet security, to mention but a few. The process of data mining involves several distinct tasks such as

- data representation;
- supervised data classification;
- association rules;
- feature selection and extraction;
- regression analysis; and
- clustering.

Data representation is a task of deciding on the representation of data. This includes the size of data, the number of features available as well as their natures and scales, and the number of clusters or classes. For example, in text data each document may be represented as "a bag of words." In this case, the words are used as features but the order of words in the document is not important.

Supervised data classification or *supervised learning* involves the supervised assignment of objects to the predefined and known classes. More precisely, the objects with known classes (labels) are used to train/learn a description of the classes. These objects constitute the *training set*. Then the supervised data classification approach is applied to label new objects from a *test set* into one or more of these classes.

Association rule mining is a process for discovering and identifying frequent patterns, correlations, associations, or causal structures in a data set. For instance, given a set of transactions, association rule mining finds the rules which enable us to predict an occurrence of a specific item based on the existence of the other items in the transaction.

Feature selection is a process of selecting the most important, informative and relevant features of data to be used in clustering or supervised data classification tasks. Feature selection methods aim to identify the most informative features which have strong discriminatory or predictive significance, to remove some uninformative or noisy features and to reduce the dimension of the problem under consideration. *Feature extraction* or *feature combination* process transforms or combines the original set of features in order to find a much smaller set of new features with higher quality of data. Both feature selection and feature extraction methods aim to improve the performance of clustering and supervised data classification algorithms.

Regression analysis is a predictive modelling technique to approximate a relationship between a dependent variable (target) and one or more independent variables (predictors). The most commonly used regression models are the linear and logistic regressions but there are also many other regression models available.

Regression analysis was initially developed in the field of statistics. However, recently it has been widely used for predicting and forecasting purposes, and has become an important tool in data mining.

Cluster analysis or *clustering* deals with the problem of organizing a collection of objects into clusters based on their similarity. Clustering involves the identification of subsets of data that are similar. Each subset intuitively corresponds to objects that are more similar to each other than to objects in other subsets. Clustering is carried out in an *unsupervised* way by trying to find these subsets without having any predefined knowledge of the clusters. This means that identifying clusters is *data driven* rather than *data informed*. Cluster analysis has been referred by a number of different names over the years: Q-analysis, typology, grouping, clumping, numerical taxonomy, and unsupervised data classification [155, 295].

Clustering is among the most useful approaches for pattern recognition, image processing, decision making, data mining, and KDD. As a tool, it has a wide range of applications in many fields like biomedical sciences, cybersecurity, signal analysis, life science taxonomy, remote sensing, demography and social sciences, geology and anthropology, economics, finance and planning.

Most clustering algorithms are either hierarchical or partitional. *Hierarchical clustering algorithms* yield a dendrogram representing the nested grouping of patterns and similarity levels at which groupings change [156, 174]. *Partitional clustering algorithms* find a partition of objects that optimizes some predefined clustering criterion [156, 223]. Partitional clustering can be divided into two subclasses: *hard partitional clustering*, where each object belongs to only one cluster and *soft (fuzzy) partitional clustering*, where each object may belong to more than one cluster. In what follows we are mainly considering hard partitional clustering.

1.2 Notations and Definitions

Throughout this book, we consider a *data set* A as a finite set of points given in an n-dimensional space $\mathbb{R}^n$. More precisely, the data set A is presented as

$$A = \{\boldsymbol{a}_1, \ldots, \boldsymbol{a}_m\}, \quad \boldsymbol{a}_i \in \mathbb{R}^n, \ i = 1, \ldots, m.$$

The data points $\boldsymbol{a}_i$, $i = 1, \ldots, m$ are called *instances (observations, objects)* and each instance has n *attributes (features)*. We say that the *dimensionality* of data (or the data set A) is n, and it is distinct from the *size* of the data set, that is m. For simplicity, the notation $\boldsymbol{a} \in A$ will sometimes be used for a data point.

Unconstrained hard clustering deals with the problem of partitioning the points of the set A into a given number k of disjoint subsets—*clusters*—A^j, $j = 1, \ldots, k$ such that the partition satisfies the following criteria:

$$
\begin{aligned}
&1.\ A^j \neq \emptyset, \quad j = 1, \ldots, k;\\
&2.\ A^j \bigcap A^q = \emptyset \quad \text{for all} \quad j, q = 1, \ldots, k,\ j \neq q; \quad \text{and} \qquad (1.1)\\
&3.\ A = \bigcup_{j=1}^{k} A^j.
\end{aligned}
$$

In addition, no condition is imposed on clusters A^j, $j = 1, \ldots, k$. These criteria mean that all clusters are non-empty—that is, $m_j \geq 1$, where m_j is the number of points in the jth cluster—each data point belongs only to one cluster, and uniting all the clusters reproduces the whole data set A.

The *number of clusters* k is an important parameter for any clustering algorithm. If $k = m$, then each cluster contains exactly one object from the set A. This partition is trivial and does not require a solution of any clustering problem. Therefore, we will always assume that the number k is significantly less than the size of the data set m.

Another important notion in clustering is a *cluster representative*. Each cluster A^j is identified by its representative. The cluster representative is a simple set and it is also known as a cluster profile, prototype, classification vector, and cluster label. It is an item that summarizes and represents the objects in the cluster. In some sense it should be close to every object in the cluster where closeness is defined using a similarity measure. The cluster representative can be, for example, a point—usually a center of the cluster—a sphere or a hyperplane.

In this book, we consider *cluster centers* as cluster representatives. We denote the centers by $\boldsymbol{x}_j \in \mathbb{R}^n$, $j = 1, \ldots, k$ and the collection of these centers by

$$\boldsymbol{x} = (\boldsymbol{x}_1, \ldots, \boldsymbol{x}_k) \in \mathbb{R}^{nk}.$$

The problem of finding these centers is called the *k-clustering (or k-partition) problem*.

Throughout this book we will consider clustering problems in data sets with only numeric features. Clustering problems with categorical features and also algorithms for their solutions are discussed and studied, for example, in [13, 112, 117, 129, 250].

1.3 Similarity Measures

The notion of *similarity* is fundamental in clustering. The similarity of the objects is measured by a matching or similarity function (similarity measure). The similarity measure is chosen based on the cluster representative and the types of features in a data set. The choice of this measure is one of the main factors in determining the complexity of the clustering task.

The similarity measures for clusters represented by centers (points) and for those represented by hyperplanes are different. In addition, the similarity measures for data sets containing only numeric features differ from those containing both numeric and categorical or only categorical features. Nevertheless, all clustering algorithms involve some kind of process for measuring the similarity of the objects. Any two points belonging to the same cluster are supposed to be more mutually similar than any two points from two different clusters.

Let $\boldsymbol{b},\ \boldsymbol{c} \in \mathbb{R}^n$ be any two points from the data set A and denote by $\bar{\mathbb{R}}_+ = \{r \in \mathbb{R} : r \geq 0\}$. A function $d : A \times A \to \bar{\mathbb{R}}_+$ is called the *similarity function* or *similarity measure* if it satisfies the following conditions:

- $d(\boldsymbol{b}, \boldsymbol{c}) = 0$ if and only if $\boldsymbol{b} = \boldsymbol{c}$ and
- $d(\boldsymbol{b}, \boldsymbol{c}) = d(\boldsymbol{c}, \boldsymbol{b})$ for all $\boldsymbol{b},\ \boldsymbol{c} \in A$.

The first condition means that the point $\boldsymbol{b}$ is similar to itself (this means that there is no any dissimilarity of a data point to itself), while the second condition states that if the point $\boldsymbol{b}$ is similar to $\boldsymbol{c}$ then the point $\boldsymbol{c}$ is similar to $\boldsymbol{b}$, that is, similarity is symmetric.

The use of different similarity measures may help to identify different cluster structures of a data set. This in turn may lead to a significant improvement in the decision making process. Furthermore, different cluster representatives, similarity measures, and the number of clusters should be used in different data sets to obtain meaningful results. For example, data given in Fig. 1.1 is best approximated using two ball-shaped clusters and their centroids ($\boldsymbol{x}_1$ and $\boldsymbol{x}_2$). On the other hand, in data given in Fig. 1.2 there are three clusters best approximated by three hyperplanes (red lines).

When a data set contains only numeric attributes various *distance functions* can be used to define the similarity measures. Nevertheless, in these cases it is more convenient to use the notion of *dissimilarity* that is dual to similarity. Indeed, the lower the value of dissimilarity, the more similar the two objects are.

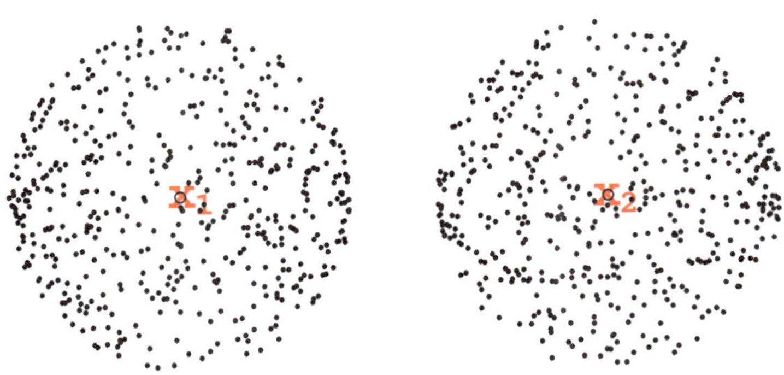

Fig. 1.1 Two ball-shaped clusters

Fig. 1.2 Clusters described by three hyperplanes

In what follows, we consider data sets with only numeric attributes and, as already mentioned, use centers of clusters as their representatives. Therefore, the similarity measure for two objects is equivalent to the distance between these objects.

We use the general Minkowski norms to define similarity measures for two points $\boldsymbol{b}, \boldsymbol{c} \in \mathbb{R}^n$:

$$d_p(\boldsymbol{b}, \boldsymbol{c}) = \left(\sum_{i=1}^{n} |b_i - c_i|^p \right)^{1/p}. \tag{1.2}$$

Note that any $p \in (0, \infty)$ can be taken here, but only for $p \geq 1$ this measure is a distance function. The most commonly used distance functions are those corresponding to values $p = 1, 2$ and $p = \infty$. In addition, any other L_p-norm with $p \geq 1$ could be considered to define similarity measures in clustering. However, in these cases the clustering problem might become very complex to be solved efficiently.

For $p = 1$, we get the similarity measure based on the L_1-*norm* (also known as the *city block* or the *Manhattan norm*):

$$d_1(\boldsymbol{b}, \boldsymbol{c}) = \sum_{i=1}^{n} |b_i - c_i|. \tag{1.3}$$

In the case of $p = 2$, the Minkowski norm (1.2) yields the well-known Euclidean distance. In cluster analysis, we usually use the squared form of this norm—the *squared Euclidean distance*:

$$d_2(\boldsymbol{b}, \boldsymbol{c}) = \sum_{i=1}^{n} (b_i - c_i)^2. \tag{1.4}$$

We also say that this similarity measure is based on the L_2-*norm*. This measure can be generalized as

$$d_2(\boldsymbol{b}, \boldsymbol{c}) = (\boldsymbol{b} - \boldsymbol{c})^T H (\boldsymbol{b} - \boldsymbol{c}),$$

where H is an $n \times n$ positive definite matrix. In the case of the squared Euclidean distance, the cluster centers are called *centroids*.

Finally, the similarity measure is based on the L_∞-*norm* (also known as the *Chebyshev norm*), when $p = \infty$ is given by

$$d_\infty(\boldsymbol{b}, \boldsymbol{c}) = \max_{i=1,\dots,n} |b_i - c_i|. \tag{1.5}$$

It should be noted that, in general, similarity measures need not to satisfy the triangular inequality (known also as the Minkowski inequality) indicating that they are not always distance functions. Particularly, the similarity measure d_2 does not satisfy this inequality, while the similarity measures d_1 and d_∞ do.

Clustering problems with the similarity measure defined by the squared Euclidean distance have been studied extensively over the last six decades while problems with other Minkowski norms have attracted significantly less attention. In addition to similarity measures considered in this section, there are many other ways of defining similarity. For instance, Bregman divergence and some monotonous functions can be used to generalize the above considered similarity measures [295]. In [105], a similarity measure (relation) is defined as an equivalence relation. This implies that such a similarity measure has a transitivity property which many other similarity measures do not have.

1.4 Types of Clustering Algorithms

There are various types of clustering algorithms available. According to [3, 136, 158, 272, 273], a proper clustering algorithm should

- produce clusters which are unlikely to be altered drastically when additional objects are incorporated;
- be stable in the sense that small changes in the features of the objects do not lead to a significant change in clustering;
- be independent on the initial ordering of the objects;
- be able to filter the possible noise and define outliers;
- be scalable, i.e., able to be extended to large scale data sets with large number of attributes;
- be able to define clusters of different shapes; and
- require the minimal knowledge about data to determine the parameters.

In addition, the result of clustering should be interpretable and usable. Evidently, most clustering algorithms do not satisfy all these requirements. For instance, the shape of the clusters found by partitional clustering algorithms usually depends on the similarity measure while the hierarchical clustering algorithms do not scale well. Furthermore, most clustering algorithms need a number of empirically determined parameters—most importantly the number of clusters. This number is not known a priori for many data sets and it should be provided by a user.

As already mentioned, most clustering algorithms can be grouped into two main classes: partitional and hierarchical. More generally, the clustering algorithms can be classified into the following categories: [3, 136]

- *Partitional clustering algorithms:* a partitioning algorithm constructs partitions of data, where each partition represents a cluster. The most popular representatives of this class of clustering algorithms are the k-means algorithm and its variations: for instance, kernel k-means, weighted k-means, and genetic k-means.
- *Hierarchical clustering algorithms:* a hierarchical algorithm creates a hierarchical decomposition of a data set. Based on the decomposition formed, the algorithm can be classified as being either *agglomerative* or *divisive*. More precisely, a hierarchical clustering algorithm produces a *dendrogram* presenting a nested grouping of data and similarity levels at which the clusters change [156, 174]. Most hierarchical clustering algorithms are variants of the single link [269], complete link [177], and minimum variance [221, 291] algorithms. In addition, it is worth of mentioning the first hierarchical clustering algorithm BIRCH (Balanced Iterative Reducing and Clustering using Hierarchies) [313] and the most well-known hierarchical clustering algorithm CURE (Clustering Using REpresentatives) [128]. With CURE it is possible to find cluster shapes other than hyperspheres. This allows one to avoid the tendency of finding clusters with similar sizes.
- *Density-based clustering algorithms:* a density-based clustering algorithm considers clusters as regions in which the density of data points exceeds a predefined threshold value [5, 100]. The general idea of the density-based algorithms is to continue growing a given cluster as long as the density (number of data points) in the "neighborhood" exceeds the threshold. Unlike the most other clustering algorithms, density-based clustering algorithms can be used to create clusters of arbitrary shape. The algorithm DBSCAN (Density-Based Spatial Clustering of Applications with Noise) [100], the KNNCLUST (k-NN density-based clustering) [282] and its stochastic extension [58] are well-known representatives of these algorithms. The performance evaluation of various density-based clustering algorithms can be found in [5].
- *Grid-based clustering algorithms:* in the grid-clustering approach the data space is divided into a finite number of cells that form a grid structure. Then all of the clustering operations are performed on the grid structure. For instance, the genetic-guided clustering algorithm [66] belongs to this class.
- *Fuzzy and probabilistic model-based clustering algorithms:* these algorithms allow a data point to belong to one or more clusters. Representatives of this class are the fuzzy c-means algorithm, the expectation-maximization clustering algorithm, and the algorithms based on Bernoulli and Gaussian mixture models.

1.5 Applications of Clustering

Clustering algorithms allow us to find hidden patterns in data without using any prior information. Therefore, the application of clustering algorithms benefits many areas of human activities. These application areas include—but are not limited to—medicine and bioinformatics, cybersecurity, text mining, and image processing, as well as economy, finance, and marketing. In addition, clustering can be combined with other tools of data analysis and machine learning to obtain more versatile approaches for KDD. Next, we describe some of these applications in more details.

Medicine and *bioinformatics* are among the most important areas that apply clustering. Drug–target interactions such as drug sensitivity and resistance are studied, for example, in [110, 222, 235] to obtain more effective drugs and individual drug therapies for leukemia patients. Recently, microarray technology is developed in biological studies to measure expression levels of thousands of genes simultaneously. This makes it possible to investigate gene activities considering the whole genome. Gene clustering is used, for example, in discovering groups of correlated genes potentially coregulated or associated with the diseases. Many clustering algorithms including those based on hierarchical clustering, biclustering and mixture models as well as the k-means algorithm and the self-organizing map have been used to solve clustering problems in gene expression data sets [67, 68, 97, 230, 280, 308]. In addition, various algorithms have been applied for clustering microarray data sets with a large number of genes to group patients with different diseases (e.g., different types of cancer) [21, 96].

In *economics*, cluster analysis is applied to compare economic developments of municipalities—or countries, or other regions and to identify common factors (e.g., unemployment rates, age structure, rate of change in the number of permanent inhabitants, and the number of individual entrepreneurs) that have an influence on economic development. In particular, clustering can be applied for comparing socio-economic development of different areas and thus, it can help decision makers to identify the regions with the largest need for stimulating their development [53]. Another interesting application of clustering techniques in economy is the *study of economic resilience* of regions under crises [71]. The use of clustering in this area allowed to identify patterns of economic resilience in Australian regions by industry categories. Applying the clustering algorithm on census data from 2001, 2006, and 2011 helped to evaluate the impact of two major shocks the "13-year drought" and the "global financial crisis" on four functional groups of regions in Australia. The most widely applied method for clustering in economy is probably the agglomerative hierarchical clustering. Nevertheless, the partitional methods like k-means and k-medoids, and fuzzy clustering methods like fuzzy c-means are also commonly used [247].

Cluster analysis has been widely applied in *marketing research* and, especially, in *market segmentation* [176, 239]. Segmenting a market means dividing its potential customers into separate groups, where customers in the same group are similar with respect to a given set of characteristics and customers belonging to different groups are dissimilar with respect to the same set of characteristics. The goal of using cluster analysis in marketing is to accurately segment customers in order to achieve more effective customer marketing via personalization. As companies collect more data—and more detailed data—about customers, the advantages of targeting specific groups of homogeneous customers compared to using a mass marketing approach will continue to grow. The most commonly used clustering algorithms for market segmentation are the k-means and Ward's minimum variance methods [87, 239]. In addition, many other clustering algorithms have been applied in this area, such as support vector clustering [149], self-organizing feature map [188], genetic clustering algorithm [283], and artificial neural networks (ANN) [147].

Further, *financial data analysis* is becoming increasingly important in the business market [192]. Banking and financial institutes are applying various data mining techniques, including clustering, to enhance the performance of their businesses. As companies collect more data from daily operations, they expect to extract more useful knowledge—for instance, the user credit category and confidence of expected return—that helps them to make a reasonable decision for new customer's requests.

Due to enormous growth of the available text data and number of documents, the *text mining* has become a mandatory practice. Here, the clustering approach allows us to find groups of similar documents each corresponding to one or more topics [83]. The k-means algorithm and its variations have been applied to solve this type of clustering problems. Generally, in text mining data points (documents) are normalized and clustering is carried out on the unit sphere. In this case, the k-means algorithm is also called the *spherical k-means algorithm* [83, 268]. Another approach for clustering of patterns represented by sentences is presented in [202]. In this approach, the similarity between patterns is expressed in terms of the distance between their corresponding sentences. Each unlabelled data point is assigned to the cluster of its k-nearest labelled neighbors as long as the average distance to the k neighbors is below a given threshold. Efficient document clustering algorithms help to improve the performance of search engines by pre-clustering the entire corpus and the postretrieval document browsing technique. Modification of these algorithms by assigning weights to documents improves the accuracy of clustering solutions [6].

Machine learning techniques are widely used in *cybersecurity* to detect malware attacks [225]. For example, *phishing* is one of the most malicious attacks that has enabled attackers to masquerade as legitimate users of organizations, such as banks, to scam money and private information from victims. Phishing is so widespread that combating the phishing attacks could overwhelm the victim organization. Therefore, it is important to group the phishing attacks to formulate effective defence mechanisms. Different clustering algorithms have been applied to identify, analyze, and group phishing emails. They include k-means, multi-start modified global k-means, the incremental nonsmooth optimization based clustering algorithm, and difference of convex optimization based clustering algorithms [260].

In addition, cluster analysis can be used in conjunction with other approaches of machine learning. For example, *clusterwise linear regression* (CLR) is a combination of clustering and regression analysis [31, 34, 35, 81, 271]. The CLR technique is used to obtain clusters by their specific regression coefficients in a linear regression model. The CLR has been used, for instance, in rainfall prediction [37], forecasting of PM10 in Earth's atmosphere [238], and consumer benefit segmentation [292].

Chapter 2
Theory of Nonsmooth Optimization

2.1 Introduction

Nonsmooth optimization (NSO) refers to the general problem of minimizing (or maximizing) functions with discontinuous gradients. These types of problems appear in many applied fields, for example, in image denoising, optimal control, data mining, economics, computational chemistry and physics. Since the classical theory of optimization presumes certain differentiability and strong regularity assumptions for the functions to be optimized it cannot be directly utilized.

This chapter contains necessary information on theoretical NSO which is used to model clustering problems, to obtain optimality conditions and to design algorithms for these problems. We first introduce some notations and basic concepts from smooth analysis. Then we generalize the concepts of differential calculus for nonsmooth convex functions [249], define subgradients and subdifferentials and present some basic results for convex problems. These concepts and results are further extended to optimization problems with *locally Lipschitz continuous* (LLC) functions [70]. We define the so-called ε-subdifferentials that approximate the subdifferentials. The notions of quasidifferential [79] and discrete gradients [18, 28] are introduced as additional tools to handle both the nonsmoothness and the nonconvexity. In addition, we formulate necessary and sufficient optimality conditions for optimization problems with LLC functions. Finally, we consider some special classes of nonconvex nonsmooth functions—DC functions, piecewise partially separable functions, and max-functions—and show how their structures can be utilized in NSO. The proofs of theorems, lemmas, and propositions in this chapter are omitted since they can be found, for example, in [32].

Our notations are fairly standard: all the vectors $\boldsymbol{x}$ are considered as column vectors and, correspondingly, all the transposed vectors $\boldsymbol{x}^T$ are considered as row vectors. We denote by $\boldsymbol{x}^T\boldsymbol{y}$ the usual *inner product* and by $\|\boldsymbol{x}\|$ the *norm in the n-dimensional real Euclidean space* $\mathbb{R}^n$. In other words

A. M. Bagirov et al., *Partitional Clustering via Nonsmooth Optimization*, Unsupervised and Semi-Supervised Learning,
https://doi.org/10.1007/978-3-030-37826-4_2

$$x^T y = \sum_{i=1}^{n} x_i y_i \quad \text{and} \quad \|x\| = \left(x^T x\right)^{\frac{1}{2}},$$

where x and y are in $\mathbb{R}^n$ and $x_i, y_i \in \mathbb{R}$ are the ith components of the vectors x and y, respectively.

We denote by $[x, y]$ the *closed line-segment* joining x and y

$$[x, y] = \left\{ z \in \mathbb{R}^n \ : z = \lambda x + (1 - \lambda) y \ \text{ for } \ 0 \leq \lambda \leq 1 \right\},$$

and by (x, y) the corresponding *open line-segment*.

An *open (closed) ball* with the center $x \in \mathbb{R}^n$ and the radius $r > 0$ is denoted by $B(x; r)$ $(\bar{B}(x; r))$. That is,

$$B(x; r) = \{y \in \mathbb{R}^n \ : \ \|y - x\| < r\}, \quad \text{and}$$
$$\bar{B}(x; r) = \{y \in \mathbb{R}^n \ : \ \|y - x\| \leq r\}.$$

We also denote by S_1 the sphere of the unit ball

$$S_1 = \left\{ y \in \mathbb{R}^n \ : \ \|y\| = 1 \right\}.$$

2.2 Preliminaries

In this section, we describe some notations, concepts, and basic results from convex and smooth analysis.

2.2.1 Convex Sets

A set $S \subset \mathbb{R}^n$ is said to be *convex* if

$$\lambda x + (1 - \lambda) y \in S,$$

whenever x and y are in S and $\lambda \in [0, 1]$. This means that the set is convex if the closed line-segment $[x, y]$ is entirely contained in S whenever its endpoints x and y are in S. Particularly, the empty set, the unit ball $\bar{B}(x; 1)$ and the whole Euclidean space $\mathbb{R}^n$ are convex.

Let $\mathcal{I} = \{i \ : \ i = 1, \ldots, m\}$. If the sets $S_i \subset \mathbb{R}^n$ are convex for $i \in \mathcal{I}$, then their intersection $\cap_{i \in \mathcal{I}} S_i$ is convex. If, in addition, each S_i is non-empty and $\mu_i \in \mathbb{R}$, then the set $\sum_{i \in \mathcal{I}} \mu_i S_i$ is convex. Figure 2.1 illustrates some convex and nonconvex sets.

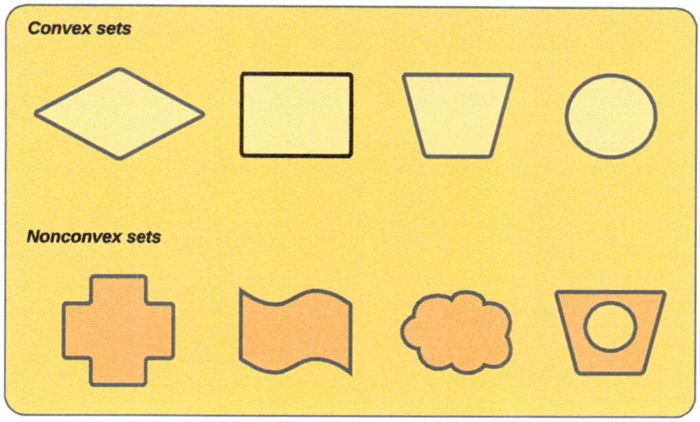

Fig. 2.1 Convex and nonconvex sets

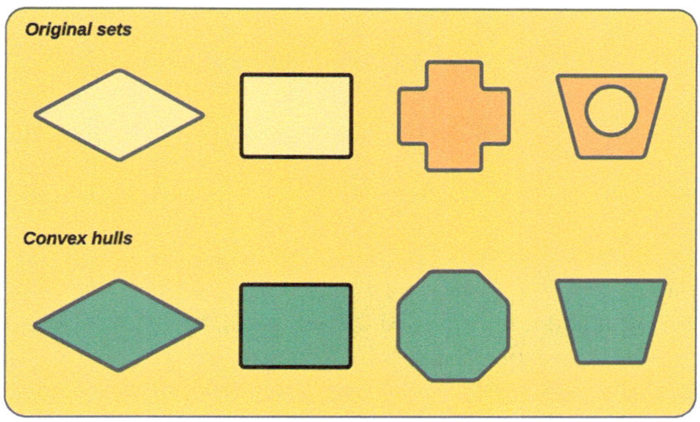

Fig. 2.2 Convex hulls of convex and nonconvex sets

A linear combination $\sum_{i \in \mathcal{I}} \lambda_i \boldsymbol{x}_i$ is called a *convex combination* of elements $\boldsymbol{x}_1, \ldots, \boldsymbol{x}_m \in \mathbb{R}^n$ if each $\lambda_i \geq 0$ and $\sum_{i \in \mathcal{I}} \lambda_i = 1$. For any set $S \subset \mathbb{R}^n$, we define the *convex hull* of S as a set of all possible convex combinations of its elements

$$\operatorname{conv} S = \left\{ \boldsymbol{x} \in \mathbb{R}^n \ : \boldsymbol{x} = \sum_{i \in \mathcal{I}} \lambda_i \boldsymbol{x}_i, \ \sum_{i \in \mathcal{I}} \lambda_i = 1, \ \boldsymbol{x}_i \in S, \ \lambda_i \geq 0 \right\}.$$

The conv S is the smallest convex set containing S. Further, S is convex if and only if $S = \operatorname{conv} S$. Figure 2.2 illustrates the convex hulls of some convex and nonconvex sets.

2.2.2 Separating Hyperplanes

Every non-zero vector $\boldsymbol{v} \in \mathbb{R}^n$ and a scalar $\alpha \in \mathbb{R}$ define a unique *hyperplane*

$$H(\boldsymbol{v}, \alpha) = \{\boldsymbol{x} \in \mathbb{R}^n : \boldsymbol{v}^T\boldsymbol{x} = \alpha\}.$$

A hyperplane divides the whole space $\mathbb{R}^n$ into two closed (or open) *halfspaces*

$$H^+(\boldsymbol{v}, \alpha) = \{\boldsymbol{x} \in \mathbb{R}^n : \boldsymbol{v}^T\boldsymbol{x} \geq \alpha\}, \quad \text{and}$$
$$H^-(\boldsymbol{v}, \alpha) = \{\boldsymbol{x} \in \mathbb{R}^n : \boldsymbol{v}^T\boldsymbol{x} \leq \alpha\}.$$

If $n = 1$, the hyperplane $H(v, \alpha) = \{x \in \mathbb{R} : vx = \alpha\}$, $v \neq 0$ is the singleton $\{\alpha/v\}$, and the halfspaces are $H^+(v, \alpha) = [\alpha/v, \infty)$ and $H^-(v, \alpha) = (-\infty, \alpha/v]$. The hyperplane is a line when $n = 2$ and is a plane when $n = 3$.

Next, we define the *separating hyperplane*: let $S_1, S_2 \subset \mathbb{R}^n$ be non-empty sets. A hyperplane $H(\boldsymbol{v}, \alpha)$ with $\boldsymbol{v} \neq \boldsymbol{0}$ *separates* S_1 and S_2 if $S_1 \subseteq H^+(\boldsymbol{v}, \alpha)$ and $S_2 \subseteq H^-(\boldsymbol{v}, \alpha)$, in other words

$$\boldsymbol{v}^T\boldsymbol{x} \geq 0 \quad \text{for all} \quad \boldsymbol{x} \in S_1, \quad \text{and}$$
$$\boldsymbol{v}^T\boldsymbol{x} \leq 0 \quad \text{for all} \quad \boldsymbol{x} \in S_2.$$

The separation is *strict* if $S_1 \cap H(\boldsymbol{v}, \alpha) = \emptyset$ and $S_2 \cap H(\boldsymbol{v}, \alpha) = \emptyset$.

Let $S \subset \mathbb{R}^n$ be a non-empty, closed convex set and $\boldsymbol{x}^* \notin S$. Then there exists a hyperplane $H(\boldsymbol{v}, \alpha)$ separating S and $\{\boldsymbol{x}^*\}$. In addition, two convex sets $S_1, S_2 \subset \mathbb{R}^n$ such that $S_1 \cap S_2 = \emptyset$ can always be separated by a hyperplane. For the strict separation, S_1 and S_2 need to be closed and also at least one of them should to be bounded. Figure 2.3 illustrates the separating hyperplane of some convex sets. Note that the separating hyperplane in Fig. 2.3 is not unique and there exist infinitely many hyperplanes separating these two convex sets.

Fig. 2.3 Separating hyperplane

2.2.3 Continuous, Lipschitz Continuous, and Convex Functions

A function $f : \mathbb{R}^n \to \mathbb{R}$ is said to be *upper semicontinuous* at $\boldsymbol{x} \in \mathbb{R}^n$ if for every sequence $\{\boldsymbol{x}_h\}$ converging to $\boldsymbol{x}$ the following holds:

$$\limsup_{h\to\infty} f(\boldsymbol{x}_h) \le f(\boldsymbol{x}),$$

and *lower semicontinuous* if

$$f(\boldsymbol{x}) \le \liminf_{h\to\infty} f(\boldsymbol{x}_h).$$

A both upper and lower semicontinuous function is *continuous*. Upper and lower semicontinuous as well as continuous functions are illustrated in Figs. 2.4, 2.5, and 2.6, respectively.

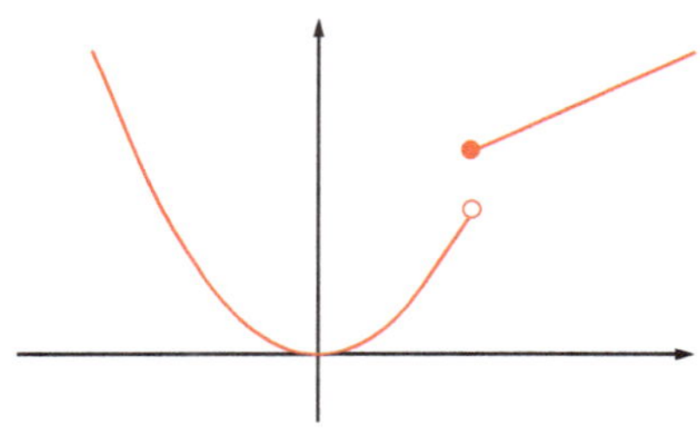

Fig. 2.4 Upper semicontinuous function

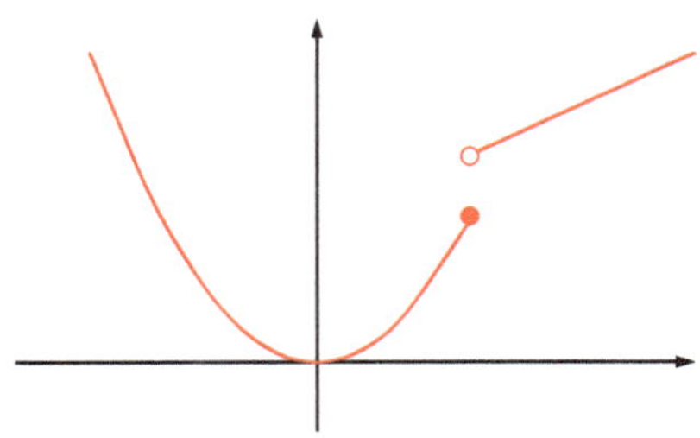

Fig. 2.5 Lower semicontinuous function

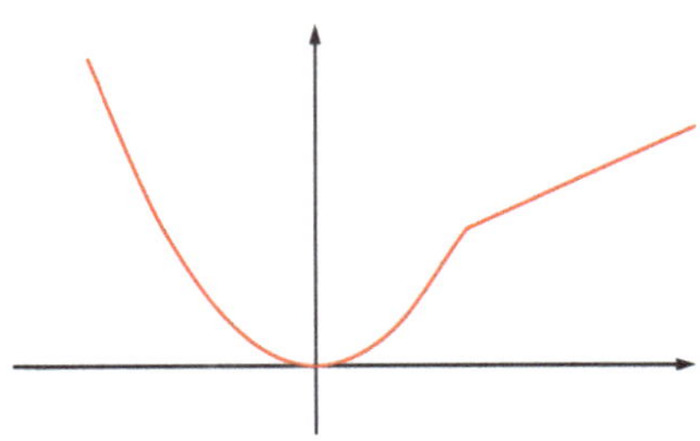

Fig. 2.6 Continuous function

A function $f : \mathbb{R}^n \to \mathbb{R}$ is called *Lipschitz continuous* if

$$|f(\boldsymbol{x}) - f(\boldsymbol{y})| \le L\|\boldsymbol{x} - \boldsymbol{y}\| \quad \text{for all} \quad \boldsymbol{x}, \boldsymbol{y} \in \mathbb{R}^n,$$

where $L > 0$ is a constant independent of $\boldsymbol{x}$ and $\boldsymbol{y}$.

A function $f : \mathbb{R}^n \to \mathbb{R}$ is called *locally Lipschitz continuous (LLC)* at $\boldsymbol{x} \in \mathbb{R}^n$ with a constant $L > 0$ if there exists a positive number ε such that

$$|f(\boldsymbol{y}) - f(\boldsymbol{z})| \le L\|\boldsymbol{y} - \boldsymbol{z}\| \quad \text{for all} \quad \boldsymbol{y}, \boldsymbol{z} \in B(\boldsymbol{x}; \varepsilon).$$

A function $f : \mathbb{R}^n \to \mathbb{R}$ is said to be *convex* if

$$f(\lambda\boldsymbol{x} + (1 - \lambda)\boldsymbol{y}) \le \lambda f(\boldsymbol{x}) + (1 - \lambda) f(\boldsymbol{y}) \tag{2.1}$$

for all $\boldsymbol{x}, \boldsymbol{y} \in \mathbb{R}^n$ and $\lambda \in [0, 1]$. If a strict inequality holds in (2.1) for all $\boldsymbol{x}, \boldsymbol{y} \in \mathbb{R}^n$ such that $\boldsymbol{x} \ne \boldsymbol{y}$ and $\lambda \in (0, 1)$, then the function f is said to be *strictly convex.* A function $f : \mathbb{R}^n \to \mathbb{R}$ is *(strictly) concave* if $-f$ is (strictly) convex. Further, a function f that is not convex is called *nonconvex.* The convex function $f : \mathbb{R}^n \to \mathbb{R}$ is LLC at any $\boldsymbol{x} \in \mathbb{R}^n$.

A function $f : \mathbb{R}^n \to \mathbb{R}$ is *positively homogeneous* if

$$f(\lambda\boldsymbol{x}) = \lambda f(\boldsymbol{x})$$

for all $\lambda \ge 0$ and *subadditive* if

$$f(\boldsymbol{x} + \boldsymbol{y}) \le f(\boldsymbol{x}) + f(\boldsymbol{y})$$

for all $\boldsymbol{x}$ and $\boldsymbol{y}$ in $\mathbb{R}^n$. A function that is both positively homogeneous and subadditive is always convex.

A function $f : \mathbb{R}^n \to \mathbb{R}$ is said to be *differentiable* at $\boldsymbol{x} \in \mathbb{R}^n$ if there exists a vector $\nabla f(\boldsymbol{x}) \in \mathbb{R}^n$ and a function $\varrho : \mathbb{R}^n \to \mathbb{R}$ such that for all $\boldsymbol{d} \in \mathbb{R}^n$

$$f(\boldsymbol{x} + \boldsymbol{d}) = f(\boldsymbol{x}) + \nabla f(\boldsymbol{x})^T\boldsymbol{d} + \|\boldsymbol{d}\|\varrho(\boldsymbol{d}),$$

and $\varrho(\boldsymbol{d}) \to 0$ whenever $\|\boldsymbol{d}\| \to 0$. The vector $\nabla f(\boldsymbol{x})$ is called the *gradient vector* of the function f at $\boldsymbol{x}$ and it is given by the formula

$$\nabla f(\boldsymbol{x}) = \Big(\frac{\partial f(\boldsymbol{x})}{\partial x_1}, \dots, \frac{\partial f(\boldsymbol{x})}{\partial x_n}\Big)^T.$$

Here, the components $\partial f(\boldsymbol{x})/\partial x_j$ for $j = 1, \dots, n$, are called *partial derivatives* of the function f. If the function is differentiable and all the partial derivatives are continuous, then the function is said to be *continuously differentiable* or *smooth* ($f \in C^1(\mathbb{R}^n)$). A smooth function is always LLC. Functions with discontinuous

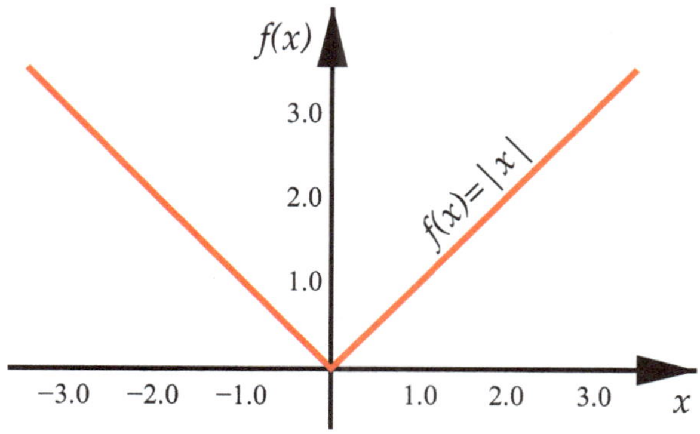

Fig. 2.7 Absolute value function

gradients are called *nonsmooth*. The simplest example of a nonsmooth function is the absolute value function $f(x) = |x|,\ x \in \mathbb{R}$ (see Fig. 2.7).

Continuous differentiability of a function $f : \mathbb{R}^n \to \mathbb{R}$ at $\boldsymbol{x}$ implies that for any $\varepsilon > 0$ there exists $\delta > 0$ such that

$$\Big|\frac{f(\boldsymbol{z} + t\boldsymbol{y}) - f(\boldsymbol{z})}{t} - \nabla f(\boldsymbol{x})^T \boldsymbol{y}\Big| < \varepsilon$$

whenever $0 < t < \delta$ and $\boldsymbol{z} \in B(\boldsymbol{x}; \delta)$.

The limit

$$f'(\boldsymbol{x}; \boldsymbol{d}) = \lim_{t \downarrow 0} \frac{f(\boldsymbol{x} + t\boldsymbol{d}) - f(\boldsymbol{x})}{t}$$

(if it exists) is called the *directional derivative* of f at $\boldsymbol{x} \in \mathbb{R}^n$ in the direction $\boldsymbol{d} \in \mathbb{R}^n$. The directional derivative can be used as an approximation for the change of function values: thus, we can use it to detect directions where function values increase, decrease, or do not change.

As a function of $\boldsymbol{d}$, the directional derivative $f'(\boldsymbol{x}; \boldsymbol{d})$ is positively homogeneous and subadditive. If at a point $\boldsymbol{x} \in \mathbb{R}^n$ the directional derivative $f'(\boldsymbol{x}; \boldsymbol{d})$ exists in every direction $\boldsymbol{d} \in \mathbb{R}^n$, then the function f is called *directionally differentiable* at $\boldsymbol{x}$. Both the differentiable and finite valued convex functions are directionally differentiable. For a differentiable function f we have

$$f'(\boldsymbol{x}; \boldsymbol{d}) = \nabla f(\boldsymbol{x})^T \boldsymbol{d},$$

and, if f is also convex, then for all $\boldsymbol{y} \in \mathbb{R}^n$ we have

$$f(\boldsymbol{y}) \geq f(\boldsymbol{x}) + f'(\boldsymbol{x}, \boldsymbol{y} - \boldsymbol{x}),$$

or

$$f(\boldsymbol{y}) \geq f(\boldsymbol{x}) + \nabla f(\boldsymbol{x})^T (\boldsymbol{y} - \boldsymbol{x}).$$

A function $f : \mathbb{R}^n \to \mathbb{R}$ is said to be *twice differentiable* at $\boldsymbol{x} \in \mathbb{R}^n$ if there exists a vector $\nabla f(\boldsymbol{x}) \in \mathbb{R}^n$, a symmetric matrix $\nabla^2 f(\boldsymbol{x}) \in \mathbb{R}^{n\times n}$, and a function $\varrho : \mathbb{R}^n \to \mathbb{R}$ such that for all $\boldsymbol{d} \in \mathbb{R}^n$

$$f(\boldsymbol{x}+\boldsymbol{d}) = f(\boldsymbol{x}) + \nabla f(\boldsymbol{x})^T\boldsymbol{d} + \frac{1}{2}\boldsymbol{d}^T\nabla^2 f(\boldsymbol{x})\boldsymbol{d} + \|\boldsymbol{d}\|^2\varrho(\boldsymbol{d}),$$

where $\varrho(\boldsymbol{d}) \to 0$ whenever $\|\boldsymbol{d}\| \to 0$. The matrix $\nabla^2 f(\boldsymbol{x})$ is called the *Hessian matrix* of the function f at $\boldsymbol{x}$ and its entries are *second order partial derivatives* of f, that is,

$$\nabla^2 f(\boldsymbol{x}) = \begin{bmatrix} \frac{\partial^2 f(\boldsymbol{x})}{\partial x_1^2} & \cdots & \frac{\partial^2 f(\boldsymbol{x})}{\partial x_1 \partial x_n} \\ \vdots & \ddots & \vdots \\ \frac{\partial^2 f(\boldsymbol{x})}{\partial x_n \partial x_1} & \cdots & \frac{\partial^2 f(\boldsymbol{x})}{\partial x_n^2} \end{bmatrix}.$$

If the function is twice differentiable and all the second order partial derivatives are continuous, then the function is said to be *twice continuously differentiable* ($f \in C^2(\mathbb{R}^n)$). Finally, we denote by $C^\infty(\mathbb{R}^n)$ the class of *infinitely continuously differentiable* functions.

2.3 Concepts of Nonsmooth Analysis

The theory of nonsmooth analysis originates from convex analysis. Therefore, we start this section by providing some important definitions and results for (not necessarily differentiable) convex functions. We define the *subgradient* and the *subdifferential* of a convex function [249], and then generalize these concepts to the class of nonconvex LLC functions [70]. In addition, we describe rules for the calculation of subgradients and subdifferentials of different classes of functions. Finally, we recall the notion of the *quasidifferential* [78].

The aim of this section is not to give the detailed description of nonsmooth analysis—for that we refer, for example, to [32, 70, 207, 249]—but rather to collect some basic definitions and results necessary in the subsequent chapters of this book.

2.3.1 Subdifferentials of Convex Functions

Definition 2.1 The *subdifferential of a convex function* $f : \mathbb{R}^n \to \mathbb{R}$ at $\boldsymbol{x} \in \mathbb{R}^n$ is the set

$$\partial_c f(\boldsymbol{x}) = \left\{ \boldsymbol{\xi} \in \mathbb{R}^n \ :\ f(\boldsymbol{y}) \ge f(\boldsymbol{x}) + \boldsymbol{\xi}^T(\boldsymbol{y}-\boldsymbol{x}) \text{ for all } \boldsymbol{y} \in \mathbb{R}^n \right\}.$$

Each vector $\boldsymbol{\xi} \in \partial_c f(\boldsymbol{x})$ is called a *subgradient* of f at $\boldsymbol{x}$.

The subdifferential $\partial_c f(\boldsymbol{x})$ is a non-empty, convex, and compact set such that $\partial_c f(\boldsymbol{x}) \subseteq B(\boldsymbol{0}; L)$, where $L > 0$ is the Lipschitz constant of f at $\boldsymbol{x}$. Note the similarity between Definition 2.1 and the smooth differential theory: if $f : \mathbb{R}^n \to \mathbb{R}$ is both convex and differentiable, then for all $\boldsymbol{y} \in \mathbb{R}^n$ we have

$$f(\boldsymbol{y}) \geq f(\boldsymbol{x}) + \nabla f(\boldsymbol{x})^T(\boldsymbol{y} - \boldsymbol{x}),$$

and the subdifferential is a singleton

$$\partial_c f(\boldsymbol{x}) = \{\nabla f(\boldsymbol{x})\}. \tag{2.2}$$

As mentioned above a finite valued convex function is directionally differentiable everywhere, that is, the directional derivative $f'(\boldsymbol{x}; \boldsymbol{d})$ exists in every direction $\boldsymbol{d} \in \mathbb{R}^n$ for all $\boldsymbol{x} \in \mathbb{R}^n$. In addition, we have

$$f'(\boldsymbol{x}; \boldsymbol{d}) = \inf_{t>0} \frac{f(\boldsymbol{x} + t\boldsymbol{d}) - f(\boldsymbol{x})}{t} \quad \text{for all} \quad \boldsymbol{d} \in \mathbb{R}^n.$$

The subdifferential $\partial_c f(\boldsymbol{x})$ can be represented with the aid of the directional derivative $f'(\boldsymbol{x}; \boldsymbol{d})$ and vice versa. Suppose that $f : \mathbb{R}^n \to \mathbb{R}$ is a convex function, then for all $\boldsymbol{x} \in \mathbb{R}^n$ we have

$$\partial_c f(\boldsymbol{x}) = \{\boldsymbol{\xi} \in \mathbb{R}^n : f'(\boldsymbol{x}, \boldsymbol{d}) \geq \boldsymbol{\xi}^T \boldsymbol{d} \text{ for all } \boldsymbol{d} \in \mathbb{R}^n\}, \quad \text{and} \tag{2.3}$$

$$f'(\boldsymbol{x}; \boldsymbol{d}) = \max \{\boldsymbol{\xi}^T \boldsymbol{d} : \boldsymbol{\xi} \in \partial_c f(\boldsymbol{x})\} \text{ for all } \boldsymbol{d} \in \mathbb{R}^n. \tag{2.4}$$

Further, any convex function can be presented by using its subgradients. This result is very useful when developing numerical methods for optimization.

Theorem 2.1 *If $f : \mathbb{R}^n \to \mathbb{R}$ is convex, then for all $\boldsymbol{y} \in \mathbb{R}^n$*

$$f(\boldsymbol{y}) = \sup \{f(\boldsymbol{x}) + \boldsymbol{\xi}^T(\boldsymbol{y} - \boldsymbol{x}) : \boldsymbol{x} \in \mathbb{R}^n, \ \boldsymbol{\xi} \in \partial_c f(\boldsymbol{x})\}.$$

The subdifferential $\partial f_c(\boldsymbol{x})$ is sometimes approximated with a larger set—the so-called ε-subdifferential—which is an extension of the subdifferential.

Definition 2.2 Let $\varepsilon \geq 0$. The *ε-subdifferential* of a convex function $f : \mathbb{R}^n \to \mathbb{R}$ at $\boldsymbol{x} \in \mathbb{R}^n$ is the set

$$\partial_\varepsilon f(\boldsymbol{x}) = \{\boldsymbol{\xi} \in \mathbb{R}^n : f(\boldsymbol{y}) \geq f(\boldsymbol{x}) + \boldsymbol{\xi}^T(\boldsymbol{y} - \boldsymbol{x}) - \varepsilon \text{ for all } \boldsymbol{y} \in \mathbb{R}^n\}.$$

Each element $\boldsymbol{\xi} \in \partial_\varepsilon f(\boldsymbol{x})$ is called an *ε-subgradient* of f at $\boldsymbol{x}$.

The geometrical interpretation of the ε-subdifferential is as follows. A subgradient $\boldsymbol{\xi} \in \partial f(\bar{\boldsymbol{x}})$ at some point $\bar{\boldsymbol{x}} \in \mathbb{R}^n$ belongs to $\partial_\varepsilon f(\boldsymbol{x})$ if the affine function

$u(\boldsymbol{y}) = f(\bar{\boldsymbol{x}}) + \boldsymbol{\xi}^T(\boldsymbol{y} - \bar{\boldsymbol{x}})$ giving the tangent line to the graph of the function f at $\bar{\boldsymbol{x}}$ satisfies the inequality $u(\boldsymbol{x}) \geq f(\boldsymbol{x}) - \varepsilon$.

Similarly to the subdifferential $\partial_c f(\boldsymbol{x})$, the ε-subdifferential $\partial_\varepsilon f(\boldsymbol{x})$ is a nonempty, convex, and compact set such that $\partial_\varepsilon f(\boldsymbol{x}) \subseteq B(\boldsymbol{0}; L)$, where $L > 0$ is the Lipschitz constant of f at $\boldsymbol{x}$. In addition, it has the following properties:

- $\partial_0 f(\boldsymbol{x}) = \partial_c f(\boldsymbol{x})$; and
- if $\varepsilon_1 \leq \varepsilon_2$, then $\partial_{\varepsilon_1} f(\boldsymbol{x}) \subseteq \partial_{\varepsilon_2} f(\boldsymbol{x})$.

We conclude this subsection by noting that the ε-subdifferential contains in a compressed form the subgradient information in the whole neighborhood of a point $\boldsymbol{x} \in \mathbb{R}^n$. This information is utilized, for instance, in bundle methods to be described in Sect. 3.3.

Theorem 2.2 *Let $f : \mathbb{R}^n \to \mathbb{R}$ be a convex function with the Lipschitz constant L at $\boldsymbol{x}$. Then for any $\varepsilon \geq 0$ we have*

$$\partial_c f(\boldsymbol{y}) \subseteq \partial_\varepsilon f(\boldsymbol{x}) \quad \textit{for all} \quad \boldsymbol{y} \in B\big(\boldsymbol{x}; \tfrac{\varepsilon}{2L}\big).$$

2.3.2 Nonconvex Analysis

The directional derivatives need not to exist for all LLC functions. Therefore, the definitions given in the previous subsection are not always applicable to nonconvex LLC functions. There are various alternatives available to generalize the concept of subdifferential to the nonconvex case (see, e.g., [77, 78, 220, 236]). We use here the *generalized directional derivative* and *subdifferential* by Clarke [70], since elements of this subdifferential can be efficiently estimated or calculated as will be shown in Theorem 2.4.

Definition 2.3 [70] Let $f : \mathbb{R}^n \to \mathbb{R}$ be a LLC at $\boldsymbol{x} \in \mathbb{R}^n$. The *generalized directional derivative* of f at $\boldsymbol{x}$ in the direction $\boldsymbol{d} \in \mathbb{R}^n$ is defined by

$$f^\circ(\boldsymbol{x}; \boldsymbol{d}) = \limsup_{\substack{\boldsymbol{y} \to \boldsymbol{x} \\ t \downarrow 0}} \frac{f(\boldsymbol{y} + t\boldsymbol{d}) - f(\boldsymbol{y})}{t}.$$

Note that the generalized directional derivative always exists for LLC functions and, as a function of $\boldsymbol{d}$, it is positively homogeneous and subadditive on $\mathbb{R}^n$. Therefore, we can now define the subdifferential for nonconvex LLC functions analogously to the equation (2.3) in convex case with the directional derivative replaced by the generalized directional derivative. In what follows we sometimes refer to this subdifferential as *Clarke subdifferential*.

Definition 2.4 [70] Let $f : \mathbb{R}^n \to \mathbb{R}$ be a LLC function at a point $\boldsymbol{x} \in \mathbb{R}^n$. Then the *subdifferential* of f at $\boldsymbol{x}$ is the set

$$\partial f(\boldsymbol{x}) = \left\{\boldsymbol{\xi} \in \mathbb{R}^n \;:\; f^\circ(\boldsymbol{x}; \boldsymbol{d}) \geq \boldsymbol{\xi}^T\boldsymbol{d} \text{ for all } \boldsymbol{d} \in \mathbb{R}^n\right\}.$$

Each vector $\boldsymbol{\xi} \in \partial f(\boldsymbol{x})$ is called a *subgradient* of f at $\boldsymbol{x}$.

Similar to the subdifferential in convex case the subdifferential $\partial f(\boldsymbol{x})$ is a nonempty, convex, and compact set such that $\partial f(\boldsymbol{x}) \subseteq B(\boldsymbol{0}; L)$, where $L > 0$ is the Lipschitz constant of f at $\boldsymbol{x}$. In addition, the generalized directional derivative can be calculated using the subdifferential $\partial f(\boldsymbol{x})$ as follows:

$$f^\circ(\boldsymbol{x}; \boldsymbol{d}) = \max\left\{\boldsymbol{\xi}^T\boldsymbol{d} \;:\; \boldsymbol{\xi} \in \partial f(\boldsymbol{x})\right\} \quad \text{for all} \quad \boldsymbol{d} \in \mathbb{R}^n.$$

The subdifferential for LLC functions is a generalization of the subdifferential for convex functions: if $f : \mathbb{R}^n \to \mathbb{R}$ is a convex function, then $f^\circ(\boldsymbol{x}; \boldsymbol{d}) = f'(\boldsymbol{x}; \boldsymbol{d})$ for all $\boldsymbol{d} \in \mathbb{R}^n$, and $\partial f(\boldsymbol{x}) = \partial_c f(\boldsymbol{x})$. Thus, we can omit the index c and use $\partial f(\boldsymbol{x})$ to denote the subdifferential of a convex function as well.

The subdifferential of LLC functions is also a generalization of the classical derivative: if $f : \mathbb{R}^n \to \mathbb{R}$ is both LLC and differentiable at $\boldsymbol{x} \in \mathbb{R}^n$, then $\nabla f(\boldsymbol{x}) \in \partial f(\boldsymbol{x})$. If, in addition, $f : \mathbb{R}^n \to \mathbb{R}$ is smooth at $\boldsymbol{x} \in \mathbb{R}^n$, then $\partial f(\boldsymbol{x}) = \left\{\nabla f(\boldsymbol{x})\right\}$.

The LLC function $f : \mathbb{R}^n \to \mathbb{R}$ is *strictly differentiable* at a point $\boldsymbol{x} \in \mathbb{R}^n$ if the subdifferential $\partial f(\boldsymbol{x})$ is a singleton. The function f is called strictly differentiable on $\mathbb{R}^n$ if it is strictly differentiable at any $\boldsymbol{x} \in \mathbb{R}^n$.

The next theorem shows that the subdifferentials of LLC functions, including subdifferentials of convex functions, are upper semicontinuous.

Theorem 2.3 *[32] Let $f : \mathbb{R}^n \to \mathbb{R}$ be LLC. Then the subdifferential mapping $\boldsymbol{x} \mapsto \partial f(\boldsymbol{x})$ is upper semicontinuous, that is at a point $\boldsymbol{x} \in \mathbb{R}^n$ for any $\varepsilon > 0$ there exists $\delta > 0$ such that*

$$\partial f(\boldsymbol{y}) \subset \partial f(\boldsymbol{x}) + B(\boldsymbol{0}; \varepsilon) \tag{2.5}$$

for all $\boldsymbol{y} \in B(\boldsymbol{x}; \delta)$. Equivalently, for any sequences $\{\boldsymbol{x}_h\}$ and $\{\boldsymbol{\xi}_h\}$ such that $\boldsymbol{x}_h \in \mathbb{R}^n$, $\boldsymbol{\xi}_h \in \partial f(\boldsymbol{x}_h)$ if $\boldsymbol{x}_h \to \boldsymbol{x}$, $\boldsymbol{\xi}_h \to \boldsymbol{\xi}$ as $h \to \infty$ then $\boldsymbol{\xi} \in \partial f(\boldsymbol{x})$.

Next, we recall a result that is essential for calculating subgradients in practice. By Rademacher's Theorem [101] a function which is Lipschitz continuous on a set $U \subseteq \mathbb{R}^n$ is differentiable almost everywhere on U. This means that the gradient exists almost everywhere and the subdifferential can be constructed as a convex hull of all possible limits of gradients at points $\boldsymbol{x}_i$ converging to $\boldsymbol{x}$. Let

$$\Omega_f = \left\{\boldsymbol{x} \in \mathbb{R}^n \;:\; f \text{ is not differentiable at the point } \boldsymbol{x}\right\}$$

be the set of points where f is not differentiable.

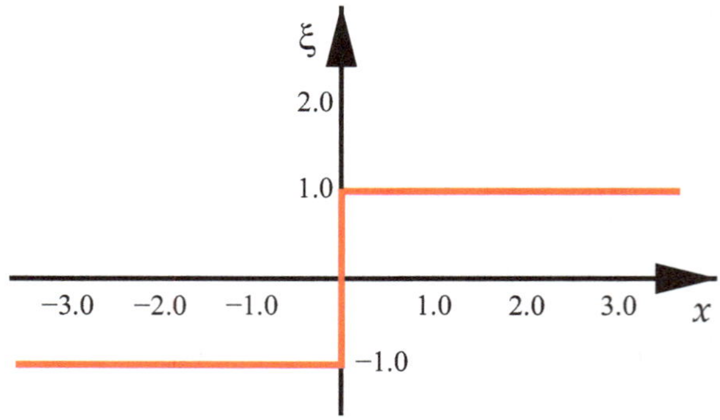

Fig. 2.8 Subdifferential of the absolute value function

Theorem 2.4 *Let $f : \mathbb{R}^n \to \mathbb{R}$ be LLC at $\boldsymbol{x} \in \mathbb{R}^n$. Then*

$$\partial f(\boldsymbol{x}) = \operatorname{conv}\Big\{\boldsymbol{\xi} \in \mathbb{R}^n \ : \ \textit{there exists } \{\boldsymbol{x}_h\} \subset \mathbb{R}^n \setminus \Omega_f \textit{ such that}$$
$$\boldsymbol{x}_h \to \boldsymbol{x} \textit{ and } \nabla f(\boldsymbol{x}_h) \to \boldsymbol{\xi}\Big\}.$$

As an example of Theorem 2.4, let us consider the absolute value function $f(x) = |x|$, $x \in \mathbb{R}$ (see Fig. 2.7). The function is differentiable everywhere but at $x = 0$, and its gradient is given by

$$\nabla f(x) = \begin{cases} 1, & \text{for } x > 0, \\ -1, & \text{for } x < 0. \end{cases}$$

The subdifferential of this function at $x = 0$ is given by (see also Fig. 2.8)

$$\partial f(0) = \operatorname{conv}\{-1, 1\} = [-1, 1].$$

The Goldstein ε-subdifferential, introduced in the next definition, is the approximation of the subdifferential of an LLC function.

Definition 2.5 Let a function $f : \mathbb{R}^n \to \mathbb{R}$ be LLC at $\boldsymbol{x} \in \mathbb{R}^n$ and let $\varepsilon \geq 0$. The *Goldstein ε-subdifferential* of f is the set

$$\partial_\varepsilon^G f(\boldsymbol{x}) = \operatorname{cl}\operatorname{conv}\big\{\partial f(\boldsymbol{y}) \ : \ \boldsymbol{y} \in B(\boldsymbol{x}; \varepsilon)\big\}.$$

Each element $\boldsymbol{\xi} \in \partial_\varepsilon^G f(\boldsymbol{x})$ is called an *ε-subgradient* of the function f at $\boldsymbol{x}$.

Similar to other subdifferential mappings presented so far, the Goldstein ε-subdifferential $\partial_\varepsilon^G f(\boldsymbol{x})$ is a non-empty, convex, and compact set such that $\partial_\varepsilon^G f(\boldsymbol{x}) \subseteq B(\boldsymbol{0}; L)$. It also has the similar properties as the ε-subdifferential for convex functions. More precisely,

- $\partial_0^G f(\boldsymbol{x}) = \partial f(\boldsymbol{x})$; and
- if $\varepsilon_1 \leq \varepsilon_2$, then $\partial_{\varepsilon_1}^G f(\boldsymbol{x}) \subseteq \partial_{\varepsilon_2}^G f(\boldsymbol{x})$.

In addition, as a corollary to Theorem 2.4 we have

$$\partial_\varepsilon^G f(\boldsymbol{x}) = \operatorname{cl}\operatorname{conv}\Big\{\boldsymbol{\xi} \in \mathbb{R}^n \;:\; \text{there exists } \{\boldsymbol{y}_h\} \subset \mathbb{R}^n \setminus \Omega_f \text{ such that } \boldsymbol{y}_h \to \boldsymbol{y},\; \nabla f(\boldsymbol{y}_h) \to \boldsymbol{\xi},\; \text{and } \boldsymbol{y} \in B(\boldsymbol{x};\varepsilon)\Big\}.$$

As in the convex case, the Goldstein ε-subdifferential contains in a compressed form the subgradient information from the whole neighborhood of $\boldsymbol{x}$.

Theorem 2.5 *Let $f : \mathbb{R}^n \to \mathbb{R}$ be LLC at $\boldsymbol{x} \in \mathbb{R}^n$. Then for any $\varepsilon \geq 0$ we have*

$$\partial f(\boldsymbol{y}) \subseteq \partial_\varepsilon^G f(\boldsymbol{x}) \quad \textit{for all} \quad \boldsymbol{y} \in B(\boldsymbol{x};\varepsilon).$$

The relationship between the ε-subdifferential and the Goldstein ε-subdifferential for convex functions is shown in the next theorem.

Theorem 2.6 *Let $f : \mathbb{R}^n \to \mathbb{R}$ be a convex function with the Lipschitz constant L at $\boldsymbol{x}$. Then for all $\varepsilon \geq 0$ we have*

$$\partial_\varepsilon^G f(\boldsymbol{x}) \subseteq \partial_{2L\varepsilon} f(\boldsymbol{x}).$$

The mapping $\boldsymbol{x} \mapsto \partial_\varepsilon^G f(\boldsymbol{x})$ is upper semicontinuous at any $\boldsymbol{x} \in \mathbb{R}^n$.

We conclude this subsection by giving definitions of semismooth and weakly semismooth functions. A function $f : \mathbb{R}^n \to \mathbb{R}$ is called *semismooth* at $\boldsymbol{x} \in \mathbb{R}^n$ if it is LLC at $\boldsymbol{x}$ and for every $\boldsymbol{d} \in \mathbb{R}^n$ the limit

$$\lim_{\substack{\boldsymbol{\xi} \in \partial f(\boldsymbol{x}+t\boldsymbol{d}'),\\ \boldsymbol{d}' \to \boldsymbol{d},\, t \downarrow 0}} \boldsymbol{\xi}^T \boldsymbol{d}'$$

exists. Further, a function $f : \mathbb{R}^n \to \mathbb{R}$ is *weakly semismooth* at $\boldsymbol{x} \in \mathbb{R}^n$ if it is LLC at $\boldsymbol{x}$ and the limit

$$\lim_{\substack{\boldsymbol{\xi} \in \partial f(\boldsymbol{x}+t\boldsymbol{d}),\\ t \downarrow 0}} \boldsymbol{\xi}^T \boldsymbol{d} \tag{2.6}$$

exists for every $\boldsymbol{d} \in \mathbb{R}^n$. Evidently, the semismoothness implies the weak semismoothness. The class of semismooth functions is fairly broad and it contains, for instance, convex, concave, max- and min-type functions. The semismooth function f is directionally differentiable and

$$f'(\boldsymbol{x},\boldsymbol{d}) = \lim_{\substack{\boldsymbol{\xi} \in \partial f(\boldsymbol{x}+t\boldsymbol{d}'),\\ \boldsymbol{d}' \to \boldsymbol{d},\, t \downarrow 0}} \boldsymbol{\xi}^T \boldsymbol{d}.$$

2.3.3 Subdifferential Calculus

In this subsection, we give some rules for calculation of subdifferentials. We consider LLC functions but differentiation rules for the smooth and convex functions can be obtained from these results as special cases. For general LLC functions we restrict ourselves with inclusions instead of equalities in differentiation rules. In order to obtain equalities we need the following regularity property.

Definition 2.6 A function $f : \mathbb{R}^n \to \mathbb{R}$ is said to be *subdifferentially regular* at $\boldsymbol{x} \in \mathbb{R}^n$ if it is LLC at $\boldsymbol{x}$, and for all $\boldsymbol{d} \in \mathbb{R}^n$ the directional derivative $f'(\boldsymbol{x}; \boldsymbol{d})$ exists and

$$f'(\boldsymbol{x}; \boldsymbol{d}) = f^\circ(\boldsymbol{x}; \boldsymbol{d}). \tag{2.7}$$

The equality (2.7) is not satisfied in general even if $f'(\boldsymbol{x}; \boldsymbol{d})$ exists. A simple example is the concave function $f(x) = -|x|$: it has the directional derivative $f'(0; 1) = -1$, but the generalized directional derivative $f^\circ(0; 1) = 1$. Nevertheless, there are some simple rules when the function $f : \mathbb{R}^n \to \mathbb{R}$ is subdifferentially regular at $\boldsymbol{x}$:

- f is smooth at $\boldsymbol{x}$;
- f is convex; or
- $f = \sum_{i \in \mathcal{I}} \lambda_i f_i$, where $\lambda_i > 0$ and f_i is subdifferentially regular at $\boldsymbol{x}$ for each $i \in \mathcal{I}$.

In Sect. 2.3.1, we noted that for a differentiable convex function the subdifferential is a singleton. Now we generalize this result to the broader class of functions: if $f : \mathbb{R}^n \to \mathbb{R}$ is both differentiable and subdifferentially regular at $\boldsymbol{x}$, then the subdifferential $\partial f(\boldsymbol{x})$ is a singleton

$$\partial f(\boldsymbol{x}) = \{\nabla f(\boldsymbol{x})\}.$$

Next, we present differentiation rules for LLC functions. The proofs of these results can be found in [32].

Theorem 2.7 *If the function $f : \mathbb{R}^n \to \mathbb{R}$ is LLC at $\boldsymbol{x}$, then for all $\lambda \in \mathbb{R}$*

$$\partial(\lambda f)(\boldsymbol{x}) = \lambda\, \partial f(\boldsymbol{x}). \tag{2.8}$$

Theorem 2.8 (Sum) *Let $f_i : \mathbb{R}^n \to \mathbb{R}$ be LLC at $\boldsymbol{x}$ and $\lambda_i \in \mathbb{R}$ for all $i \in \mathcal{I}$. Then the function*

$$f(\boldsymbol{x}) := \sum_{i \in \mathcal{I}} \lambda_i f_i(\boldsymbol{x})$$

is LLC at $\boldsymbol{x}$ and

$$\partial f(\boldsymbol{x}) \subseteq \sum_{i\in\mathcal{I}} \lambda_i \partial f_i(\boldsymbol{x}). \tag{2.9}$$

In addition, if each f_i is subdifferentially regular at $\boldsymbol{x}$ and each $\lambda_i \geq 0$, then f is subdifferentially regular at $\boldsymbol{x}$ and equality holds in (2.9).

Next, we generalize two main results of differential calculus, namely the *mean-value theorem* and the *chain rule*.

Theorem 2.9 (Mean-Value Theorem) *Let the function $f : \mathbb{R}^n \to \mathbb{R}$ be Lipschitz continuous on an open set $U \subseteq \mathbb{R}^n$. Let $\boldsymbol{x}, \boldsymbol{y} \in \mathbb{R}^n$ be such that $\boldsymbol{x} \neq \boldsymbol{y}$ and the line-segment $[\boldsymbol{x}, \boldsymbol{y}] \subset U$. Then there exists a point $\boldsymbol{z} \in (\boldsymbol{x}, \boldsymbol{y})$ such that*

$$f(\boldsymbol{y}) - f(\boldsymbol{x}) \in \partial f(\boldsymbol{z})^T(\boldsymbol{y} - \boldsymbol{x}),$$

where

$$\partial f(\boldsymbol{z})^T(\boldsymbol{y} - \boldsymbol{x}) = \Big\{\boldsymbol{u} \in \mathbb{R}^n : \exists\, \boldsymbol{\xi} \in \partial f(\boldsymbol{z}) \text{ such that } \boldsymbol{u} = \boldsymbol{\xi}^T(\boldsymbol{y} - \boldsymbol{x})\Big\}.$$

Theorem 2.10 (Chain Rule) *Let $f : \mathbb{R}^n \to \mathbb{R}$ be such that $f = g \circ \boldsymbol{h}$, where $\boldsymbol{h} : \mathbb{R}^n \to \mathbb{R}^m$ is LLC at $\boldsymbol{x} \in \mathbb{R}^n$ and $g : \mathbb{R}^m \to \mathbb{R}$ is LLC at $\boldsymbol{h}(\boldsymbol{x}) \in \mathbb{R}^m$. Then f is LLC at $\boldsymbol{x}$ and*

$$\partial f(\boldsymbol{x}) \subseteq \operatorname{conv}\Big\{\partial \boldsymbol{h}(\boldsymbol{x})^T \partial g\big(\boldsymbol{h}(\boldsymbol{x})\big)\Big\}. \tag{2.10}$$

There are several possibilities to achieve equality in (2.10) as the following theorem shows. Recall the set $\mathcal{I} = \{i : i = 1, \ldots, m\}$.

Theorem 2.11 *Let $f : \mathbb{R}^n \to \mathbb{R}$ be such that the assumptions of Theorem 2.10 are satisfied. Then*

(i) *if g is subdifferentially regular at $\boldsymbol{h}(\boldsymbol{x})$, each h_i is subdifferentially regular at $\boldsymbol{x}$ and for any $\boldsymbol{\alpha} \in \partial g\big(\boldsymbol{h}(\boldsymbol{x})\big)$ we have $\alpha_i \geq 0$ for all $i \in \mathcal{I}$, then f is subdifferentially regular at $\boldsymbol{x}$ and*

$$\partial f(\boldsymbol{x}) = \operatorname{conv}\Big\{\partial \boldsymbol{h}(\boldsymbol{x})^T \partial g\big(\boldsymbol{h}(\boldsymbol{x})\big)\Big\};$$

(ii) *if g is subdifferentially regular at $\boldsymbol{h}(\boldsymbol{x})$ and h_i is smooth at $\boldsymbol{x}$ for all $i \in \mathcal{I}$, then f is subdifferentially regular at $\boldsymbol{x}$ and*

$$\partial f(\boldsymbol{x}) = \nabla \boldsymbol{h}(\boldsymbol{x})^T \partial g\big(\boldsymbol{h}(\boldsymbol{x})\big); \text{ and}$$

(iii) *if $m = 1$ and g is smooth at $h(\boldsymbol{x})$, then f is subdifferentially regular at $\boldsymbol{x}$ and*

$$\partial f(\boldsymbol{x}) = g'\big(h(\boldsymbol{x})\big)\partial h(\boldsymbol{x}).$$

Based on the chain rule, we can prove the generalization of the classical differentiation rules for products and quotients of functions.

Theorem 2.12 (Products) *Let f_1 and f_2 be LLC at $\boldsymbol{x} \in \mathbb{R}^n$. Then the function $f_1 f_2$ is LLC at $\boldsymbol{x}$ and*

$$\partial(f_1 f_2)(\boldsymbol{x}) \subseteq f_2(\boldsymbol{x})\partial f_1(\boldsymbol{x}) + f_1(\boldsymbol{x})\partial f_2(\boldsymbol{x}). \tag{2.11}$$

In addition, if $f_1(\boldsymbol{x}), f_2(\boldsymbol{x}) \geq 0$ and f_1, f_2 are both subdifferentially regular at $\boldsymbol{x}$, then the function $f_1 f_2$ is subdifferentially regular at $\boldsymbol{x}$ and equality holds in (2.11).

Theorem 2.13 (Quotients) *Let f_1 and f_2 be LLC at $\boldsymbol{x} \in \mathbb{R}^n$ and $f_2(\boldsymbol{x}) \neq 0$. Then the function f_1/f_2 is LLC at $\boldsymbol{x}$ and*

$$\partial\Big(\frac{f_1}{f_2}\Big)(\boldsymbol{x}) \subseteq \frac{f_2(\boldsymbol{x})\partial f_1(\boldsymbol{x}) - f_1(\boldsymbol{x})\partial f_2(\boldsymbol{x})}{\big(f_2(\boldsymbol{x})\big)^2}. \tag{2.12}$$

In addition, if $f_1(\boldsymbol{x}) \geq 0$, $f_2(\boldsymbol{x}) > 0$ and f_1, f_2 are both subdifferentially regular at $\boldsymbol{x}$, then the function f_1/f_2 is subdifferentially regular at $\boldsymbol{x}$ and equality holds in (2.12).

Finally, we give a result for a class of functions which are frequently encountered in NSO and which, indeed, are used also in nonsmooth formulations of clustering problems, namely the max-functions. The problem of minimizing max-function is called the *minimax problem*.

Theorem 2.14 (Max-Function) *Let $f_i : \mathbb{R}^n \to \mathbb{R}$ be LLC at $\boldsymbol{x}$ for all $i \in \mathcal{I}$. Then the function*

$$f(\boldsymbol{x}) := \max\big\{f_i(\boldsymbol{x}) : i \in \mathcal{I}\big\}$$

is LLC at $\boldsymbol{x}$ and

$$\partial f(\boldsymbol{x}) \subseteq \operatorname{conv}\big\{\partial f_i(\boldsymbol{x}) : i \in \mathcal{I}(\boldsymbol{x})\big\}, \tag{2.13}$$

where

$$\mathcal{I}(\boldsymbol{x}) := \big\{i \in \mathcal{I} :\ f_i(\boldsymbol{x}) = f(\boldsymbol{x})\big\}.$$

In addition, if f_i is subdifferentially regular at $\boldsymbol{x}$ for all $i \in \mathcal{I}$, then f is also subdifferentially regular at $\boldsymbol{x}$ and equality holds in (2.13). Moreover,

$$f'(\boldsymbol{x}, \boldsymbol{d}) = \max_{i \in \mathcal{I}(\boldsymbol{x})} f_i'(\boldsymbol{x}, \boldsymbol{d}), \quad \boldsymbol{d} \in \mathbb{R}^n.$$

2.3.4 Quasidifferentials

As already mentioned, there are other generalizations of the subdifferential for nonconvex nonsmooth functions different from the Clarke subdifferential. In this subsection, we briefly recall the notion of quasidifferential. We omit the proofs since they can be found in [79].

Definition 2.7 A function $f : X \to \mathbb{R}$ is called *quasidifferentiable* at $\boldsymbol{x} \in X \subseteq \mathbb{R}^n$ if it is directionally differentiable at $\boldsymbol{x}$ and there exists a pair of compact convex sets $[\underline{\partial} f(\boldsymbol{x}), \overline{\partial} f(\boldsymbol{x})]$ such that the directional derivative $f'(\boldsymbol{x}; \boldsymbol{d})$ of the function f at $\boldsymbol{x}$ in the direction $\boldsymbol{d} \in \mathbb{R}^n$ can be represented in the form

$$f'(\boldsymbol{x}; \boldsymbol{d}) = \max\left\{\boldsymbol{\xi}^T \boldsymbol{d} \,:\, \boldsymbol{\xi} \in \underline{\partial} f(\boldsymbol{x})\right\} + \min\left\{\boldsymbol{v}^T \boldsymbol{d} \,:\, \boldsymbol{v} \in \overline{\partial} f(\boldsymbol{x})\right\}. \tag{2.14}$$

The set $\underline{\partial} f(\boldsymbol{x})$ is called the *subdifferential* of the function f at $\boldsymbol{x}$ and the $\overline{\partial} f(\boldsymbol{x})$ is called the *superdifferential* of the function f at $\boldsymbol{x}$. The pair $\mathcal{D} f(\boldsymbol{x}) = [\underline{\partial} f(\boldsymbol{x}), \overline{\partial} f(\boldsymbol{x})]$ is called the *quasidifferential* of the function f at $\boldsymbol{x}$.

The simplest examples of quasidifferentiable functions include a convex function f_1 and a concave function f_2 with the quasidifferentials given at a point $\boldsymbol{x} \in \mathbb{R}^n$

$$\begin{aligned} \mathcal{D} f_1(\boldsymbol{x}) &= \left[\partial f_1(\boldsymbol{x}), \{\boldsymbol{0}\}\right], \quad \text{and} \\ \mathcal{D} f_2(\boldsymbol{x}) &= \left[\{\boldsymbol{0}\}, \partial f_2(\boldsymbol{x})\right], \end{aligned}$$

respectively.

The quasidifferential mapping enjoys the full calculus in a sense that the equalities can be used instead of inclusions (cf. subdifferentially regular functions with Clarke subdifferential in Sect. 2.3.3). The following rules can be used to determine when the function $f : \mathbb{R}^n \to \mathbb{R}$ is quasidifferentiable at $\boldsymbol{x} \in \mathbb{R}^n$ and to compute quasidifferentials.

Theorem 2.15 (Sum and Product) *Let functions $f_1, f_2 : \mathbb{R}^n \to \mathbb{R}$ be quasidifferentiable at a point $\boldsymbol{x}$. Then*

(i) *the function $f = f_1 + f_2$ is quasidifferentiable at $\boldsymbol{x}$ and*

$$\mathcal{D} f(\boldsymbol{x}) = \mathcal{D} f_1(\boldsymbol{x}) + \mathcal{D} f_2(\boldsymbol{x}).$$

In other words, if $[\underline{\partial} f_1(\boldsymbol{x}), \overline{\partial} f_1(\boldsymbol{x})]$ and $[\underline{\partial} f_2(\boldsymbol{x}), \overline{\partial} f_2(\boldsymbol{x})]$ are quasidifferentials of the functions f_1 and f_2 at $\boldsymbol{x}$, respectively, then

$$\begin{aligned} \underline{\partial} f(\boldsymbol{x}) &= \underline{\partial} f_1(\boldsymbol{x}) + \underline{\partial} f_2(\boldsymbol{x}), \quad \textit{and} \\ \overline{\partial} f(\boldsymbol{x}) &= \overline{\partial} f_1(\boldsymbol{x}) + \overline{\partial} f_2(\boldsymbol{x}). \end{aligned}$$

(ii) *the function* $f = f_1 \cdot f_2$ *is quasidifferentiable at* $\boldsymbol{x}$, *and*

$$\mathcal{D}f(\boldsymbol{x}) = f_1(\boldsymbol{x})\mathcal{D}f_2(\boldsymbol{x}) + f_2(\boldsymbol{x})\mathcal{D}f_1(\boldsymbol{x}),$$

where

$$\underline{\partial} f(\boldsymbol{x}) = \begin{cases} f_1(\boldsymbol{x})\underline{\partial} f_2(\boldsymbol{x}) + f_2(\boldsymbol{x})\underline{\partial} f_1(\boldsymbol{x}), & \text{if} \quad f_1(\boldsymbol{x}) \geq 0,\ f_2(\boldsymbol{x}) \geq 0, \\ f_1(\boldsymbol{x})\overline{\partial} f_2(\boldsymbol{x}) + f_2(\boldsymbol{x})\underline{\partial} f_1(\boldsymbol{x}), & \text{if} \quad f_1(\boldsymbol{x}) \leq 0,\ f_2(\boldsymbol{x}) \geq 0, \\ f_1(\boldsymbol{x})\overline{\partial} f_2(\boldsymbol{x}) + f_2(\boldsymbol{x})\overline{\partial} f_1(\boldsymbol{x}), & \text{if} \quad f_1(\boldsymbol{x}) \leq 0,\ f_2(\boldsymbol{x}) \leq 0, \\ f_1(\boldsymbol{x})\underline{\partial} f_2(\boldsymbol{x}) + f_2(\boldsymbol{x})\overline{\partial} f_1(\boldsymbol{x}), & \text{if} \quad f_1(\boldsymbol{x}) \geq 0,\ f_2(\boldsymbol{x}) \leq 0, \end{cases}$$

and

$$\overline{\partial} f(\boldsymbol{x}) = \begin{cases} f_1(\boldsymbol{x})\overline{\partial} f_2(\boldsymbol{x}) + f_2(\boldsymbol{x})\overline{\partial} f_1(\boldsymbol{x}), & \text{if} \quad f_1(\boldsymbol{x}) \geq 0,\ f_2(\boldsymbol{x}) \geq 0, \\ f_1(\boldsymbol{x})\underline{\partial} f_2(\boldsymbol{x}) + f_2(\boldsymbol{x})\overline{\partial} f_1(\boldsymbol{x}), & \text{if} \quad f_1(\boldsymbol{x}) \leq 0,\ f_2(\boldsymbol{x}) \geq 0, \\ f_1(\boldsymbol{x})\underline{\partial} f_2(\boldsymbol{x}) + f_2(\boldsymbol{x})\underline{\partial} f_1(\boldsymbol{x}), & \text{if} \quad f_1(\boldsymbol{x}) \leq 0,\ f_2(\boldsymbol{x}) \leq 0, \\ f_1(\boldsymbol{x})\overline{\partial} f_2(\boldsymbol{x}) + f_2(\boldsymbol{x})\underline{\partial} f_1(\boldsymbol{x}), & \text{if} \quad f_1(\boldsymbol{x}) \geq 0,\ f_2(\boldsymbol{x}) \leq 0. \end{cases}$$

Theorem 2.16 (Max- and Min-Functions) *Let functions* $\varphi_i,\ i \in \mathcal{I}$ *be defined on an open set* $X \subset \mathbb{R}^n$, *be quasidifferentiable at a point* $\boldsymbol{x} \in X$ *and*

$$f_1(\boldsymbol{x}) = \max_{i \in \mathcal{I}} \varphi_i(\boldsymbol{x}), \quad \text{and}$$
$$f_2(\boldsymbol{x}) = \min_{i \in \mathcal{I}} \varphi_i(\boldsymbol{x}).$$

Then the functions f_1 *and* f_2 *are quasidifferentiable at* $\boldsymbol{x}$ *and*

$$\mathcal{D}f_1(\boldsymbol{x}) = \big[\underline{\partial} f_1(\boldsymbol{x}), \overline{\partial} f_1(\boldsymbol{x})\big], \quad \text{and}$$
$$\mathcal{D}f_2(\boldsymbol{x}) = \big[\underline{\partial} f_2(\boldsymbol{x}), \overline{\partial} f_2(\boldsymbol{x})\big].$$

Here

$$\underline{\partial} f_1(\boldsymbol{x}) = \operatorname{conv} \bigcup_{j \in \mathcal{M}(\boldsymbol{x})} \left(\underline{\partial}\varphi_j(\boldsymbol{x}) - \sum_{i \in \mathcal{M}(\boldsymbol{x}), i \neq j} \overline{\partial}\varphi_i(\boldsymbol{x}) \right),$$

$$\overline{\partial} f_1(\boldsymbol{x}) = \sum_{j \in \mathcal{M}(\boldsymbol{x})} \overline{\partial}\varphi_j(\boldsymbol{x}),$$

$$\underline{\partial} f_2(\boldsymbol{x}) = \sum_{j \in \mathcal{N}(\boldsymbol{x})} \underline{\partial}\varphi_j(\boldsymbol{x}), \quad \text{and}$$

$$\overline{\partial} f_2(\boldsymbol{x}) = \operatorname{conv} \bigcup_{j \in \mathcal{N}(\boldsymbol{x})} \left(\overline{\partial}\varphi_j(\boldsymbol{x}) - \sum_{i \in \mathcal{N}(\boldsymbol{x}), i \neq j} \underline{\partial}\varphi_i(\boldsymbol{x}) \right).$$

Here, $\left[\underline{\partial}\varphi_j(\boldsymbol{x}), \overline{\partial}\varphi_j(\boldsymbol{x})\right]$ *is a quasidifferential of the function* φ_j *at the point* $\boldsymbol{x}$*, and*

$$\mathcal{M}(\boldsymbol{x}) = \left\{i \in \mathcal{I} : \varphi_i(\boldsymbol{x}) = f_1(\boldsymbol{x})\right\}, \quad \textit{and}$$
$$\mathcal{N}(\boldsymbol{x}) = \left\{i \in \mathcal{I} : \varphi_i(\boldsymbol{x}) = f_2(\boldsymbol{x})\right\}.$$

By Applying Theorems 2.15 and 2.16 we obtain quasidifferentials for some important functions frequently encountered in NSO like DC functions (see Sect. 2.7) and functions represented as a sum of maximum or minimum functions.

It is worth of noting that the quasidifferential mapping is not uniquely defined: if $\mathcal{D}f(\boldsymbol{x}) = \left[\underline{\partial}f(\boldsymbol{x}), \overline{\partial}f(\boldsymbol{x})\right]$ is a quasidifferential of the function f at a point $\boldsymbol{x}$, then for any compact convex set $U \subset \mathbb{R}^n$ the pair $\mathcal{D}f(\boldsymbol{x}) = \left[\underline{\partial}f(\boldsymbol{x}) + U, \overline{\partial}f(\boldsymbol{x}) - U\right]$ is also a quasidifferential of f at $\boldsymbol{x}$.

For the relationship between the Clarke subdifferential and the quasidifferential we refer to [79].

2.4 Optimality Conditions

In this section, we first define global and local minima of functions. After that, we generalize the classical first order optimality conditions for unconstrained NSO problems using both the Clarke subdifferential and quasidifferential. At the end of this section, we define the notion of a descent direction and show how to find it for a LLC function.

We consider an unconstrained NSO problem of the form

$$\begin{cases} \text{minimize} & f(\boldsymbol{x}) \\ \text{subject to} & \boldsymbol{x} \in \mathbb{R}^n, \end{cases} \tag{2.15}$$

where the objective function $f : \mathbb{R}^n \to \mathbb{R}$ is LLC at $\boldsymbol{x}$ for all $\boldsymbol{x} \in \mathbb{R}^n$.

Definition 2.8 A point $\boldsymbol{x} \in \mathbb{R}^n$ is a *global minimizer* of f if

$$f(\boldsymbol{x}) \leq f(\boldsymbol{y}) \quad \text{for all} \quad \boldsymbol{y} \in \mathbb{R}^n.$$

In practice, this means that the global minimizer gives the smallest value—*global minimum*—for the problem (2.15) on the whole space $\mathbb{R}^n$.

Definition 2.9 A point $\boldsymbol{x} \in \mathbb{R}^n$ is a *local minimizer* of f if there exists $\varepsilon > 0$ such that

$$f(\boldsymbol{x}) \leq f(\boldsymbol{y}) \quad \text{for all} \quad \boldsymbol{y} \in B(\boldsymbol{x}; \varepsilon).$$

The local minimizer $\boldsymbol{x}$ is guaranteed to be the smallest value—*local minimum*—for the problem (2.15) in some neighborhood of the point $\boldsymbol{x}$.

The necessary conditions for a LLC function to attain its local minimum in an unconstrained case are given in the next theorem. For a convex function these conditions are also sufficient for global optimality.

Theorem 2.17 *Let $f : \mathbb{R}^n \to \mathbb{R}$ be a LLC function at $\boldsymbol{x} \in \mathbb{R}^n$. If $\boldsymbol{x}$ is the local minimizer of the function f, then*

(i) $f^\circ(\boldsymbol{x}; \boldsymbol{d}) \geq 0$ *for all $\boldsymbol{d} \in \mathbb{R}^n$; and*
(ii) $\boldsymbol{0} \in \partial f(\boldsymbol{x})$.

Theorem 2.18 *If $f : \mathbb{R}^n \to \mathbb{R}$ is a convex function, then the following conditions are equivalent:*

(i) *the function f attains its global minimum value at $\boldsymbol{x}$;*
(ii) $f'(\boldsymbol{x}; \boldsymbol{d}) \geq 0$ *for all $\boldsymbol{d} \in \mathbb{R}^n$; and*
(iii) $\boldsymbol{0} \in \partial f(\boldsymbol{x})$.

Definition 2.10 A point $\boldsymbol{x} \in \mathbb{R}^n$ satisfying the condition $\boldsymbol{0} \in \partial f(\boldsymbol{x})$ is called a *(Clarke) stationary point* of f.

The necessary optimality condition can be formulated also with the help of the Goldstein ε-subdifferential.

Theorem 2.19 *Let $f : \mathbb{R}^n \to \mathbb{R}$ be a LLC function at $\boldsymbol{x} \in \mathbb{R}^n$. If f attains its local minimum value at $\boldsymbol{x}$, then for all $\varepsilon \geq 0$ and $\boldsymbol{y} \in B(\boldsymbol{x}; \varepsilon)$ we have*

$$\boldsymbol{0} \in \partial_\varepsilon^G f(\boldsymbol{y}).$$

In addition, the quasidifferential (see Definition 2.7 in Sect. 2.3.4) can be used to formulate optimality conditions.

Theorem 2.20 *[79] Let $f : \mathbb{R}^n \to \mathbb{R}$ be a quasidifferentiable function. If a point $\boldsymbol{x} \in \mathbb{R}^n$ is the local minimizer of the function f, then*

$$-\overline{\partial} f(\boldsymbol{x}) \subseteq \underline{\partial} f(\boldsymbol{x}).$$

If, in addition, f is LLC at $\boldsymbol{x}$ and

$$-\overline{\partial} f(\boldsymbol{x}) \subset \operatorname{int} \underline{\partial} f(\boldsymbol{x}),$$

then the point $\boldsymbol{x}$ is a strict local minimizer *of f on $\mathbb{R}^n$.*

Unless the optimal solution of the problem (2.15) has been found, an essential part of most iterative optimization methods is to find a direction such that the objective function values decrease along that direction. We complete this section by defining a *descent direction* for an objective and show how to find it for a LLC function.

Definition 2.11 The direction $\boldsymbol{d} \in \mathbb{R}^n$ is called a *descent direction* for $f : \mathbb{R}^n \to \mathbb{R}$ at $\boldsymbol{x} \in \mathbb{R}^n$, if there exists $\varepsilon > 0$ such that for all $t \in (0, \varepsilon]$ we have

$$f(\boldsymbol{x} + t\boldsymbol{d}) < f(\boldsymbol{x}).$$

Theorem 2.21 *Let $f : \mathbb{R}^n \to \mathbb{R}$ be a LLC function at $\boldsymbol{x} \in \mathbb{R}^n$. The direction $\boldsymbol{d} \in \mathbb{R}^n$ is a descent direction for f if any of the following holds:*

(i) $f^\circ(\boldsymbol{x}; \boldsymbol{d}) < 0$;
(ii) $\boldsymbol{\xi}^T\boldsymbol{d} < 0$ *for all* $\boldsymbol{\xi} \in \partial f(\boldsymbol{x})$; *or*
(iii) $\boldsymbol{\xi}^T\boldsymbol{d} < 0$ *for all* $\boldsymbol{\xi} \in \partial_\varepsilon^G f(\boldsymbol{x})$.

As a consequence, from the above results we obtain that at a non-stationary point a descent direction always exists.

2.5 Discrete Gradient

In practice, the computation of subdifferential or even one subgradient is not always an easy task. In this section, we introduce the notion of *discrete gradient* for a LLC function [28]. A discrete gradient is an approximation of the subgradient at a given point and only function values are used to compute it.

Let $f : \mathbb{R}^n \to \mathbb{R}$ be a LLC function and

$$G = \left\{\boldsymbol{w} \in \mathbb{R}^n \ : \ \boldsymbol{w} = (w_1, \ldots, w_n), \ |w_j| = 1, \ j = 1, \ldots, n\right\}$$

be a set of all vertices of the unit hypercube in $\mathbb{R}^n$. Given $\boldsymbol{w} \in G$ and $\alpha \in (0, 1]$, define the sequence of n vectors

$$\boldsymbol{w}_j \equiv \boldsymbol{w}_j(\alpha) = \left(\alpha w_1, \alpha^2 w_2, \ldots, \alpha^j w_j, 0, \ldots, 0\right), \quad j = 1, \ldots, n.$$

Take any vector $\boldsymbol{g}$ from the sphere of the unit ball S_1, any vector $\boldsymbol{w} \in G$ and positive numbers $\lambda > 0, \ \alpha \in (0, 1]$. Consider the points

$$\boldsymbol{x}^0 = \boldsymbol{x} + \lambda \boldsymbol{g}, \quad \boldsymbol{x}^j = \boldsymbol{x}^0 + \lambda \boldsymbol{w}_j(\alpha), \quad j = 1, \ldots, n. \tag{2.16}$$

For $\boldsymbol{g} \in S_1$, compute

$$\bar{c} = \max\left\{|g_i|, \ i = 1, \ldots, n\right\},$$

and define the set

$$\mathcal{I}(\boldsymbol{g}) = \left\{i \in \{1, \ldots, n\} : \ |g_i| = \bar{c}\right\}. \tag{2.17}$$

It is clear that $|g_i| \geq 1/n$ for all $i \in \mathcal{I}(\boldsymbol{g})$.

Definition 2.12 The *discrete gradient* of the function $f : \mathbb{R}^n \to \mathbb{R}$ at the point $\boldsymbol{x} \in \mathbb{R}^n$ is the vector

$$\boldsymbol{\Gamma}^i(\boldsymbol{x}, \boldsymbol{g}, \boldsymbol{w}, \lambda, \alpha) = (\Gamma_1^i, \dots, \Gamma_n^i) \in \mathbb{R}^n, \quad \boldsymbol{g} \in S_1, \ i \in \mathcal{I}(\boldsymbol{g}),$$

with the following coordinates:

$$\Gamma_j^i = (\lambda \alpha^j w_j)^{-1}\big(f(\boldsymbol{x}^j) - f(\boldsymbol{x}^{j-1})\big), \quad j = 1, \dots, n, \ j \neq i, \quad \text{and}$$

$$\Gamma_i^i = (\lambda g_i)^{-1}\left(f(\boldsymbol{x} + \lambda \boldsymbol{g}) - f(\boldsymbol{x}) - \lambda \sum_{j=1, j\neq i}^{n} \Gamma_j^i g_j\right).$$

It follows from Definition 2.12 that

$$f(\boldsymbol{x} + \lambda \boldsymbol{g}) - f(\boldsymbol{x}) = \lambda \boldsymbol{\Gamma}^i(\boldsymbol{x}, \boldsymbol{g}, \boldsymbol{w}, \lambda, \alpha)^T \boldsymbol{g} \tag{2.18}$$

for all $\boldsymbol{g} \in S_1, \ \boldsymbol{w} \in G, \ \lambda > 0, \ \alpha > 0$.

For any $\boldsymbol{x} \in \mathbb{R}^n$, $\boldsymbol{g} \in S_1$, $i \in \mathcal{I}(\boldsymbol{g})$, $\boldsymbol{w} \in G$, $\lambda > 0$, and $\alpha > 0$ we have

$$\|\boldsymbol{\Gamma}^i\| \leq C(n)L, \quad \text{where} \quad C(n) = (n^2 + 2n^{3/2} - 2n^{1/2})^{1/2},$$

and L is a Lipschitz constant of the function f at $\boldsymbol{x}$. Furthermore, the set

$$D(\boldsymbol{x}, \lambda, \alpha) = \operatorname{cl}\operatorname{conv}\Big\{\boldsymbol{v} \in \mathbb{R}^n : \text{there exist } \boldsymbol{g} \in S_1, \ \boldsymbol{w} \in G \ \text{such that } \boldsymbol{v} = \boldsymbol{\Gamma}^i(\boldsymbol{x}, \boldsymbol{g}, \boldsymbol{w}, \lambda, \alpha)\Big\}$$

is an approximation to the subdifferential $\partial f(\boldsymbol{x})$ for a sufficiently small $\lambda > 0$ as stated in the following Theorem [28]:

Theorem 2.22 *Let $f : \mathbb{R}^n \to \mathbb{R}$ be a semismooth function at $\boldsymbol{x}$. For $\lambda > 0$ and $\boldsymbol{g} \in S_1$ define*

$$o(\lambda, \boldsymbol{g}) = f(\boldsymbol{x} + \lambda \boldsymbol{g}) - f(\boldsymbol{x}) - \lambda f'(\boldsymbol{x}; \boldsymbol{g}).$$

If $\lambda^{-1} o(\lambda, \boldsymbol{g}) \to 0$ uniformly with respect to $\boldsymbol{g} \in S_1$ as $\lambda \downarrow 0$, then for any $\varepsilon > 0$ there exists $\lambda_0 > 0, \ \alpha_0 \in (0, 1]$ such that

$$D(\boldsymbol{x}, \lambda, \alpha) \subset \partial f(\boldsymbol{x}) + B(\boldsymbol{0}; \varepsilon)$$

for all $\lambda \in (0, \lambda_0)$ and $\alpha \in (0, \alpha_0)$.

The previous theorem assumes only the semismoothness of the function f. As noted above, the class of semismooth functions is fairly broad. Thus, discrete gradients can be used to approximate subdifferentials of a broad class of nonsmooth functions.

The discrete gradient is defined with respect to a given direction $\boldsymbol{g} \in S_1$. In practice, we first define a sequence of points $\boldsymbol{x}^0, \dots, \boldsymbol{x}^n$ (see (2.16)) and compute the values of the function f at these points. This means that we need to compute $n+2$ values of f including the point $\boldsymbol{x}$. The ith coordinate of the discrete gradient is defined so that it satisfies the equality (2.18). This equality can be considered as a version of the mean-value theorem (see Theorem 2.9).

We complete this section by giving the necessary optimality condition using the set $D(\boldsymbol{x}, \lambda, \alpha)$, and by showing that, if this condition is not satisfied, then the set $D(\boldsymbol{x}, \lambda, \alpha)$ can be used to compute descent directions (see Definition 2.11) for the objective function.

Theorem 2.23 *Let $\boldsymbol{x}^* \in \mathbb{R}^n$ be a local minimizer of the function f. Then there exists $\lambda_0 > 0$ such that for all $\lambda \in (0, \lambda_0)$ and for all $\alpha \in (0, 1]$*

$$\mathbf{0} \in D(\boldsymbol{x}^*, \lambda, \alpha).$$

Theorem 2.24 *Let $\boldsymbol{x} \in \mathbb{R}^n$, $\lambda > 0$, $\alpha \in (0, 1]$, and $\mathbf{0} \notin D(\boldsymbol{x}, \lambda, \alpha)$. In other words,*

$$\|\boldsymbol{v}^0\| = \min\Big\{\|\boldsymbol{v}\| \ : \ \boldsymbol{v} \in D(\boldsymbol{x}, \lambda, \alpha)\Big\} > 0.$$

Then, $\boldsymbol{g}^0 = -\|\boldsymbol{v}^0\|^{-1}\boldsymbol{v}^0$ is a descent direction of the function f at $\boldsymbol{x}$.

2.6 Piecewise Partially Separable Functions

Many practical problems involve nonsmooth functions with large number of variables. The clustering problem is among such problems. Most of the large-scale optimization problems have a special structure, which can be exploited to design efficient algorithms. In this section, we discuss one of them: *piecewise partial separability* of nonsmooth functions. In particular, we show how to calculate the discrete gradient for a piecewise partially separable function. This information can be utilized with the discrete gradient method to be described in Sect. 3.8.

2.6.1 *Piecewise Partially Separable and Chained Functions*

Let f be a scalar function defined on an open set $\mathcal{X}_0 \subseteq \mathbb{R}^n$ containing a closed set $\mathcal{X} \subseteq \mathbb{R}^n$.

Definition 2.13 The function $f : \mathcal{X}_0 \to \mathbb{R}$ is called *partially separable* if there exists a family of $n \times n$ diagonal matrices U_l, $l = 1, \dots, M$ such that the function f can be represented as

$$f(\boldsymbol{x}) = \sum_{l=1}^{M} f_l(U_l \boldsymbol{x}). \tag{2.19}$$

If $M = n$ and $\text{diag}(U_l) = \boldsymbol{e}_l$ where $\boldsymbol{e}_l$ is the lth column of the identity matrix, then the function f is called *separable.*

Without loss of generality we assume that all entries of the matrices U_l are either 0 or 1. Although, any function f can be considered as a partially separable function if we take $M = 1$ and $U_1 = I$, where I is the identity matrix, we consider only the cases where $M > 1$ and m_l—the number of non-zero elements in the diagonal of U_l—is much smaller than n, $(l = 1, \dots, M)$. In other terms, the function f is called partially separable if it can be represented as the sum of functions of a much smaller number of variables.

Definition 2.14 The function f is said to be *piecewise (partially) separable* if there exists a finite family of closed sets $\mathcal{X}_1, \dots, \mathcal{X}_m$ such that

$$\bigcup_{j=1}^{m} \mathcal{X}_j = \mathcal{X},$$

and the function f is (partially) separable on each set $\mathcal{X}_j$, $j = 1, \dots, m$.

Simple examples of piecewise separable functions are piecewise linear functions and the maximum function $f(\boldsymbol{x}) = \max_{h=1,\dots,n} x_h^2$. In addition, an interesting and important subclass of partially separable functions is the so-called *chained functions.*

Definition 2.15 The function f is said to be *k-chained*, $k \leq n$, if it can be represented as

$$f(\boldsymbol{x}) = \sum_{l=1}^{n-k+1} f_l(x_l, \dots, x_{l+k-1}), \quad \boldsymbol{x} \in \mathbb{R}^n.$$

For instance, if $k = 2$, the function f is

$$f(\boldsymbol{x}) = \sum_{l=1}^{n-1} f_l(x_l, x_{l+1}).$$

Definition 2.16 The function f is said to be *piecewise k-chained* if there exists a finite family of closed sets $\mathcal{X}_1, \dots, \mathcal{X}_m$ such that

$$\bigcup_{j=1}^{m} \mathcal{X}_j = \mathcal{X},$$

and the function f is k-chained on each set $\mathcal{X}_j,\ j = 1, \ldots, m$.

The relationship between various chained and (piecewise) separable functions are listed below [25]:

- a separable function is 1-chained;
- a piecewise separable function is piecewise 1-chained;
- any k-chained function is partially separable;
- any piecewise k-chained function is piecewise partially separable;
- all separable functions are piecewise separable with $m = 1$;
- all partially separable functions are piecewise partially separable with $m = 1$; and
- any piecewise separable function is piecewise partially separable.

2.6.2 Properties of Piecewise Partially Separable Functions

Some interesting properties of piecewise partially separable functions are as follows (see [25], for more details):

Theorem 2.25 (Scalar Multiplication, Sum, Max- and Min-Functions) *Let g and h be piecewise partially separable (piecewise k-chained, piecewise separable) continuous functions on the closed set $\mathcal{X}$. Then*

(i) $f(\boldsymbol{x}) = \alpha g(\boldsymbol{x}),\ \alpha \in \mathbb{R}$, *is piecewise partially separable (piecewise k-chained, piecewise separable);*
(ii) $f(\boldsymbol{x}) = g(\boldsymbol{x}) + h(\boldsymbol{x})$ *is piecewise partially separable (piecewise k-chained, piecewise separable); and*
(iii) $f(\boldsymbol{x}) = \max\{g(\boldsymbol{x}), h(\boldsymbol{x})\}$, $f(\boldsymbol{x}) = \min\{g(\boldsymbol{x}), h(\boldsymbol{x})\}$, *and* $f(\boldsymbol{x}) = |g(\boldsymbol{x})|$ *are piecewise partially separable (piecewise k-chained, piecewise separable).*

Next, we describe briefly the Lipschitz continuity and the directional differentiability of piecewise partially separable functions. Let the function f be partially separable and functions $f_l,\ l = 1, \ldots, M$ in (2.19) be directionally differentiable. Then, the function f is also directionally differentiable and

$$f'(\boldsymbol{x}; \boldsymbol{d}) = \sum_{l=1}^{M} f_l'(U_l\boldsymbol{x}; U_l\boldsymbol{d}), \quad \boldsymbol{x}, \boldsymbol{d} \in \mathbb{R}^n. \tag{2.20}$$

Let f be a continuous and piecewise partially separable function on the closed convex set $\mathcal{X} \subset \mathbb{R}^n$. In addition, let each function f_l be LLC on $\mathcal{X}_j,\ j = 1, \ldots, m$. Then the function f is also LLC on $\mathcal{X}$ and, therefore, also Clarke subdifferentiable (see Definition 2.4). Nevertheless, in general, piecewise partially separable functions are not subdifferentially regular (see Definition 2.6) and, thus, computation of the subdifferential or even one subgradient of such functions may not be an easy task.

2.6.3 Calculation of Discrete Gradients

As noted before, for a general nonsmooth function f we need to compute its values at $n+2$ points in order to compute one discrete gradient (see Definition 2.12). Nevertheless, for nonsmooth piecewise partially separable functions the computation of discrete gradients can be significantly simplified.

Let us consider the function

$$f(\boldsymbol{x}) = \sum_{l=1}^{M} \max_{j \in \mathcal{J}_l} \min_{k \in \mathcal{K}_j} f_{ljk}(\boldsymbol{x}), \tag{2.21}$$

where functions $f_{ljk},\ k \in \mathcal{K}_j,\ j \in \mathcal{J}_l,\ l = 1, \ldots, M$, are partially separable (see Definition 2.13) and $\mathcal{K}_j, \mathcal{J}_l$ are non-empty finite index sets. This means that there exists a family of $n \times n$ diagonal matrices U_{ljk}^t with $t = 1, \ldots, M_{ljk}$ such that

$$f_{ljk}(\boldsymbol{x}) = \sum_{t=1}^{M_{ljk}} f_{ljk}^t(U_{ljk}^t \boldsymbol{x}), \quad k \in \mathcal{K}_j,\ j \in \mathcal{J}_l,\ l = 1, \ldots, M.$$

Here, functions f_{ljk}^t are called *term functions*. By Definition 2.14, the function f is piecewise partially separable. In addition, if all functions f_{ljk} are r-chained (separable), then the function f is piecewise r-chained (piecewise separable) (see Definitions 2.14 and 2.16).

The total number of term functions is

$$N_0 = \sum_{l=1}^{M} \sum_{j \in \mathcal{J}_l} \sum_{k \in \mathcal{K}_j} M_{ljk},$$

which means that for one evaluation of the function f we have to compute term functions N_0 times. In addition, for each evaluation of the discrete gradient we need to compute $n+2$ times the function f. Therefore, the total number of computation of term functions for one evaluation of the discrete gradient is

$$N_t = (n+2)N_0.$$

This number can be significantly decreased if one exploits the piecewise partial separability of the function f. In order to compute the discrete gradient of f at a point $\boldsymbol{x} \in \mathbb{R}^n$ with respect to any direction $\boldsymbol{g} \in S_1$ (where as before S_1 is the sphere of the unit ball), we first define the sequence (cf. the equation (2.16))

$$\boldsymbol{x}^0, \ldots, \boldsymbol{x}^{i-1}, \boldsymbol{x}^{i+1}, \ldots, \boldsymbol{x}^n, \quad i \in \mathcal{I}(\boldsymbol{g}),$$

where $\mathcal{I}(\boldsymbol{g})$ is defined in (2.17). For all $q \in \{1, \dots, n\}$, $q \neq i$, the point $\boldsymbol{x}^q$ differs from $\boldsymbol{x}^{q-1}$ by one coordinate only.

At the point $\boldsymbol{x}^q$, $q \in \{1, \dots, n\}$, $q \neq i$, the function f_{ljk} can be calculated using the simplified scheme

$$f_{ljk}(\boldsymbol{x}^q) = \sum_{t \in \mathcal{Q}^q_{ljk}} f^t_{ljk}(U^t_{ljk}\boldsymbol{x}^q) + \sum_{t \in \bar{\mathcal{Q}}^q_{ljk}} f^t_{ljk}(U^t_{ljk}\boldsymbol{x}^{q-1}), \tag{2.22}$$

where

$$\mathcal{Q}^q_{ljk} = \Big\{t \in \{1, \dots, M_{ljk}\} \ : \ (U^t_{ljk})_{qq} = 1\Big\}, \quad \text{and}$$
$$\bar{\mathcal{Q}}^q_{ljk} = \Big\{t \in \{1, \dots, M_{ljk}\} \ : \ (U^t_{ljk})_{qq} = 0\Big\}.$$

This means that we compute only functions f^t_{ljk}, $t \in \mathcal{Q}^q_{ljk}$ at the point $\boldsymbol{x}^q$ and all other functions remain the same as at the point $\boldsymbol{x}^{q-1}$. Thus, in order to calculate the function f at the point $\boldsymbol{x}^q$, the term functions need to be computed only

$$N_s = \sum_{l=1}^{M} \sum_{j \in \mathcal{J}_l} \sum_{k \in \mathcal{K}_j} |\mathcal{Q}^q_{ljk}|$$

times at this point. Since $M_{ljk} = |\mathcal{Q}^q_{ljk}| + |\bar{\mathcal{Q}}^q_{ljk}|$ and usually $|\mathcal{Q}^q_{ljk}| \ll |\bar{\mathcal{Q}}^q_{ljk}|$, we can expect that $N_s \ll N_0$. For example, if all functions f_{ljk} are r-chained, then $|\mathcal{Q}^q_{ljk}| \leq r$ and $|\bar{\mathcal{Q}}^q_{ljk}| \geq n - r - 1$ and, if the functions are separable, then $|\mathcal{Q}^q_{ljk}| = 1$ and $|\bar{\mathcal{Q}}^q_{ljk}| = n - 1$.

Now, to compute a discrete gradient at the point $\boldsymbol{x} \in \mathbb{R}^n$ with respect to the direction $\boldsymbol{g} \in S_1$ we need to compute the function f at the points $\boldsymbol{x}$ and $\boldsymbol{x} + \lambda \boldsymbol{g}$ using the formula (2.21). At all the other points $\boldsymbol{x}^q$, $q = 1, \dots, n$ we can use the simplified scheme (2.22). Therefore, the total number of computation of term functions is

$$N_{ts} = 2N_0 + nN_s,$$

which is significantly less than N_t when n is large.

2.7 DC Optimization

In this section, we consider a broad subclass of nonconvex NSO: the *DC problems*. This type of problems frequently arise in practical applications. Notably, the clustering problem can be represented as a nonsmooth DC problem (see Sect. 4.4).

A function $f : \mathbb{R}^n \to \mathbb{R}$ is called a *DC function*, if there exist two convex functions $f_1, f_2 : \mathbb{R}^n \to \mathbb{R}$ such that

$$f(\boldsymbol{x}) = f_1(\boldsymbol{x}) - f_2(\boldsymbol{x}).$$

Here, the functions f_1 and f_2 are called *DC components* and the difference $f_1 - f_2$ is a *DC decomposition* of f. DC functions, defined on $\mathbb{R}^n$, are always LLC and they may be nonsmooth. Particularly, if f is nonsmooth then at least one of the DC components is also nonsmooth. In addition, DC functions are typically nonconvex. Nevertheless, DC functions have the structure that can be used to separate the convex (f_1) and the concave $(-f_2)$ behavior of the function f.

DC functions preserve the DC structure under some simple operations frequently used in optimization [284]:

Theorem 2.26 *Assume that* $f^i = f_1^i - f_2^i$, $i \in \mathcal{I}$, *where* $\mathcal{I} = \{i : i = 1, \ldots, m\}$ *are DC functions. Then*

(i) $g(\boldsymbol{x}) = \sum_{i \in \mathcal{I}} c_i f^i(\boldsymbol{x})$ *is a DC function with any* $c_i \in \mathbb{R}$*;*

(ii) $g(\boldsymbol{x}) = \prod_{i \in \mathcal{I}} f^i(\boldsymbol{x})$ *and* $h(\boldsymbol{x}) = f_1(\boldsymbol{x})/f_2(\boldsymbol{x})$, $f_2(\boldsymbol{x}) \not\equiv 0$ *are DC functions;*

(iii) $g(\boldsymbol{x}) = |f^i(\boldsymbol{x})|$ *is a DC function for* $i \in \mathcal{I}$*;*

(iv) $g(\boldsymbol{x}) = \min\{f^i(\boldsymbol{x}) : i \in \mathcal{I}\}$ *is a DC function. One of its DC decomposition* $g = g_1 - g_2$ *is given by*

$$g_1(\boldsymbol{x}) = \sum_{i \in \mathcal{I}} f_1^i(\boldsymbol{x}), \quad \textit{and}$$

$$g_2(\boldsymbol{x}) = \max_{i \in \mathcal{I}} \left\{ f_2^i(\boldsymbol{x}) + \sum_{j \in \mathcal{I},\ j \neq i} f_1^j(\boldsymbol{x}) \right\};$$

(v) $h(\boldsymbol{x}) = \max\{f^i(\boldsymbol{x}) : i \in \mathcal{I}\}$ *is a DC function. One of its DC decomposition* $h = h_1 - h_2$ *is given by*

$$h_1(\boldsymbol{x}) = \max_{i \in \mathcal{I}} \left\{ f_1^i(\boldsymbol{x}) + \sum_{j \in \mathcal{I},\ j \neq i} f_2^j(\boldsymbol{x}) \right\}, \quad \textit{and}$$

$$h_2(\boldsymbol{x}) = \sum_{i \in \mathcal{I}} f_2^i(\boldsymbol{x}).$$

Note that the DC representation of a function is not unique: from one DC representation we can easily construct new ones by adding any convex function to both DC components f_1 and f_2.

The class of DC functions is very broad: any convex or concave function as well as every twice continuously differentiable function is a DC function. In addition, as shown above simple operations can be used to obtain more complex DC structures from simple ones. Finally, every continuous function can be approximated by a DC function with any given precision [284].

An *unconstrained DC optimization problem* is formulated as

$$\begin{cases} \text{minimize} & f(\boldsymbol{x}) = f_1(\boldsymbol{x}) - f_2(\boldsymbol{x}) \\ \text{subject to} & \boldsymbol{x} \in \mathbb{R}^n. \end{cases} \tag{2.23}$$

Next, we present *necessary conditions for local optimality* of this problem.

Theorem 2.27 *Let f_1 and f_2 be convex functions. If $\boldsymbol{x}^* \in \mathbb{R}^n$ is a local minimizer of $f = f_1 - f_2$, then*

$$\partial f_2(\boldsymbol{x}^*) \subseteq \partial f_1(\boldsymbol{x}^*), \tag{2.24}$$

$$\mathbf{0} \in \partial f(\boldsymbol{x}^*), \quad \textit{and} \tag{2.25}$$

$$\partial f_1(\boldsymbol{x}^*) \cap \partial f_2(\boldsymbol{x}^*) \neq \emptyset. \tag{2.26}$$

Points satisfying (2.24) are called inf-*stationary*. This condition guarantees local optimality of $\boldsymbol{x}^*$ if f_2 is a polyhedral convex function of the form

$$f_2(\boldsymbol{x}) = \max_{i=1,\dots,m} \left\{ \boldsymbol{c}_i^T \boldsymbol{x} - b_i \right\},$$

where $\boldsymbol{c}_i \in \mathbb{R}^n$ and $b_i \in \mathbb{R}$. Nevertheless, the condition (2.24) is hard to be verified in practice since we usually do not know the whole subdifferentials of DC components f_1 and f_2.

Points satisfying (2.25) are called *Clarke stationary* (see Definition 2.10). This condition is often utilized in general nonconvex NSO. However, DC functions are not, in general, subdifferentially regular (see Definition 2.6). This means that for the subdifferential of f we have

$$\partial f(\boldsymbol{x}^*) \subseteq \partial f_1(\boldsymbol{x}^*) - \partial f_2(\boldsymbol{x}^*), \tag{2.27}$$

and therefore, the Clarke stationarity, in general, is difficult to verify using only values and subgradients of DC components.

The condition (2.26) is called *criticality*, and the points satisfying this condition are called *critical points*. The condition (2.26) is a relaxation of inf-stationarity, and it is the most commonly used optimality condition in DC optimization due to the fact that it can be easily verified. However, the major drawback is that a critical point needs not to be a local optimum or even a saddle point of the problem (2.23) [114, 161].

There exist interesting relationships between the optimality conditions (2.24), (2.25) and (2.26) (see [114]): inf-stationarity always implies Clarke stationarity and a Clarke stationarity point is always a critical point. Nevertheless, the opposite is true only under some additional assumptions. For example, if the function f_2 is differentiable at a critical point $\boldsymbol{x}^* \in \mathbb{R}^n$, then we have

$$\partial f_2(\boldsymbol{x}^*) = \{\nabla f_2(\boldsymbol{x}^*)\} \subseteq \partial f_1(\boldsymbol{x}^*), \quad \text{and}$$
$$\boldsymbol{0} \in \partial f(\boldsymbol{x}^*) = \partial f_1(\boldsymbol{x}^*) - \partial f_2(\boldsymbol{x}^*),$$

indicating also the inf-stationarity and Clarke stationarity of the point $\boldsymbol{x}^*$. If only the first DC component f_1 is differentiable at a critical point $\boldsymbol{x}^* \in \mathbb{R}^n$, then the equality holds in (2.27) and we get

$$\boldsymbol{0} \in \partial f(\boldsymbol{x}^*) = \partial f_1(\boldsymbol{x}^*) - \partial f_2(\boldsymbol{x}^*) = \{\nabla f_1(\boldsymbol{x}^*)\} - \partial f_2(\boldsymbol{x}^*),$$

which means that the Clarke stationarity is achieved. However, $\partial f_2(\boldsymbol{x}^*)$ may contain more than one subgradient and cannot be a subset of $\partial f_1(\boldsymbol{x}^*) = \{\nabla f_1(\boldsymbol{x}^*)\}$. In this case, the point $\boldsymbol{x}^*$ is not inf-stationary. The relationships between the sets of different stationary points are summarized in Fig. 2.9.

We complete this section by noting that a DC function $f = f_1 - f_2$ is quasidifferentiable and its quasidifferential at a point $\boldsymbol{x} \in \mathbb{R}^n$ is

$$\mathcal{D}f(\boldsymbol{x}) = [\partial f_1(\boldsymbol{x}), -\partial f_2(\boldsymbol{x})].$$

It is easy to notice that for unconstrained DC optimization problems the inf-stationarity follows from the necessary optimality condition using the quasidifferential (see Theorem 2.20).

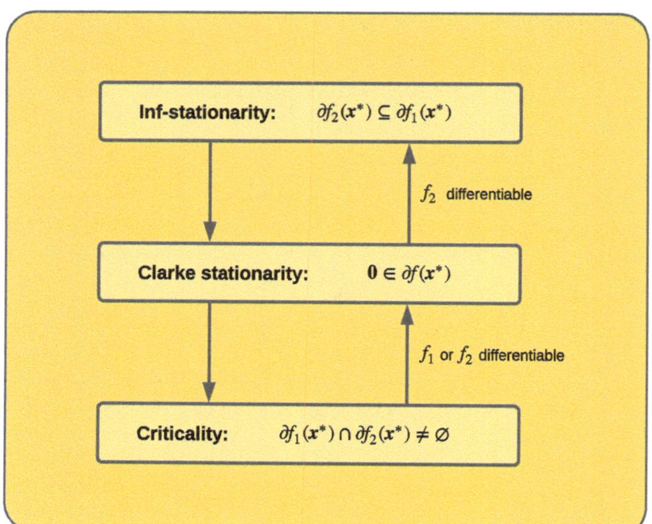

Fig. 2.9 Relationships between sets of different stationary points

2.8 Smoothing of Nonsmooth Functions

Smoothing techniques have been used in NSO since 1970s. The use of such techniques allows one to apply powerful smooth optimization algorithms for solving NSO problems. Most smoothing techniques can only be applied when the objective and/or constraint functions have some specific structures: for instance, functions represented as

- a maximum of the finite number of smooth functions;
- a minimum of the finite number of smooth functions;
- a maximum of minimum of the finite number of smooth functions; and
- a sum of maxima of minima of the finite number of smooth functions.

There are also some smoothing methods which do not require such structures. For instance, the smoothing method introduced in [227] is applicable to general convex functions and the so-called Steklov smoothing method (see for example, [98]) can be applied to any LLC function. However, the computation of smoothing function in this method is very time-consuming.

Since the objective functions in the clustering problems are represented as a sum of maxima of minima of simple smooth functions and, consequently, have a specific structure we apply smoothing techniques to exploit such a structure. In this section, first we study smoothing of functions represented as a maximum of the finite number of continuously differentiable functions. More specifically, we consider the *finite minimax problem*

$$\begin{cases} \text{minimize} & f(\boldsymbol{x}) \\ \text{subject to} & \boldsymbol{x} \in \mathbb{R}^n, \end{cases} \tag{2.28}$$

where

$$f(\boldsymbol{x}) = \max_{i \in \mathcal{I}} f_i(\boldsymbol{x}), \tag{2.29}$$

and the functions $f_i,\ i \in \mathcal{I}$ are smooth. Note that this type of problems are frequently encountered in practical applications. By Definition 2.6 and Theorem 2.14 the objective function f in the problem (2.28) is subdifferentially regular and its subdifferential at a point $\boldsymbol{x} \in \mathbb{R}^n$ can be expressed as

$$\partial f(\boldsymbol{x}) = \operatorname{conv}\left\{\nabla f_i(\boldsymbol{x}) : i \in \mathcal{I}(\boldsymbol{x})\right\},$$

where $\mathcal{I}(\boldsymbol{x}) = \left\{i \in \mathcal{I} : f_i(\boldsymbol{x}) = f(\boldsymbol{x})\right\}$.

Smoothing techniques, without loss of generality, can be divided into two classes: *local* and *global smoothing techniques*. Local smoothing techniques try to smooth the objective function f given in (2.28) in some neighborhood of the so-called kink points (points where the function is not differentiable). Such techniques have been considered, for example, in [38, 307, 310].

Smoothing techniques from the second class approximate the objective function f globally, that is, in the whole domain of the function. This class includes, in particular, the exponential [302, 305] and the hyperbolic smoothing [30, 299–301] techniques.

We will concentrate on the hyperbolic smoothing technique and will start by introducing it for a simple maximum function. In order to apply this technique to more general minimax problems (2.28), we reformulate the objective function (2.29) in the minimax problem and establish the relationship between the original minimax and the reformulated problems. We describe the main properties of the hyperbolic smoothing function. Based on these properties, an algorithm—a hyperbolic smoothing method—for solving the finite minimax problem (2.28) is given in Sect. 3.9 and the method is applied to design a clustering algorithm in Sect. 8.6.

2.8.1 *Hyperbolic Smoothing of a Simple Maximum Function*

We start with the definition of the hyperbolic smoothing function [299–301]. Let us first consider the following simple maximum function:

$$\varphi(x) = \max\{0, x\}, \quad x \in \mathbb{R}. \tag{2.30}$$

The *hyperbolic smoothing function* approximating the function (2.30) is defined as

$$\phi_\tau(x) = \frac{x + \sqrt{x^2 + \tau^2}}{2}. \tag{2.31}$$

Here, $\tau > 0$ is called a *precision* or *smoothing parameter*. The function ϕ_τ is an increasing convex C^∞-function and

$$\varphi(x) < \phi_\tau(x) \le \varphi(x) + \frac{\tau}{2}$$

for all $\tau > 0$ and $x \in \mathbb{R}$. The hyperbolic function for smoothing the function (2.30) is illustrated in Fig. 2.10, where the blue curve shows the smoothing function.

2.8.2 *Reformulation of Minimax Problem*

In order to apply the hyperbolic smoothing (2.31) to the finite maximum function (2.29), we represent it as a sum of the maximum of two functions by adding a new auxiliary variable $t \in \mathbb{R}$. Consider the function

$$F(\boldsymbol{x}, t) = t + \sum_{i \in \mathcal{I}} \max\big\{0, f_i(\boldsymbol{x}) - t\big\}. \tag{2.32}$$

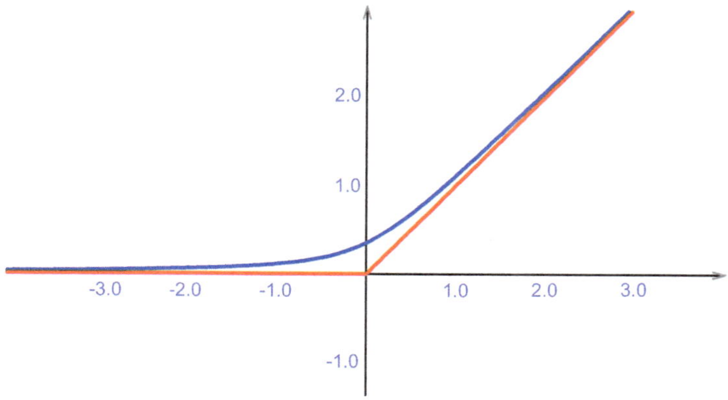

Fig. 2.10 Hyperbolic smoothing of the function (2.30)

It is clear that $f(\boldsymbol{x}) = F\big(\boldsymbol{x}, f(\boldsymbol{x})\big)$. In addition, we have

$$f(\boldsymbol{x}) = \min_{t \in \mathbb{R}} F(\boldsymbol{x}, t).$$

Let us denote by

$$\Psi_i(\boldsymbol{x}, t) = \max\big\{0, f_i(\boldsymbol{x}) - t\big\}, \quad i \in \mathcal{I}.$$

For a given point $(\boldsymbol{x}, t) \in \mathbb{R}^{n+1}$, the index set $\mathcal{I}$ can be represented as

$$\mathcal{I} = \mathcal{I}_1 \cup \mathcal{I}_2 \cup \mathcal{I}_3,$$

where

$$\begin{aligned}
\mathcal{I}_1 &\equiv \mathcal{I}_1(\boldsymbol{x}, t) = \big\{i \in \mathcal{I} : f_i(\boldsymbol{x}) - t < 0\big\}, \\
\mathcal{I}_2 &\equiv \mathcal{I}_2(\boldsymbol{x}, t) = \big\{i \in \mathcal{I} : f_i(\boldsymbol{x}) - t = 0\big\}, \quad \text{and} \\
\mathcal{I}_3 &\equiv \mathcal{I}_3(\boldsymbol{x}, t) = \big\{i \in \mathcal{I} : f_i(\boldsymbol{x}) - t > 0\big\}.
\end{aligned}$$

Since functions f_i, $i \in \mathcal{I}$ are smooth it follows from Theorem 2.14 that functions Ψ_i are subdifferentially regular, and the subdifferential $\partial \Psi_i(\boldsymbol{x}, t)$, $(\boldsymbol{x}, t) \in \mathbb{R}^{n+1}$ is given by

$$\partial \Psi_i(\boldsymbol{x}, t) = \begin{cases} \mathbf{0}_{n+1}, & \text{if } i \in \mathcal{I}_1, \\ \operatorname{conv}\big\{\mathbf{0}_{n+1}, (\nabla f_i(\boldsymbol{x}), -1)\big\}, & \text{if } i \in \mathcal{I}_2, \\ \big\{(\nabla f_i(\boldsymbol{x}), -1)\big\}, & \text{if } i \in \mathcal{I}_3, \end{cases}$$

where $\mathbf{0}_{n+1}$ is used to denote $(n+1)$-dimensional vector of zeros. In addition, applying Theorem 2.8 the subdifferential of the function F at the point $(\boldsymbol{x}, t)$ can be expressed as

$$\partial F(\boldsymbol{x}, t) = \big\{(\mathbf{0}_n, 1)\big\} + \sum_{i \in \mathcal{I}_1} \mathbf{0}_{n+1}$$
$$+ \sum_{i \in \mathcal{I}_2} \operatorname{conv}\big\{\mathbf{0}_{n+1}, (\nabla f_i(\boldsymbol{x}), -1)\big\} + \sum_{i \in \mathcal{I}_3} \big\{(\nabla f_i(\boldsymbol{x}), -1)\big\}.$$

The following propositions establish the relationship between functions f and F. Their proofs can be found in [30].

Proposition 2.1 (Stationary Points) *Assume that*

(i) *a point $\boldsymbol{x}^* \in \mathbb{R}^n$ is a stationary point of f. Then $(\boldsymbol{x}^*, t^*) \in \mathbb{R}^{n+1}$ is a stationary point of the function F where $t^* = f(\boldsymbol{x}^*)$; and*
(ii) *a point $(\boldsymbol{x}^*, t^*) \in \mathbb{R}^{n+1}$ is a stationary point of the function F. Then $\boldsymbol{x}^* \in \mathbb{R}^n$ is a stationary point of f.*

Proposition 2.2 (Local Minimizer) *Assume that*

(i) *a point $\boldsymbol{x}^* \in \mathbb{R}^n$ is a local minimizer of f. Then $(\boldsymbol{x}^*, t^*) \in \mathbb{R}^{n+1}$ is a local minimizer of the function F where $t^* = f(\boldsymbol{x}^*)$; and*
(ii) *a point $(\boldsymbol{x}^*, t^*) \in \mathbb{R}^{n+1}$ is a local minimizer of the function F. Then $\boldsymbol{x}^* \in \mathbb{R}^n$ is a local minimizer of f.*

In addition, the values of global minima of functions f and F are equal [307].

2.8.3 *Hyperbolic Smoothing of the Maximum Function*

Applying (2.31), we obtain the following hyperbolic smoothing of the function (2.32):

$$\Phi_\tau(\boldsymbol{x}, t) = t + \frac{1}{2} \sum_{i \in \mathcal{I}} \Big(f_i(\boldsymbol{x}) - t + \sqrt{(f_i(\boldsymbol{x}) - t)^2 + \tau^2} \Big), \quad \tau > 0. \tag{2.33}$$

Obviously, $\Phi_\tau(\boldsymbol{x}, t)$ is smooth and for any $\boldsymbol{x} \in \mathbb{R}^n$, $t \in \mathbb{R}$, and $\tau > 0$ we have

$$0 < \Phi_\tau(\boldsymbol{x}, t) - F(\boldsymbol{x}, t) \le \frac{m\tau}{2}.$$

The gradient of the function Φ_τ is given by

$$\nabla \Phi_\tau(\boldsymbol{x}, t) = \big(G_\tau^1(\boldsymbol{x}, t), G_\tau^2(\boldsymbol{x}, t)\big),$$

where

$$G^1_\tau(\boldsymbol{x}, t) = \frac{1}{2}\sum_{i\in\mathcal{I}}\Big(1 + \beta_{i\tau}(\boldsymbol{x}, t)\Big)\nabla f_i(\boldsymbol{x}),$$

$$G^2_\tau(\boldsymbol{x}, t) = 1 - \frac{1}{2}|\mathcal{I}| - \frac{1}{2}\sum_{i\in\mathcal{I}}\beta_{i\tau}(\boldsymbol{x}, t), \quad \text{and}$$

$$\beta_{i\tau}(\boldsymbol{x}, t) = \frac{f_i(\boldsymbol{x}) - t}{\sqrt{(f_i(\boldsymbol{x}) - t)^2 + \tau^2}}.$$

Proposition 2.3 *Assume that*

$$\boldsymbol{v} = \lim_{\tau\to 0}\nabla\Phi_\tau(\boldsymbol{x}, t).$$

Then $\boldsymbol{v} \in \partial F(\boldsymbol{x}, t)$.

Proposition 2.4 *Assume that sequences* $\{\boldsymbol{x}_h\}$, $\{t_h\}$ *and* $\{\tau_h\}$ *are given such that* $\boldsymbol{x}_h \in \mathbb{R}^n$, $t_h \in \mathbb{R}$, $t_h \geq f(\boldsymbol{x}_h)$ *and* $\tau_h > 0$, $h = 1, 2, \ldots$. *In addition, assume that* $\boldsymbol{x}_h \to \boldsymbol{x}$, $t_h \to t$, $\tau_h \to 0$ *as* $h \to \infty$ *and*

$$\boldsymbol{v} = \lim_{h\to\infty}\nabla\Phi_{\tau_h}(\boldsymbol{x}_h, t_h).$$

Then $\boldsymbol{v} \in \partial F(\boldsymbol{x}, t)$.

Proposition 2.5 *Suppose that functions* f_i, $i \in \mathcal{I}$ *are smooth and their gradients* ∇f_i *are LLC. Then the gradient* $\nabla\Phi_\tau$ *is also LLC for any given* $\tau > 0$.

2.8.4 Hyperbolic Smoothing of the Minimum Function

Hyperbolic smoothing for minimum functions can be defined similarly to that for maximum functions. For the function

$$\bar{\varphi}(x) = \min\{0, x\}, \quad x \in \mathbb{R},$$

we have

$$\bar{\varphi}(x) = -\max\{0, -x\},$$

and therefore, the hyperbolic smoothing function $\bar{\phi}_\tau$ for $\bar{\varphi}$ is given by

$$\bar{\phi}_\tau(x) = \frac{x - \sqrt{x^2 + \tau^2}}{2}. \tag{2.34}$$

Now consider the minimum function

$$\bar{f}(\boldsymbol{x}) = \min_{j \in \mathcal{J}} \bar{f}_j(\boldsymbol{x}), \quad \mathcal{J} = \{1, \ldots, q\},$$

where functions $\bar{f}_j, \ j \in \mathcal{J}$ are smooth. Then

$$\bar{f}(\boldsymbol{x}) = -\max_{j \in \mathcal{J}} \big(- \bar{f}_j(\boldsymbol{x}) \big).$$

Applying (2.31) we define the hyperbolic smoothing function $\bar{\Phi}_\tau$ with $\tau > 0$ for the function $\bar{f}$ as

$$\bar{\Phi}_\tau(\boldsymbol{x}, t) = t + \frac{1}{2} \sum_{j \in \mathcal{J}} \Big(\bar{f}_j(\boldsymbol{x}) - t - \sqrt{(\bar{f}_j(\boldsymbol{x}) - t)^2 + \tau^2} \Big).$$

Chapter 3
Nonsmooth Optimization Methods

3.1 Introduction

Consider the following unconstrained minimization problem

$$\begin{cases} \text{minimize} & f(\boldsymbol{x}) \\ \text{subject to} & \boldsymbol{x} \in \mathbb{R}^n. \end{cases} \tag{3.1}$$

Here, the objective function $f : \mathbb{R}^n \to \mathbb{R}$ is, in general, LLC. The basic scheme of various iterative (smooth) optimization algorithms for solving the problem (3.1) is given in Fig. 3.1. For example, the well-known *steepest descent method* for smooth optimization finds the descent direction by using the formula

$$\boldsymbol{d}_h = -\nabla f(\boldsymbol{x}_h),$$

(h is the iteration counter) and the step size t_h by minimizing the function $f(\boldsymbol{x}_h + t\boldsymbol{d}_h)$ subject to $t > 0$. In addition, the steepest descent method—as well as several other smooth optimization methods—terminates when the norm $\|\nabla f(\boldsymbol{x}_h)\|$ of the gradient of the objective is small enough.

All this works fine as long as the objective function is continuously differentiable. However, in many practical applications, including clustering problems, we can only assume that the objective function $f : \mathbb{R}^n \to \mathbb{R}$ is LLC: it is not necessarily differentiable nor convex. Moreover, in practical applications, we do not usually know the whole subdifferential $\partial f(\boldsymbol{x})$ of the nonsmooth function but only one arbitrary element $\boldsymbol{\xi} \in \partial f(\boldsymbol{x})$. These facts may cause difficulties in almost all steps of an iterative algorithm as listed below.

A. M. Bagirov et al., *Partitional Clustering via Nonsmooth Optimization*, Unsupervised and Semi-Supervised Learning,
https://doi.org/10.1007/978-3-030-37826-4_3

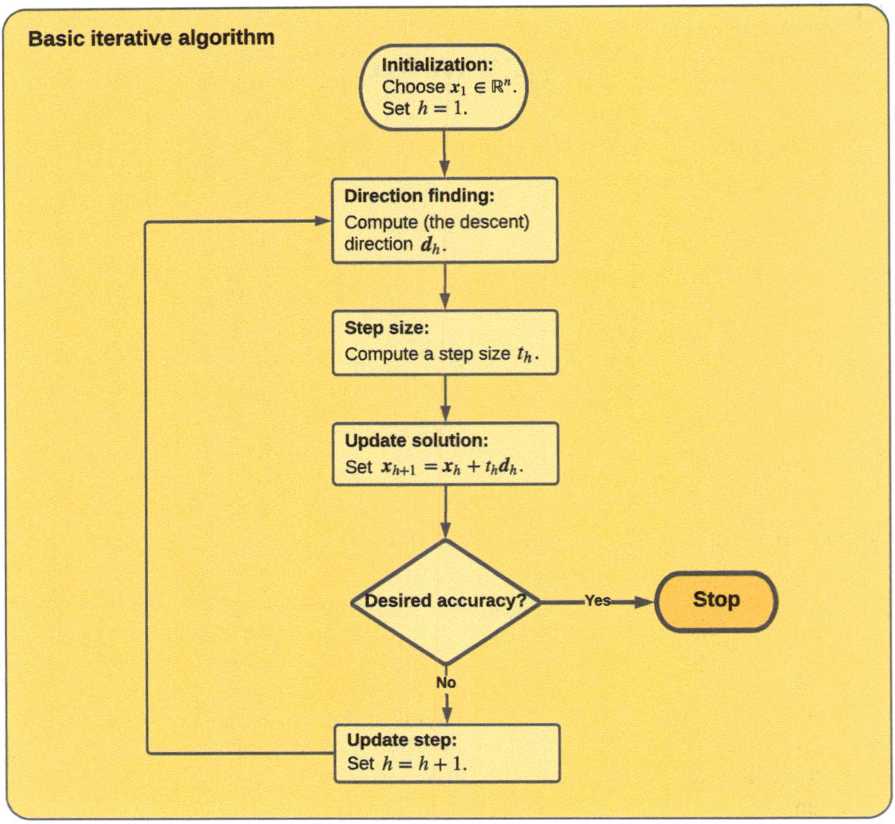

Fig. 3.1 Basic iterative algorithm

- *Finding a descent direction and step sizes*: for a smooth function a descent direction can be obtained as the opposite direction of the gradient $\nabla f(\boldsymbol{x})$. However, for a nonsmooth function the gradient does not exist at every point and, as already noted, its generalization, the subdifferential $\partial f(\boldsymbol{x})$, is normally not fully known. On the other hand, a direction opposite to an arbitrary subgradient $\boldsymbol{\xi} \in \partial f(\boldsymbol{x})$ does not need to be a descent direction. This also makes the computation of the step size t challenging.
- *Necessary optimality condition and a stopping criterion*: for a smooth function a necessary condition for a local minimum is that $\nabla f(\boldsymbol{x}) = 0$. By continuity the norm $\|\nabla f(\boldsymbol{x})\|$ becomes small when we are close to an optimal point providing a good stopping criterion (i.e., $\|\nabla f(\boldsymbol{x})\| < \varepsilon$) for algorithms. For nonsmooth functions the gradient usually does not exist at optimal points and the above mentioned property is no longer true when we replace the gradient with an arbitrary subgradient.

- *Approximation of subgradients*: in practical applications it is common to approximate the gradient by finite difference estimates. However, this approach is valid only for smooth functions, since for them we have $f'(\boldsymbol{x}; \boldsymbol{d}) = \nabla f(\boldsymbol{x})^T \boldsymbol{d}$. In the nonsmooth case, the mapping $\boldsymbol{x} \mapsto \partial f(\boldsymbol{x})$ is not continuous and, thus, the conventional finite difference estimates may give an element which does not belong to the subdifferential.
- *Numerical difficulties due to the nonsmoothness*: the discontinuity of the mapping $\boldsymbol{x} \mapsto \partial f(\boldsymbol{x})$ also means that a small variation on $\boldsymbol{x}_h$ may cause large variation on $\partial f(\boldsymbol{x}_h)$. Therefore, an algorithm may perform very differently in different platforms due to rounding errors that yield different sequences $\{\boldsymbol{x}_h\}$. This makes the numerical comparison of different algorithms—or even the same algorithm in different platforms—a very delicate task.

All these facts imply that careful theoretical bases and special tools are needed to solve NSO problems.

In iterative optimization methods—whether the problem is smooth or nonsmooth—we try to generate a sequence $\{\boldsymbol{x}_h\}$ that converges to a minimum point $\boldsymbol{x}^*$ of the objective function, that is, $\boldsymbol{x}_h \to \boldsymbol{x}^*$ whenever $h \to \infty$. If an iterative method converges to a (local) minimum $\boldsymbol{x}^*$ from any arbitrary starting point $\boldsymbol{x}_1$, it is said to be *globally convergent*. If it converges to a (local) minimum in some neighborhood of $\boldsymbol{x}^*$, it is said to be *locally convergent*. It is worth of noting that a globally convergent method does not necessarily attempt to find the global minimum of the objective function.

Recall that the objective function f in the problem (3.1) is LLC. In addition, unless said otherwise, we assume that at every point $\boldsymbol{x}$ both the value of the objective $f(\boldsymbol{x})$ and one arbitrary subgradient $\boldsymbol{\xi}$ from the subdifferential $\partial f(\boldsymbol{x})$ can be evaluated.

3.2 Subgradient Method

The idea behind the *subgradient methods* is to generalize the gradient method (e.g., the steepest descent method) by replacing the gradient with an arbitrary subgradient. Therefore, the search direction in the standard subgradient method is given by

$$\boldsymbol{d}_h = -\boldsymbol{\xi}_h,$$

where $\boldsymbol{\xi}_h$ is any subgradient from the subdifferential $\partial f(\boldsymbol{x}_h)$. The subgradient method uses the same search direction as the steepest descent method when the objective function is continuously differentiable. In this case, the function f has only one subgradient at any point $\boldsymbol{x}_h$ which is the gradient vector $\nabla f(\boldsymbol{x}_h)$ itself. However, for a nonsmooth objective we take an arbitrary subgradient from the subdifferential (since the whole subdifferential is usually unknown) and that causes

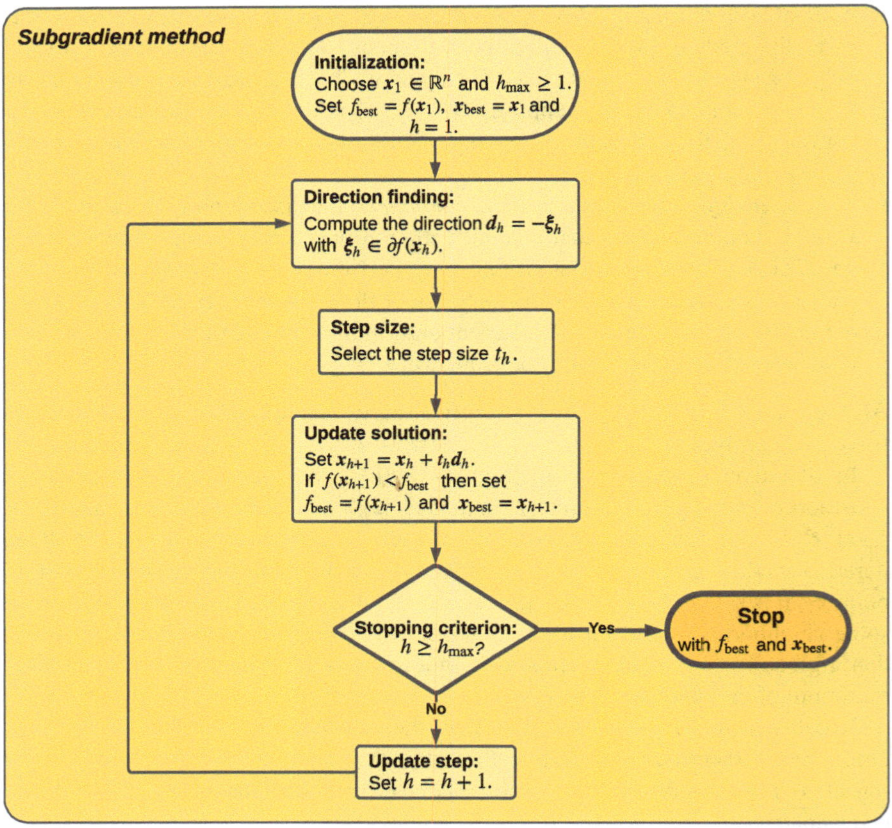

Fig. 3.2 Subgradient method

some challenges: first, a non-descent search direction may occur and, thus, the standard line search operations cannot be applied.

One option is to select step sizes a priori. Various step size rules have been developed (see, e.g., [46]). Due to possible non-descent search direction one needs to keep track of the lowest (best) value $f_h^{\text{best}} = \min\{f_{h-1}^{\text{best}}, f_h\}$ and the corresponding point x_h^{best} obtained so far.

The norm of an arbitrary subgradient $\|\xi_h\|$ need not to be small even if $\mathbf{0} \in \partial f(x_h)$ and, thus, there is no implementable subgradient based stopping criterion in the subgradient method. Typically, the maximum number of iterations $h_{\max}$ is chosen and the computation terminates when $h > h_{\max}$. Figure 3.2 illustrates the standard subgradient method.

The standard subgradient method is proved to be globally convergent if the objective function is convex and step sizes satisfy

$$t_h \geq 0, \quad \lim_{h \to \infty} t_h = 0 \quad \text{and} \quad \sum_{j=1}^{\infty} t_j = \infty.$$

Due to the simple structure and low storage requirements of subgradient methods, they are widely used and are among popular methods of NSO. Nevertheless, the rate of convergence of the standard subgradient method is not even linear. To overcome this drawback the variable metric ideas were adopted to the subgradient context in [267] by introducing two space dilation methods (the *ellipsoid method* and the *r-algorithm*). In addition, two *adaptive variable metric methods*, differing in the step size control, were derived in [285]. An extensive overview of various subgradient methods can be found in [267].

3.3 Proximal Bundle Method

As noted in the previous section, the whole subdifferential of a nonsmooth objective function usually is not possible to calculate and the use of a single arbitrary subgradient may lead to difficulties in almost all steps of an iterative optimization algorithm.

The basic idea behind various versions of bundle methods is to approximate the whole subdifferential of the objective with a bundle of subgradients. In addition, a null step is used if the current search direction is not "good enough." This allows us to obtain more information about the local behavior of the objective function than if one uses a single arbitrary subgradient. Furthermore, the sequence $\{f_h\}$ is non-increasing.

To date, bundle methods are regarded as the most effective and reliable methods for NSO. In this section, we briefly introduce the most frequently used one—the *proximal bundle method* (PBM). For more details on this method and various other bundle methods, we refer, e.g., to [145, 169, 180, 206, 207, 258, 297]. The flowchart of the simplified version of the PBM is given in Fig. 3.3.

As already mentioned, bundle methods approximate the subdifferential of the objective. This is done by gathering subgradients from previous iterations into the bundle. The subgradient information is used to construct a (convex) piecewise linear local approximation, the so-called *cutting-plane model*, for the objective. Suppose that at the hth iteration of the algorithm we have the current iteration point $\boldsymbol{x}_h$, some auxiliary points $\boldsymbol{y}_j \in \mathbb{R}^n$ (from previous iterations), and the corresponding subgradients $\boldsymbol{\xi}_j \in \partial f(\boldsymbol{y}_j)$ for $j \in \mathcal{J}_h$ available. Here, the index set $\mathcal{J}_h$ is a nonempty subset of $\{1, \ldots, h\}$. We will discuss later on how to choose it. The cutting-plane model is defined as

$$\hat{f}_h(\boldsymbol{x}) = \max_{j \in \mathcal{J}_h} \left\{ f(\boldsymbol{y}_j) + \boldsymbol{\xi}_j^T (\boldsymbol{x} - \boldsymbol{y}_j) \right\}.$$

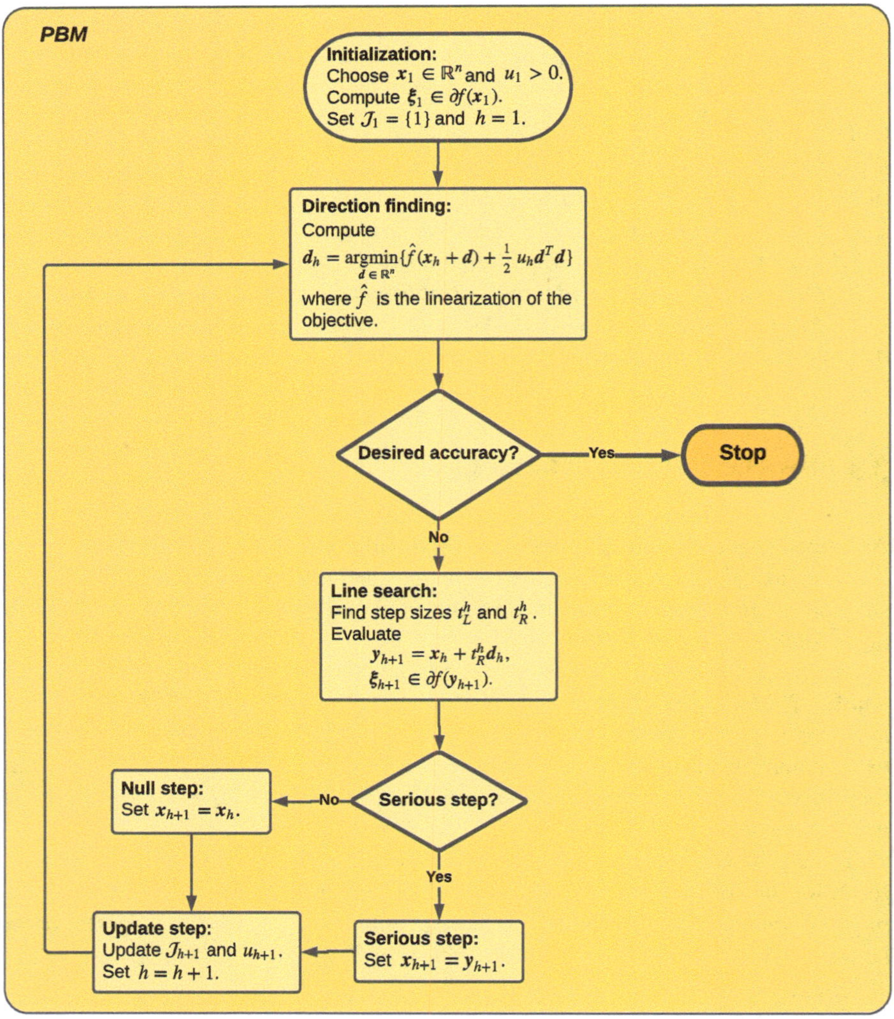

Fig. 3.3 Proximal bundle method

If f is convex, then the cutting-plane model $\hat{f}_h$ underestimates f everywhere. However, if f is nonconvex, then the cutting-plane model is not guaranteed to be an underestimate of the objective even locally. The most common way to deal with this challenge is to do some downward shifting to the model (see, e.g., [179, 207] and the limited memory bundle method in the next section). More recently, the model itself is designed to capture the nonconvex behavior of the objective function. This is the case, for example, in the DC optimization methods to be described in Sects. 3.5 and 3.6.

A descent direction for the cutting-plane model $\hat{f}_h$ and, therefore, hopefully also for the original objective f is determined by solving a *quadratic programming direction finding problem*

$$\boldsymbol{d}_h = \underset{\boldsymbol{d}\in\mathbb{R}^n}{\operatorname{argmin}} \left\{ \hat{f}_h(\boldsymbol{x}_h + \boldsymbol{d}) + \frac{1}{2} u_h \boldsymbol{d}^T \boldsymbol{d} \right\}. \tag{3.2}$$

Here, the *stabilizing term* $\frac{1}{2}u_h\boldsymbol{d}^T\boldsymbol{d}$ guarantees the existence of the solution $\boldsymbol{d}_h$ and keeps the approximation local enough. The weighting parameter $u_h > 0$ improves the convergence rate and intends to accumulate second order information about the curvature of f around $\boldsymbol{x}_h$.

Although the direction $\boldsymbol{d}_h$ is a descent direction for the model $\hat{f}_h$, it is not necessarily the descent direction for the objective f, or the decrease in function values along this direction may not be sufficient. In order to determine the step size along the search direction $\boldsymbol{d}_h$, we use the *line search procedure*. There exist several line search procedures suitable for bundle methods. Here, we discuss one of them that is used (with slight modifications) also for the limited memory bundle method to be described in the next section. The flowchart of the procedure is given in Fig. 3.4.

Assume that $\varepsilon_L \in (0, 1)$ and $\varepsilon_R \in (\varepsilon_L, 1)$ are some fixed line search parameters. We search for the two step sizes $t_R^h \in (0, 1]$ and $t_L^h \in [0, t_R^h]$ such that exactly one of the possibilities—a serious step or a null step—occurs. A necessary condition for a *serious step* is to have

$$t_L^h = t_R^h > 0 \quad \text{and} \quad f(\boldsymbol{x}_h + t_R^h \boldsymbol{d}_h) \le f(\boldsymbol{x}_h) + \varepsilon_L t_R^h v_h, \tag{3.3}$$

where v_h is the predicted amount of descent

$$v_h = \hat{f}_h(\boldsymbol{x}_h + \boldsymbol{d}_h) - f(\boldsymbol{x}_h) < 0.$$

If the condition (3.3) is satisfied, we set $\boldsymbol{x}_{h+1} = \boldsymbol{y}_{h+1} = \boldsymbol{x}_h + t_R^h \boldsymbol{d}_h$ and we call this step a serious step.

If the current search direction is not good enough, that is, the value of the objective at the new auxiliary point $\boldsymbol{y}_{h+1} = \boldsymbol{x}_h + t_R^h \boldsymbol{d}_h$ is not decreased enough, the *null step* occurs. In this case, we have

$$t_R^h > t_L^h = 0 \quad \text{and} \quad -\beta_{h+1}^h + \boldsymbol{\xi}_{h+1}^T \boldsymbol{d}_h \ge \varepsilon_R v_h, \tag{3.4}$$

where $\boldsymbol{\xi}_{h+1} \in \partial f(\boldsymbol{y}_{h+1})$, and β_j^h, $j \in \mathcal{J}_h$, is the *subgradient locality measure* [193, 213] given by

$$\beta_j^h = \max \left\{ \left| f(\boldsymbol{x}_h) - f(\boldsymbol{y}_j) - \boldsymbol{\xi}_j^T(\boldsymbol{x}_h - \boldsymbol{y}_j) \right|, \gamma \left\| \boldsymbol{x}_h - \boldsymbol{y}_j \right\|^2 \right\} \tag{3.5}$$

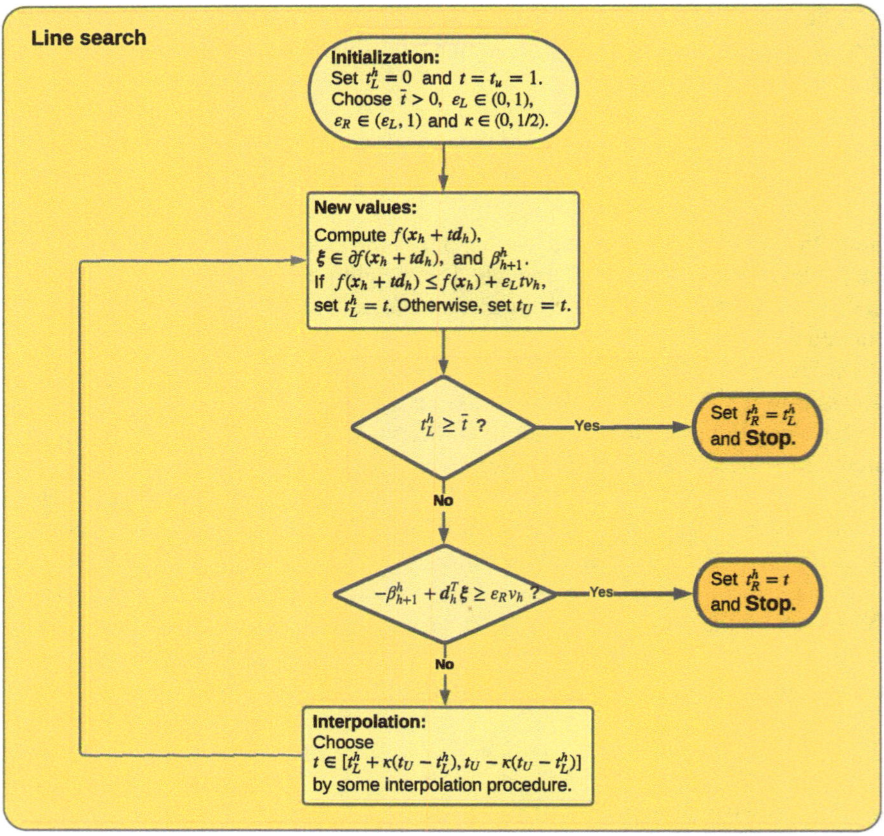

Fig. 3.4 Line search procedure

and $\gamma \geq 0$. If the objective function is convex, we can set $\gamma = 0$ and the subgradient locality measure reverts to the *linearization error* that is used to evaluate the accuracy of the piecewise linear model. In the case of a null step, we set $\boldsymbol{x}_{h+1} = \boldsymbol{x}_h$.

Having a null step means that there exists discontinuity in the gradient of f. Then the requirement (3.4) ensures that $\boldsymbol{x}_h$ and $\boldsymbol{y}_{h+1}$ lie on the opposite sides of this discontinuity and the new subgradient $\boldsymbol{\xi}_{h+1} \in \partial f(\boldsymbol{y}_{h+1})$ forces a remarkable modification to the next direction finding problem. Under the weak semismoothness assumption (see the equation (2.6)) the line search procedure is guaranteed to find the step sizes t_L^h and t_R^h such that exactly one of the two possibilities—a serious step or a null step—occurs.

The PBM is terminated if

$$v_h \geq -\varepsilon,$$

where $\varepsilon > 0$ is a final accuracy tolerance supplied by the user.

Next, we discuss how to update the weighting parameter u_h and the index set $\mathcal{J}_h$. The minimal requirements for the weights $u_h > 0$ are that they lie in a compact interval and the condition $u_{h+1} \geq u_h$ is valid at null steps. Therefore, the simplest strategy for updating the weighting parameter u_h is to keep it fixed. That is $u_h = \bar{u}$ for some fixed $\bar{u}$. However, this choice may lead to some difficulties: if $\bar{u}$ is very large, we will have small $|v_h|$ and $\|\boldsymbol{d}_h\|$ and, thus, almost all steps are serious but with very small improvement to function values. On the other hand, if $\bar{u}$ is very small, we will have large $|v_h|$ and $\|\boldsymbol{d}_h\|$ and every serious step is followed by many null steps. To avoid these difficulties, the weight u_h can be kept as a variable which is updated when necessary. Different updating rules are given, for instance, in [180].

As mentioned at the beginning of this section, the index set $\mathcal{J}_h$ must be a nonempty subset of $\{1, \ldots, h\}$. Since, in practice, the choice $\mathcal{J}_h = \{1, \ldots, h\}$ would cause serious difficulties with storage and computations after a large number of iterations, the size of the set has to be bounded. That is, we set $|\mathcal{J}_h| \leq m_\xi$, where m_ξ is a predefined *size of the bundle*. Usually the index set $\mathcal{J}_h$ is chosen as follows:

$$\mathcal{J}_h = \begin{cases} \{1, \ldots, h\}, & \text{if} \quad h \leq m_\xi, \\ \mathcal{J}_{h-1} \cup \{h\} \setminus \{h - m_\xi\}, & \text{if} \quad h > m_\xi. \end{cases}$$

The PBM is proved to be globally convergent under the weak semismoothness assumption (see the Eq. (2.6)). Furthermore, if $\mathcal{J}_h \neq \{1, \ldots, h\}$, the global convergence of the method can be guaranteed by using the *subgradient aggregation strategy*, which accumulates information from previous iterations [179]. The convergence rate of the PBM is linear for convex functions [248] and for piecewise linear problems it achieves a finite convergence [258].

3.4 Limited Memory Bundle Method

There exist a vast variety of practical problems involving nonsmooth functions with a large number of variables. Nevertheless, most NSO methods were designed to solve only small- or medium sized problems. For instance, in the PBM explained above, the computational demand often significantly increases even for relatively small problems. One reason for this is that the PBM needs a relatively large bundle ($m_\xi \approx n$) and, therefore, the size of the quadratic programming direction finding problem (3.2) increases as the number of variables increases.

The standard subgradient methods (Sect. 3.2) are applicable to solve large-scale problems due to their low storage requirements. However, it is known that they converge very slowly for such problems.

In this section, we consider the *limited memory bundle method* (LMBM) [131–133, 168] for solving general, possibly nonconvex, large-scale NSO problems. The method is a hybrid of the *variable metric bundle method* [203, 289] for small- and

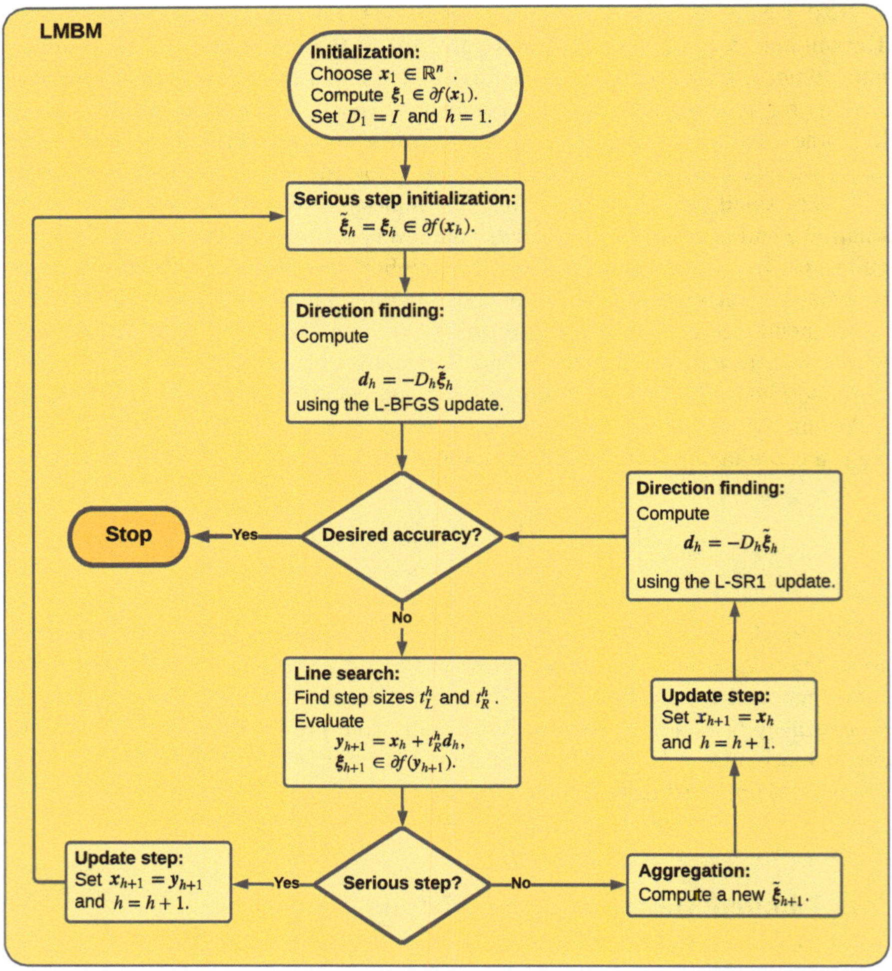

Fig. 3.5 Limited memory bundle method

medium sized NSO and the *limited memory variable metric method* (see e.g. [56]) for smooth large-scale optimization.

The flowchart of the LMBM is presented in Fig. 3.5. The LMBM is characterized by the usage of null steps together with the aggregation of subgradients. Moreover, the search direction is calculated by the limited memory variable metric approach (the limited memory BFGS (L-BFGS) update after a serious step and the limited memory SR1 (L-SR1) update, otherwise). Thus, the LMBM avoids solving the time-consuming quadratic programming direction finding problem (3.2) appearing in standard bundle methods as well as storing and manipulating large matrices as is the case in the variable metric bundle methods. The usage of null steps gives sufficient information about the nonsmooth objective whenever the current search direction is not good enough. In addition, unlike the standard subgradient methods it enables non-increasing iterates. Finally, a simple aggregation of subgradients that uses only three subgradients guarantees the convergence of the aggregate subgradients to zero and makes it possible to evaluate a stopping criterion.

These improvements make the LMBM suitable for solving large-scale NSO problems. Namely, the number of operations needed for the calculation of the search direction and the aggregate values is only linearly dependent on the number of variables while, for example, with the variable metric bundle method this dependence is quadratic. In what follows, we describe different components of the LMBM in more details.

The fundamental idea in the LMBM is to calculate the search direction using the classical limited memory variable metric scheme $\boldsymbol{d}_h = -D_h\nabla f(\boldsymbol{x}_h)$, where D_h is the limited memory variable metric update that represents the approximation of the inverse of the Hessian matrix. However, since for a nonsmooth objective f the gradient $\nabla f(\boldsymbol{x}_h)$ does not need to exist, we use (an aggregate) subgradient $\tilde{\boldsymbol{\xi}}_h$ instead of the usual gradient. Therefore, the *search direction* is given by

$$\boldsymbol{d}_h = -D_h\tilde{\boldsymbol{\xi}}_h. \tag{3.6}$$

The role of the matrix D_h, which is not formed explicitly, is to accumulate information about previous subgradients.

As noted in the previous sections, the search direction computed using an arbitrary subgradient is not necessarily a descent one. This is true even if we replace it with a little bit better approximation—an aggregate subgradient $\tilde{\boldsymbol{\xi}}_h$ (to be described later). Thus, similarly to the PBM we need to use the *line search procedure* that is able to generate a null step if the current search direction is not good enough. A new iteration point $\boldsymbol{x}_{h+1}$ and a new auxiliary point $\boldsymbol{y}_{h+1}$ are given as

$$\begin{aligned} \boldsymbol{x}_{h+1} &= \boldsymbol{x}_h + t_L^h\boldsymbol{d}_h, \quad \text{and} \\ \boldsymbol{y}_{h+1} &= \boldsymbol{x}_h + t_R^h\boldsymbol{d}_h, \quad \text{for} h \geq 1 \end{aligned}$$

with $\boldsymbol{y}_1 = \boldsymbol{x}_1$, where $t_R^h \in (0, t_{\max}]$ and $t_L^h \in [0, t_R^h]$ are step sizes, and $t_{\max} > 1$ is the upper bound for the step size.

A necessary condition for a serious step to be taken is to have

$$t_R^h = t_L^h > 0 \quad \text{and} \quad f(\boldsymbol{y}_{h+1}) \leq f(\boldsymbol{x}_h) - \varepsilon_L t_R^h w_h, \tag{3.7}$$

where $\varepsilon_L \in (0, 1/2)$ is a line search parameter, and $w_h > 0$ represents the desirable amount of descent of f at $\boldsymbol{x}_h$. If the condition (3.7) is satisfied, we set $\boldsymbol{x}_{h+1} = \boldsymbol{y}_{h+1}$ and a serious step is taken. Otherwise, a null step occurs. In null steps, we have

$$t_R^h > t_L^h = 0 \quad \text{and} \quad -\beta_{h+1} + \boldsymbol{\xi}_{h+1}^T\boldsymbol{d}_h \geq -\varepsilon_R w_h, \tag{3.8}$$

where $\varepsilon_R \in (\varepsilon_L, 1/2)$ is a line search parameter, $\boldsymbol{\xi}_{h+1} \in \partial f(\boldsymbol{y}_{h+1})$, and β_{h+1} is the subgradient locality measure ($\beta_{h+1} = \beta^h_{h+1}$ in (3.5)). In this case, we set $\boldsymbol{x}_{h+1} = \boldsymbol{x}_h$ but information about the objective function is increased since we store the auxiliary point $\boldsymbol{y}_{h+1}$ and the corresponding auxiliary subgradient $\boldsymbol{\xi}_{h+1} \in \partial f(\boldsymbol{y}_{h+1})$, and we use them to compute new aggregate values and the limited memory update matrix.

The LMBM uses the subgradient $\boldsymbol{\xi}_h \in \partial f(\boldsymbol{x}_h)$ after the serious step and the aggregate subgradient $\tilde{\boldsymbol{\xi}}_h$ after the null step to find search directions (i.e., we set $\tilde{\boldsymbol{\xi}}_h = \boldsymbol{\xi}_h$ if the previous step was a serious step, see Fig. 3.5). The *aggregation procedure* used in the LMBM differs significantly from that usually used in bundle methods (see, e.g., [179]). In the LMBM, we use the procedure similar to the variable metric bundle methods [289], where only three subgradients and two locality measures are needed. We proceed by solving the following problem:

$$\begin{cases} \text{minimize} & \varphi(\lambda_1, \lambda_2, \lambda_3) \\ \text{subject to} & \lambda_i \geq 0 \quad \text{for all} \quad i \in \{1, 2, 3\}, \\ & \sum_{i=1}^{3} \lambda_i = 1, \end{cases}$$

where

$$\begin{aligned} \varphi(\lambda_1, \lambda_2, \lambda_3) = {} & \left[\lambda_1 \boldsymbol{\xi}_m + \lambda_2 \boldsymbol{\xi}_{h+1} + \lambda_3 \tilde{\boldsymbol{\xi}}_h\right]^T D_h \left[\lambda_1 \boldsymbol{\xi}_m + \lambda_2 \boldsymbol{\xi}_{h+1} + \lambda_3 \tilde{\boldsymbol{\xi}}_h\right] \\ & + 2\left(\lambda_2 \beta_{h+1} + \lambda_3 \tilde{\beta}_h\right). \end{aligned} \tag{3.9}$$

Here, $\boldsymbol{\xi}_m \in \partial f(\boldsymbol{x}_h)$ is the current subgradient (m denotes the index of the iteration after the latest serious step, i.e., $\boldsymbol{x}_h = \boldsymbol{x}_m$), $\tilde{\boldsymbol{\xi}}_h$ is the current aggregate subgradient from the previous iteration, and as before $\boldsymbol{\xi}_{h+1} \in \partial f(\boldsymbol{y}_{h+1})$. In addition, β_{h+1} is the current subgradient locality measure and $\tilde{\beta}_h$ is the current aggregate subgradient locality measure ($\tilde{\beta}_1 = 0$). The optimal values λ^h_i, $i \in \{1, 2, 3\}$, are easy to calculate (see [289]).

The new *aggregate subgradient* $\tilde{\boldsymbol{\xi}}_{h+1}$ and the *aggregate subgradient locality measure* $\tilde{\beta}_{h+1}$ are computed as

$$\tilde{\boldsymbol{\xi}}_{h+1} = \lambda^h_1 \boldsymbol{\xi}_m + \lambda^h_2 \boldsymbol{\xi}_{h+1} + \lambda^h_3 \tilde{\boldsymbol{\xi}}_h \quad \text{and} \quad \tilde{\beta}_{h+1} = \lambda^h_2 \beta_{h+1} + \lambda^h_3 \tilde{\beta}_h. \tag{3.10}$$

This simple aggregation procedure gives us a possibility to retain the global convergence without solving the complicated quadratic programming direction finding problem (3.2) appearing in standard bundle methods. Moreover, only one trial point $\boldsymbol{y}_{h+1}$ and the corresponding subgradient $\boldsymbol{\xi}_{h+1} \in \partial f(\boldsymbol{y}_{h+1})$ need to be stored instead of $n+3$ subgradients typically stored in bundle methods. Finally, it is worth of noting that the aggregate values need to be computed only if the last step was a null step. Otherwise, we set $\tilde{\boldsymbol{\xi}}_{h+1} = \boldsymbol{\xi}_{h+1}$ and $\tilde{\beta}_{h+1} = 0$ (see Fig. 3.5).

The idea in the limited memory matrix updating is that instead of storing and manipulating large $n \times n$ matrices D_h, one stores a small number of the so-called *correction vectors* obtained at the previous iterations of the algorithm and uses these vectors to implicitly define the variable metric matrices. In the smooth case, these correction vectors are given by $\boldsymbol{s}_h = \boldsymbol{x}_{h+1} - \boldsymbol{x}_h$ and $\boldsymbol{u}_h = \nabla f(\boldsymbol{x}_{h+1}) - \nabla f(\boldsymbol{x}_h)$. In the LMBM, the correction vectors are slightly modified from those used for smooth optimization: since we may have $\boldsymbol{x}_{h+1} = \boldsymbol{x}_h$ due to the usage of null steps we use the auxiliary point $\boldsymbol{y}_{h+1}$ instead of $\boldsymbol{x}_{h+1}$ when updating $\boldsymbol{s}_h$. In addition, since the gradient need not to exist for nonsmooth objective the correction vectors $\boldsymbol{u}_h$ are computed using subgradients, that is, the vectors are given by $\boldsymbol{s}_h = \boldsymbol{y}_{h+1} - \boldsymbol{x}_h$ and $\boldsymbol{u}_h = \boldsymbol{\xi}_{h+1} - \boldsymbol{\xi}_m$.

Let us denote by $\hat{m}_c$ the user-specified maximum number of stored correction vectors ($3 \leq \hat{m}_c$) and by $\hat{m}_h = \min\{h-1, \hat{m}_c\}$ the current number of stored correction vectors. Then $n \times \hat{m}_h$ dimensional *correction matrices* S_h and U_h are defined by

$$\begin{aligned} S_h &= \left[\boldsymbol{s}_{h-\hat{m}_h} \, \ldots \, \boldsymbol{s}_{h-1}\right] \quad \text{and} \\ U_h &= \left[\boldsymbol{u}_{h-\hat{m}_h} \, \ldots \, \boldsymbol{u}_{h-1}\right]. \end{aligned} \tag{3.11}$$

The oldest correction vectors are deleted to make room for new ones when the available storage space is used up. Thus, except for the first few iterations, we always have the $\hat{m}_c$ most recent correction pairs $(\boldsymbol{s}_i, \boldsymbol{u}_i)$ available.

In the LMBM, both the L-BFGS and the L-SR1 update formulas [56] are used for the calculation of the search direction and the aggregate values. In the case of a null step, the LMBM uses the L-SR1 update formula, since with this formula it is possible to preserve the boundedness and some other properties of generated matrices that are needed to guarantee the global convergence of the method. The *inverse L-SR1 update* is defined by

$$D_h = \vartheta_h I - \left(\vartheta_h U_h - S_h\right)\left(\vartheta_h U_h^T U_h - R_h - R_h^T + C_h\right)^{-1}\left(\vartheta_h U_h - S_h\right)^T,$$

where R_h is an upper triangular matrix of the order $\hat{m}_h$ given by

$$(R_h)_{ij} = \begin{cases} \boldsymbol{s}_{h-\hat{m}_h-1+i}^T \boldsymbol{u}_{h-\hat{m}_h-1+j}, & \text{if } i \leq j, \\ 0, & \text{otherwise.} \end{cases}$$

The matrix C_h is diagonal with the order $\hat{m}_h$ such that

$$C_h = \operatorname{diag}\left[\boldsymbol{s}_{h-\hat{m}_h}^T \boldsymbol{u}_{h-\hat{m}_h}, \ldots, \boldsymbol{s}_{h-1}^T \boldsymbol{u}_{h-1}\right],$$

and ϑ_h is a positive scaling parameter.

Since these additional properties are not required after a serious step the more efficient L-BFGS update is employed. The *inverse L-BFGS update* is defined by

$$D_h = \vartheta_h I + \begin{bmatrix} S_h & \vartheta_h U_h \end{bmatrix} \begin{bmatrix} (R_h^{-1})^T C_h + \vartheta_h U_h^T U_h R_h^{-1} & -(R_h^{-1})^T \\ -R_h^{-1} & 0 \end{bmatrix} \begin{bmatrix} S_h^T \\ \vartheta_h U_h^T \end{bmatrix}.$$

The similar representations for the direct L-BFGS and L-SR1 updates can be given; however, the implementation of the LMBM only needs the inverse update formulas.

Note that we never compute the matrix D_h, but only the product $D_h \boldsymbol{v}$, where $\boldsymbol{v}$ is equal to $\boldsymbol{\xi}_m$, $\boldsymbol{\xi}_{h+1}$, or $\tilde{\boldsymbol{\xi}}_h$. This way the number of operations needed for the calculation of the search direction and the aggregate values is only linearly dependent on the number of variables and we do not need to store large $n \times n$ matrices.

The final task to be considered is to formulate the stopping criterion in the algorithm. As noted before, the norm of an arbitrary subgradient does not need to be small when we are close to an optimal point. In the LMBM, the aggregate subgradient $\tilde{\boldsymbol{\xi}}_h$ provides a better approximation than an arbitrary subgradient does, however, still the direct test $\|\tilde{\boldsymbol{\xi}}_h\| < \varepsilon$, for some $\varepsilon > 0$, is too uncertain as a stopping criterion. In the LMBM, the term $\tilde{\boldsymbol{\xi}}_h^T D_h \tilde{\boldsymbol{\xi}}_h = -\tilde{\boldsymbol{\xi}}_h^T \boldsymbol{d}_h$ and the aggregate subgradient locality measure $\tilde{\beta}_h$ are used to improve the accuracy of the sole norm $\|\tilde{\boldsymbol{\xi}}_h\|$. Therefore, the LMBM uses the value

$$w_h = -\tilde{\boldsymbol{\xi}}_h^T \boldsymbol{d}_h + 2\tilde{\beta}_h > 0 \tag{3.12}$$

as a *stopping parameter*, and it stops if $w_h \le \varepsilon$ for some user specified $\varepsilon > 0$. In addition, the parameter w_h is used during the line search procedure to represent the desirable amount of descent (see, Eq. (3.7)).

The LMBM is proved to be globally convergent for LLC objective functions under the weak semismoothness assumption [133]. The basic schema of the convergence proof is given in the next subsection.

3.4.1 *Convergence of the* LMBM

In this subsection, we recall some technical details of the convergence properties of the LMBM. A more detailed description with the proofs of Lemmas is given in [133]. The following assumptions are used in the proofs:

Assumption 3.1 *The objective function* $f : \mathbb{R}^n \to \mathbb{R}$ *is LLC.*

Assumption 3.2 *The objective function* $f : \mathbb{R}^n \to \mathbb{R}$ *is weakly semismooth (see Eq.* (2.6)*).*

Assumption 3.3 *The level set* $\text{lev}_{f(\boldsymbol{x}_1)} f = \{\boldsymbol{x} \in \mathbb{R}^n : f(\boldsymbol{x}) \le f(\boldsymbol{x}_1)\}$ *is bounded for any starting point* $\boldsymbol{x}_1 \in \mathbb{R}^n$.

The optimality condition $\boldsymbol{0} \in \partial f(\boldsymbol{x})$ is sufficient when f is convex. Since the convexity of the function f is not assumed, we can only prove that the LMBM converges to a stationary point. In order to do this, we assume that the final accuracy tolerance ε is equal to zero. The convergence proof of the LMBM can be divided into the following three tasks:

- *finite convergence of the line search procedure:* we omit the proof, since it is true for most bundle-type methods and consequently is not a specific property of the LMBM;
- *termination of the* LMBM *after a finite number of iterations:* if the LMBM terminates after a finite number of iterations, say at iteration h, then the point $\boldsymbol{x}_h$ is a Clarke stationary point of the objective; and
- *infinite number of iterations:* if the sequence $\{\boldsymbol{x}_h\}$ generated by the LMBM is infinite, then its every accumulation point $\bar{\boldsymbol{x}}$ is a Clarke stationary point of the objective.

Proposition 3.1 *Each execution of the line search procedure is finite.*

Proof See [289]. □

Remark 3.1 The sequence $\{\boldsymbol{x}_h\}$, generated by the LMBM, is bounded according to Assumption 3.3. The monotonicity of the sequence $\{f_h\}$ follows from the condition (3.7), being satisfied for serious steps and the fact that $\boldsymbol{x}_{h+1} = \boldsymbol{x}_h$ for null steps. The sequence $\{\boldsymbol{y}_h\}$ is also bounded, since $\boldsymbol{x}_{h+1} = \boldsymbol{y}_{h+1}$ for serious steps and $\|\boldsymbol{y}_{h+1} - \boldsymbol{x}_{h+1}\| \leq C t_{\max}$ for null steps with a predefined $C > 0$ (see [133] for more details). The local boundedness and the upper semicontinuity of the subdifferential imply the boundedness of subgradients $\boldsymbol{\xi}_h$ as well as their convex combinations.

Lemma 3.1 *Suppose that the* LMBM *is not terminated before the hth iteration. Then, there exist numbers $\lambda^{h,j} \geq 0$ for $j = 1, \ldots, h$ and $\tilde{\sigma}_h \geq 0$ such that*

$$(\tilde{\boldsymbol{\xi}}_h, \tilde{\sigma}_h) = \sum_{j=1}^{h} \lambda^{h,j} (\boldsymbol{\xi}_j, \|\boldsymbol{y}_j - \boldsymbol{x}_h\|), \quad \sum_{j=1}^{h} \lambda^{h,j} = 1 \quad \textit{and} \quad \tilde{\beta}_h \geq \gamma \tilde{\sigma}_h^2.$$

Lemma 3.2 *Let $\bar{\boldsymbol{x}} \in \mathbb{R}^n$ be given and suppose that there exist vectors $\bar{\boldsymbol{g}}$, $\bar{\boldsymbol{\xi}}_i$, $\bar{\boldsymbol{y}}_i$, and numbers $\bar{\lambda}_i \geq 0$ for $i = 1, \ldots, l$, $l \geq 1$, such that*

$$\begin{aligned} &(\bar{\boldsymbol{g}}, 0) = \sum_{i=1}^{l} \bar{\lambda}_i (\bar{\boldsymbol{\xi}}_i, \|\bar{\boldsymbol{y}}_i - \bar{\boldsymbol{x}}\|), \\ &\bar{\boldsymbol{\xi}}_i \in \partial f(\bar{\boldsymbol{y}}_i), \quad i = 1, \ldots, l, \quad \textit{and} \\ &\sum_{i=1}^{l} \bar{\lambda}_i = 1. \end{aligned} \tag{3.13}$$

Then $\bar{\boldsymbol{g}} \in \partial f(\bar{\boldsymbol{x}})$.

Proposition 3.2 *If the* LMBM *terminates at the hth iteration, then the point* $\boldsymbol{x}_h$ *is stationary for* f.

Proof If the LMBM terminates, then the condition $\varepsilon = 0$ implies that $w_h = 0$. Thus, by (3.5), (3.10), (3.12), Lemma 3.1 and the fact that we initialize $\tilde{\beta}_h = 0$ after serious steps, we have $\tilde{\boldsymbol{\xi}}_h = \mathbf{0}$ and $\tilde{\beta}_h = \tilde{\sigma}_h = 0$. Now, by Lemma 3.1 and by applying Lemma 3.2 with

$$\bar{\boldsymbol{x}} = \boldsymbol{x}_h, \quad l = h, \quad \bar{\boldsymbol{g}} = \tilde{\boldsymbol{\xi}}_h, \quad \text{and}$$

$$\bar{\boldsymbol{\xi}}_i = \boldsymbol{\xi}_i, \quad \bar{\boldsymbol{y}}_i = \boldsymbol{y}_i, \quad \bar{\lambda}_i = \lambda^{h,i} \quad \text{for } i \leq h,$$

we obtain $\mathbf{0} = \tilde{\boldsymbol{\xi}}_h \in \partial f(\boldsymbol{x}_h)$ and, thus, $\boldsymbol{x}_h$ is stationary for f. □

From now on, we suppose that the LMBM does not terminate, that is, $w_h > 0$ for all h.

Lemma 3.3 *If there exist a point* $\bar{\boldsymbol{x}} \in \mathbb{R}^n$ *and an infinite set* $\mathcal{H} \subset \{1, 2, \ldots\}$ *such that* $\{\boldsymbol{x}_h\}_{h\in\mathcal{H}} \to \bar{\boldsymbol{x}}$ *and* $\{w_h\}_{h\in\mathcal{H}} \to 0$, *then* $\mathbf{0} \in \partial f(\bar{\boldsymbol{x}})$.

Lemma 3.4 *Suppose that the number of serious steps is finite, and the last serious step occurs at the iteration* $m - 1$. *Then there exists a number* $h^* \geq m$, *such that* $w_{h+1} \leq w_h$ *for all* $h \geq h^*$. *In addition, we have* $w_h \to 0$ *as* $h \to \infty$.

Corollary 3.1 *Suppose that the number of serious steps is finite and the last serious step occurs at the iteration* $m - 1$. *Then, the point* $\boldsymbol{x}_m$ *is Clarke stationary for* f.

Proposition 3.3 *Every accumulation point* $\bar{\boldsymbol{x}}$ *of the sequence* $\{\boldsymbol{x}_h\}$ *generated by the* LMBM *is stationary for* f.

Proof Let $\bar{\boldsymbol{x}}$ be an accumulation point of $\{\boldsymbol{x}_h\}$ and let $\mathcal{H} \subset \{1, 2, \ldots\}$ be an infinite set such that $\{\boldsymbol{x}_h\}_{h\in\mathcal{H}} \to \bar{\boldsymbol{x}}$. Following Corollary 3.1, we can restrict our consideration to the case where the number of serious steps (with $t_L^h > 0$) is infinite. By Proposition 3.1, the line search procedure is terminating. In the termination with a serious step we have the condition (3.7) satisfied and either $t_L^h \geq t_{\min}$ for some $t_{\min} \in (0, 1)$ or $\beta_{k+1} > \varepsilon_A w_h$ with $\varepsilon_A \in (0, \varepsilon_R - \varepsilon_L)$ (for more details see [133]). Denote by

$$\mathcal{H}' = \left\{h : t_L^h > 0, \text{there exists} i \in \mathcal{H},\ i \leq h \text{ such that} \boldsymbol{x}_i = \boldsymbol{x}_h\right\}.$$

Obviously, $\mathcal{H}'$ is infinite and $\{\boldsymbol{x}_h\}_{h\in\mathcal{H}'} \to \bar{\boldsymbol{x}}$. The continuity of f implies that $\{f_h\}_{h\in\mathcal{H}'} \to f(\bar{\boldsymbol{x}})$ and, thus, $f_h \downarrow f(\bar{\boldsymbol{x}})$ by the monotonicity of the sequence $\{f_h\}$ obtained by (3.7) and (3.8). Now, using the fact that $t_L^h \geq 0$ for all $h \geq 1$, we obtain

$$0 \leq \varepsilon_L t_L^h w_h \leq f_h - f_{h+1} \to 0 \quad \text{for} \quad h \geq 1. \tag{3.14}$$

Thus, if the set $\mathcal{H}_1 = \{h \in \mathcal{H}' : t_L^h \geq t_{\min}\}$ is infinite, then $\{w_h\}_{h\in\mathcal{H}_1} \to 0$ and $\{\boldsymbol{x}_h\}_{h\in\mathcal{H}_1} \to \bar{\boldsymbol{x}}$ by (3.14) and, thus, by Lemma 3.3 we have $\mathbf{0} \in \partial f(\bar{\boldsymbol{x}})$.

If the set $\mathcal{H}_1$ is finite, then the set $\mathcal{H}_2 = \{h \in \mathcal{H}' : \beta_{h+1} > \varepsilon_A w_h\}$ has to be infinite. To the contrary, let us assume that

$$w_h \geq \delta > 0 \quad \text{for all} \quad h \in \mathcal{H}_2.$$

From (3.14), we have $\{t_L^h\}_{h \in \mathcal{H}_2} \to 0$. In addition, we have

$$\|\boldsymbol{x}_{h+1} - \boldsymbol{x}_h\| \leq C t_L^h,$$

with a predefined $C > 0$ (see [133] for more details) for all $h \geq 1$. Thus, we have

$$\left\{\|\boldsymbol{x}_{h+1} - \boldsymbol{x}_h\|\right\}_{h \in \mathcal{H}_2} \to 0.$$

By (3.5), (3.14), the boundedness of $\{\boldsymbol{\xi}_h\}$, and the fact that $\boldsymbol{y}_{h+1} = \boldsymbol{x}_{h+1}$ for serious steps, we obtain $\{\beta_{h+1}\}_{h \in \mathcal{H}_2} \to 0$, which is in contradiction with

$$\varepsilon_A \delta \leq \varepsilon_A w_h < \beta_{h+1} \quad \text{for} h \in \mathcal{H}_2.$$

Therefore, there exists an infinite set $\mathcal{H}_3 \subset \mathcal{H}_2$ such that

$$\{w_h\}_{h \in \mathcal{H}_3} \to 0 \quad \text{and} \quad \{\boldsymbol{x}_h\}_{h \in \mathcal{H}_3} \to \bar{\boldsymbol{x}}.$$

Then, applying Lemma 3.3 we obtain that $\boldsymbol{0} \in \partial f(\bar{\boldsymbol{x}})$. □

Remark 3.2 If $\varepsilon > 0$, then the LMBM terminates in a finite number of steps.

3.5 DC Diagonal Bundle Method

NSO is traditionally based on convex analysis and the convergence of most numerical methods is obtained under the convexity assumption. Usually, the convex model of the objective is reasonably accurate also for nonconvex problems, except regions in the domain where there exists the so-called concave behavior of the objective function. In this case the linearization error, used as an accuracy measure of the piecewise linear model, has negative values and the model is no longer an underestimate of the objective. The common way to deal with this difficulty is to do some downward shifting (e.g., to use the subgradient locality measures (3.5) instead of linearization errors), but the amount of this shifting might be more or less arbitrary.

This is the case also in the PBM and the LMBM described in the previous sections. Nevertheless, whenever the structure of a problem is known this information can be used to develop a better approach to handle nonconvexity. This is true, in particular, when the objective function has a DC structure (see Sect. 2.7). We recall that the function $f : \mathbb{R}^n \to \mathbb{R}$ is DC if it can be given in the form $f(\boldsymbol{x}) = f_1(\boldsymbol{x}) - f_2(\boldsymbol{x})$, where f_1 and f_2 are convex functions.

In this section, we present the basic ideas of the *DC diagonal bundle method* (DCD-BUNDLE) [167, 170]. Although the DCD-BUNDLE was originally developed specifically for solving clustering problems that are given in the nonsmooth DC formulation (see Sect. 4.4), it can be applied to solve any general DC optimization problem where either f_1 or f_2 is smooth or we can compute the subgradient $\boldsymbol{\xi} \in \partial f(\boldsymbol{x})$ at any $\boldsymbol{x}$. Note that the last requirement is not always easy to meet since DC functions, in general, are not subdifferentially regular (see Definition 2.6).

For simplicity, we now assume that the first DC component f_1 is smooth. This is the case, for instance, in the DC formulation of the clustering problem with the squared Euclidean distance. Furthermore, we assume that at every point $\boldsymbol{x}$ one can evaluate the values of the DC components f_1 and f_2 of the objective f, the gradient $\nabla f_1(\boldsymbol{x})$ of the first component, and one arbitrary subgradient $\boldsymbol{\xi}_2$ from the subdifferential $\partial f_2(\boldsymbol{x})$ of the second component. Due to the smoothness of f_1 we have $\nabla f_1(\boldsymbol{x}) - \boldsymbol{\xi}_2 \in \partial f(\boldsymbol{x})$ for any $\boldsymbol{\xi}_2 \in \partial f_2(\boldsymbol{x})$. Figure 3.6 illustrates the DCD-BUNDLE.

The DCD-BUNDLE combines the approach used to design the LMBM with the sparse matrix updating and different usage of metrics depending on a convex or concave behavior of the objective at the current iteration point. In addition, it utilizes the explicit DC structure of the problem. The method shares the good properties of the LMBM. More precisely, the time-consuming quadratic programming direction finding problem (3.2) appearing in the standard bundle methods need not to be solved nor the number of stored subgradients needs to grow with the dimension of the problem. Furthermore, the method uses only a few vectors to represent the diagonal variable metric approximation of the Hessian matrix and, thus, similar to the LMBM it avoids storing and manipulating large matrices as is the case in the variable metric bundle method [203, 289]. In addition, in many large-scale problems the Hessian (if it exists) is sparse. Due to the diagonal variable metric update formula, the DCD-BUNDLE can handle the large dimensionality and the sparsity of the objective. The usage of different metrics in the DCD-BUNDLE gives us a possibility to better deal with the nonconvexity of the problem than the downward shifting using subgradient locality measures.

Similar to the LMBM, the DCD-BUNDLE uses at most $\hat{m}_c$ in the recent correction vectors to compute the updates for the matrices. These correction vectors are given by

$$
\begin{aligned}
\boldsymbol{s}_h &= \boldsymbol{y}_{h+1} - \boldsymbol{x}_h, \\
\boldsymbol{u}_{1,h} &= \nabla f_1(\boldsymbol{y}_{h+1}) - \nabla f_1(\boldsymbol{x}_h), \qquad \text{and} \\
\boldsymbol{u}_{2,h} &= \boldsymbol{\xi}_{2,h+1} - \boldsymbol{\xi}_{2,m},
\end{aligned}
$$

where $\boldsymbol{x}_h$ is the current iteration point, $\boldsymbol{y}_{h+1} = \boldsymbol{x}_h + \boldsymbol{d}_h$ is a new auxiliary point with $\boldsymbol{d}_h$, the current search direction, $\boldsymbol{\xi}_{2,h+1} \in \partial f_2(\boldsymbol{y}_{h+1})$, and $\boldsymbol{\xi}_{2,m} \in \partial f_2(\boldsymbol{x}_h)$. As with the LMBM we may have $\boldsymbol{x}_{h+1} = \boldsymbol{x}_h$ due to the null steps and thus we use here the auxiliary point $\boldsymbol{y}_{h+1}$ instead of $\boldsymbol{x}_{h+1}$. In addition, we compute the subgradient differences separately for both DC components. For the smooth component f_1,

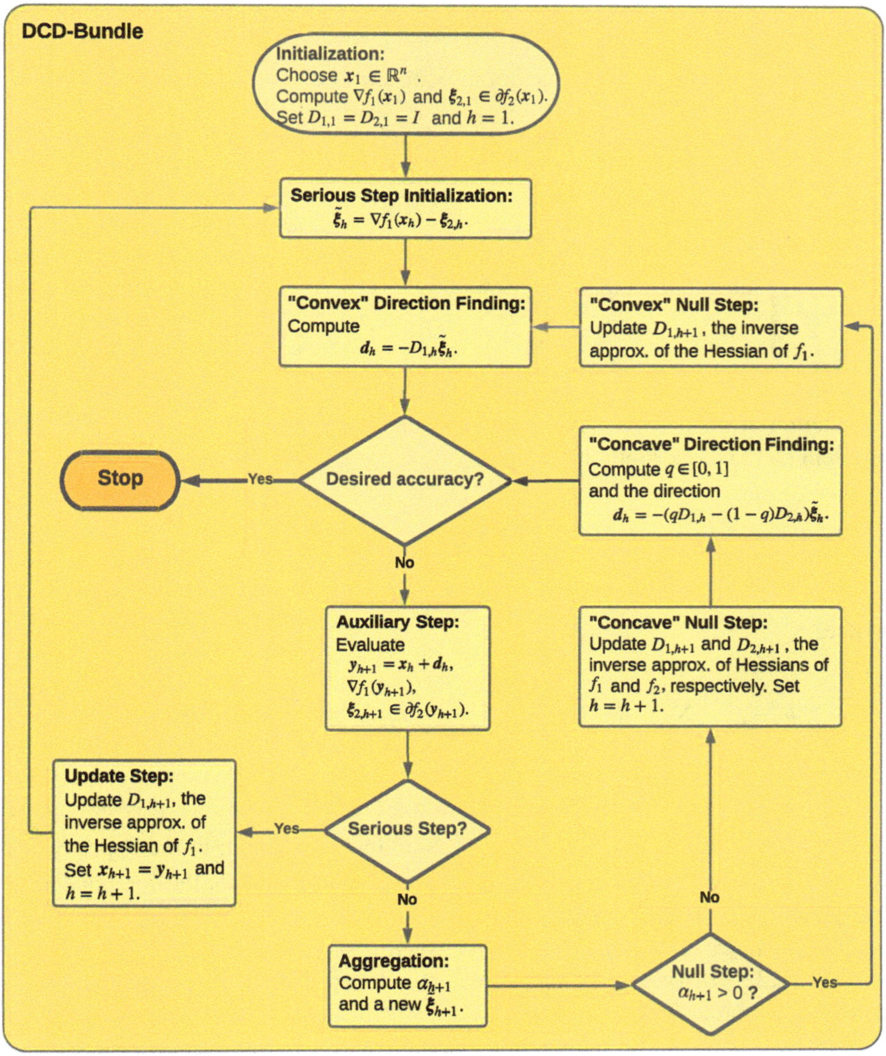

Fig. 3.6 DC diagonal bundle method

we use the gradient vectors similar to smooth variable metric methods. Since the second component f_2 is nonsmooth the correction vectors $\boldsymbol{u}_2$ are computed using subgradients. Therefore, instead of just two correction matrices S_h and U_h used in the LMBM, we have three correction matrices

$$S_h = \big[\boldsymbol{s}_{h-\hat{m}_h+1} \dots \boldsymbol{s}_h\big],$$

$$U_{1,h} = \big[\boldsymbol{u}_{1,h-\hat{m}_h+1} \dots \boldsymbol{u}_{1,h}\big], \qquad \text{and}$$

$$U_{2,h} = \left[\boldsymbol{u}_{2,h-\hat{m}_h+1} \dots \boldsymbol{u}_{2,h}\right],$$

where $\hat{m}_h = \min\{h, \hat{m}_c\}$. We use these matrices to compute separate approximations to both f_1 and f_2.

The diagonal approximation of the Hessian $B_{l,h+1}$ ($l = 1, 2$) is defined by solving the problem

$$\begin{cases} \text{minimize} & \left\| B_{l,h+1} S_h - U_{l,h} \right\|_F^2 \\ \text{subject to} & \left(B_{l,h+1}\right)_{ij} = 0, \quad \text{for} \quad i \neq j, \\ & \left(B_{l,h+1}\right)_{ii} \geq \mu, \quad \text{for} \quad i = 1, 2, \dots, n, \end{cases} \tag{3.15}$$

for some $\mu > 0$, where $\|.\|_F$ denotes the Frobenius norm of a matrix. Note that the verification of the positive definiteness is added to the problem as a constraint. The problem (3.15) has a solution

$$\left(B_{l,h+1}\right)_{ii} = \begin{cases} \boldsymbol{b}_i / Q_{ii}, & \text{if} \quad \boldsymbol{b}_i / Q_{ii} > \mu, \\ \mu, & \text{otherwise,} \end{cases}$$

where

$$\boldsymbol{b} = 2 \sum_{i=h-\hat{m}_h+1}^{\hat{m}_h} \operatorname{diag}(\boldsymbol{s}_i) \boldsymbol{u}_{l,i} \quad \text{and} \quad Q = 2 \sum_{i=h-\hat{m}_h+1}^{\hat{m}_h} \left[\operatorname{diag}(\boldsymbol{s}_i)\right]^2,$$

with $\boldsymbol{s}_i \in S$ and $\boldsymbol{u}_{l,i} \in U_l$, $l = 1, 2$. In the DCD-BUNDLE, we use the inverse of this matrix, that is, $D_{l,h} = \left(B_{l,h}\right)^{-1}$. Note that in addition to the upper bound $\mu_{\max} = \frac{1}{\mu}$, we use the lower bound $\mu_{\min}$ ($0 < \mu_{\min} < \mu_{\max}$) for the components of the matrix. We call the approximations $D_{1,h}$ and $D_{2,h}$ the "*convex approximation*" and the "*concave approximation*," respectively.

The DCD-BUNDLE uses the above mentioned diagonal approximations to compute a search direction. If the previous step was a serious step, we suppose that the convex model of the function is good enough. In this case, we use directly the "convex approximation" of the Hessian, the current subgradient of the objective function and compute the search direction by the formula

$$\boldsymbol{d}_h = -D_{1,h} \boldsymbol{\xi}_h,$$

where $\boldsymbol{\xi}_h = \nabla f_1(\boldsymbol{x}_h) - \boldsymbol{\xi}_{2,h} \in \partial f(\boldsymbol{x}_h)$ and $\boldsymbol{\xi}_{2,h} \in \partial f_2(\boldsymbol{x}_h)$. Otherwise, we use the sign of the linearization error to detect the "convex" or "concave" behavior of the objective. The *linearization error* α_{h+1} at a point $\boldsymbol{y}_{h+1}$ is defined by

$$\alpha_{h+1} = f(\boldsymbol{x}_h) - f(\boldsymbol{y}_{h+1}) + \left(\nabla f_1(\boldsymbol{y}_{h+1}) - \boldsymbol{\xi}_{2,h+1}\right)^T \boldsymbol{d}_h.$$

If $\alpha_{h+1} > 0$, we still use the "convex approximation" of the Hessian, but instead of an arbitrary subgradient $\boldsymbol{\xi}_h$ we use an *aggregate subgradient* $\tilde{\boldsymbol{\xi}}_h$. Thus, the search direction is given by

$$\boldsymbol{d}_h = -D_{1,h}\tilde{\boldsymbol{\xi}}_h.$$

Finally, in the case of negative α_{h+1}, we first compute the convex combination of the "convex" and negative "concave" approximations such that the combination still remains positive definite. Then we use this combination to compute the search direction. In other words, we compute the smallest $q_h \in [0, 1]$ such that $q_h D_{1,h} - (1 - q_h)D_{2,h}$ is positive definite. Since $D_{1,h}$ and $D_{2,h}$ are both diagonal matrices this value is very easy to compute. The search direction is computed by the formula

$$\boldsymbol{d}_h = -\big(q_h D_{1,h} - (1 - q_h)D_{2,h}\big)\tilde{\boldsymbol{\xi}}_h. \tag{3.16}$$

Next, we compute a new auxiliary point: $\boldsymbol{y}_{h+1} = \boldsymbol{x}_h + \boldsymbol{d}_h$. Note that no line search is used here (cf. the LMBM). A necessary condition for a serious step to be taken is to have

$$f(\boldsymbol{y}_{h+1}) \le f(\boldsymbol{x}_h) - \varepsilon_L w_h, \tag{3.17}$$

where $\varepsilon_L \in (0, 1/2)$ is a given descent parameter and $w_h > 0$ represents the desirable amount of descent of f at $\boldsymbol{x}_h$. If the condition (3.17) is satisfied, we set $\boldsymbol{x}_{h+1} = \boldsymbol{y}_{h+1}$ and a serious step is taken. In the case of a serious step we consider the current "convex approximation" to be good enough and continue with this metric even if the linearization error α_{h+1} was negative.

If the condition (3.17) is not satisfied, a null step occurs. In null steps, we search for a scalar $t \in (0, 1]$ such that $\boldsymbol{\xi}^t_{h+1} = \nabla f_1(\boldsymbol{x}_h + t\boldsymbol{d}_h) - \boldsymbol{\xi}^t_{2,h+1}$, with $\boldsymbol{\xi}^t_{2,h+1} \in \partial f_2(\boldsymbol{x}_h + t\boldsymbol{d}_h)$, satisfies the condition (cf. (3.8))

$$-\beta_{h+1} + (\boldsymbol{\xi}^t_{h+1})^T \boldsymbol{d}_h \ge -\varepsilon_R w_h. \tag{3.18}$$

Here, $\varepsilon_R \in (\varepsilon_L, 1/2)$ is a given parameter and β_{h+1} (with $\gamma \ge 0$) is the subgradient locality measure defined similar to bundle methods as

$$\beta_{h+1} = \max\Big\{|f(\boldsymbol{x}_h) - f(\boldsymbol{x}_h + t\boldsymbol{d}_h) + t(\boldsymbol{\xi}^t_{h+1})^T \boldsymbol{d}_h|, \gamma\|t\boldsymbol{d}_h\|^2\Big\}. \tag{3.19}$$

It has been proved in [113] that t satisfying (3.19) always exists. For the simplicity, we omitted computation of t in Fig. 3.6. In the case of a null step, we set $\boldsymbol{x}_{h+1} = \boldsymbol{x}_h$. Nevertheless, information about the objective is increased as we use the auxiliary point $\boldsymbol{y}^t_{h+1} = \boldsymbol{x}_h + t\boldsymbol{d}_h$ and the corresponding auxiliary subgradient $\boldsymbol{\xi}^t_{h+1} \in \partial f(\boldsymbol{y}^t_{h+1})$ in the computation of next aggregate values. The aggregation procedure and stopping criterion used in the DCD-BUNDLE are similar to those of the original

LMBM (see the previous section) if we replace D_h by $D_{1,h}$ and $\boldsymbol{\xi}_i$ by $\nabla f_1(\boldsymbol{y}_i) - \boldsymbol{\xi}_{2,i}^t$ $(i = m, h+1)$.

The DCD-BUNDLE finds the Clarke stationary point of the DC optimization problem (2.23). If the function f_2 is differentiable, then this point is also an inf-stationary point (see Theorem 2.27 and Fig. 2.9). In addition, the inf-stationarity of the solution can be obtained under a milder assumption:

Assumption 3.4 *If the subdifferential $\partial f_2(\boldsymbol{x})$ is not a singleton at a point $\boldsymbol{x} \in \mathbb{R}^n$, then we can always compute two subgradients $\boldsymbol{\xi}_2^1, \boldsymbol{\xi}_2^2 \in \partial f_2(\boldsymbol{x})$ such that $\boldsymbol{\xi}_2^1 \neq \boldsymbol{\xi}_2^2$.*

This assumption is satisfied, for example, for the clustering problems (see Sect. 4.4).

Figure 3.7 illustrates an *algorithm for finding* inf-*stationary points* [36]. The algorithm involves a special procedure to escape from the Clarke stationary point

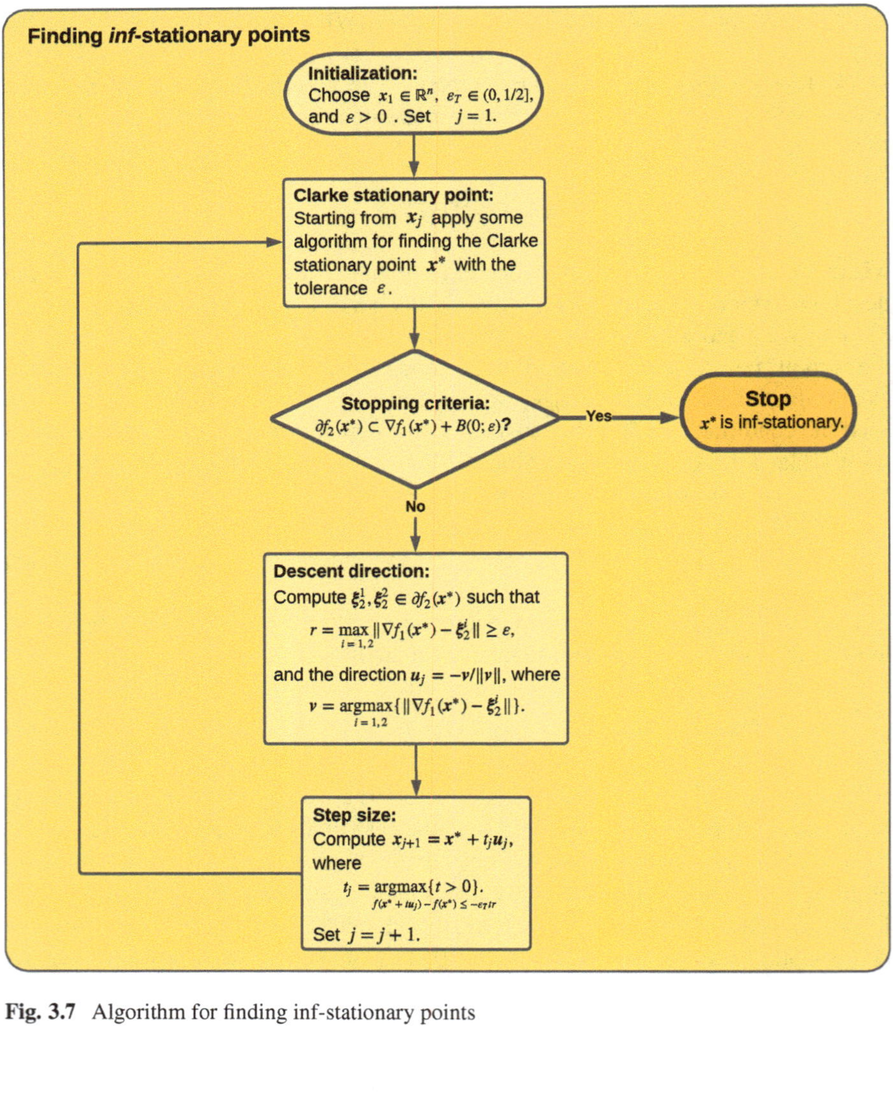

Fig. 3.7 Algorithm for finding inf-stationary points

that is not inf-stationary. The algorithm can be used together with the DCD-BUNDLE and the nonsmooth DC method (to be described in the next section).

Let us assume that $\boldsymbol{x}^*$ is a Clarke stationary point. If the subdifferential $\partial f_2(\boldsymbol{x}^*)$ is a singleton, then $\boldsymbol{x}^*$ is also an inf-stationary point. This follows from the fact that at a Clarke stationary point we have

$$\mathbf{0} \in \partial f(\boldsymbol{x}^*) = \nabla f_1(\boldsymbol{x}^*) - \boldsymbol{\xi}_2^*,$$

where $\boldsymbol{\xi}_2^*$ is the only subgradient in $\partial f_2(\boldsymbol{x}^*)$.

If $\partial f_2(\boldsymbol{x}^*) \subset \{\nabla f_1(\boldsymbol{x}^*)\} + B(\mathbf{0}; \varepsilon)$ for some sufficiently small $\varepsilon > 0$, then the point $\boldsymbol{x}^*$ can be considered as an approximate inf-stationary point. Otherwise, if the subdifferential $\partial f_2(\boldsymbol{x}^*)$ is not a singleton at $\boldsymbol{x}^*$, we can compute the subgradients $\boldsymbol{\xi}_2^1, \boldsymbol{\xi}_2^2 \in \partial f_2(\boldsymbol{x}^*)$ such that $\boldsymbol{\xi}_2^1 \neq \boldsymbol{\xi}_2^2$ and the direction $\boldsymbol{u} = -\boldsymbol{v}/\|\boldsymbol{v}\|$, where

$$\boldsymbol{v} = \underset{i=1,2}{\operatorname{argmax}} \|\nabla f_1(\boldsymbol{x}^*) - \boldsymbol{\xi}_2^i\|.$$

Then we have $f'(\boldsymbol{x}^*; \boldsymbol{u}) \leq -\|\boldsymbol{v}\|$ and the direction $\boldsymbol{u}$ is a descent direction for the objective. Therefore, we can use it to find a new starting point for the DCD-BUNDLE or the nonsmooth DC method.

3.5.1 Convergence of the DCD-BUNDLE

In this subsection, we discuss briefly about the convergence properties of the DCD-BUNDLE. More detailed proofs of Lemmas can be found in [170]. The following assumptions are needed in the proofs.

Assumption 3.5 *The objective function $f = f_1 - f_2$ is DC and f_1 is smooth.*

Assumption 3.6 *The level set $\operatorname{lev}_{f(\boldsymbol{x}_1)} f$ is bounded for every starting point $\boldsymbol{x}_1 \in \mathbb{R}^n$.*

Note that Assumption 3.6 is the same as Assumption 3.3 used in the LMBM. Assumptions 3.1 and 3.2 are also valid here since a DC function is always LLC and semismooth.

We first show that a point generated by the DCD-BUNDLE is a Clarke stationary point of the DC optimization problem (2.23). In order to do so, we assume that the optimality tolerance $\varepsilon = 0$. This part of the proof closely follows the ideas of the convergence proof of the LMBM. Nevertheless, some results are easier to prove for the DCD-BUNDLE due to the diagonal update formula and the lack of the line search during serious steps. After that, we prove that the DCD-BUNDLE combined with the algorithm for finding inf-stationary points terminates after a finite number of iterations at an inf-stationary point.

Remark 3.3 The sequence $\{\boldsymbol{x}_h\}$ generated by the DCD-BUNDLE is bounded by Assumption 3.6. The monotonicity of the sequence $\{f_h\}$ is obtained due to the condition (3.17) being satisfied at serious steps and the fact that $\boldsymbol{x}_{h+1} = \boldsymbol{x}_h$ for null steps. The matrices D_1 and D_2 are bounded since all their components are in a closed interval $[\mu_{\min}, \mu_{\max}]$. Therefore, the search direction $\boldsymbol{d}_h$ and the sequence $\{\boldsymbol{y}_h\}$ are also bounded. The boundedness of subgradients $\boldsymbol{\xi}_h$ and their convex combinations follow from the local boundedness and the upper semicontinuity of the subdifferential.

The next lemma and proposition are similar to Lemma 3.1 and Proposition 3.2, respectively, given in the proof of the LMBM.

Lemma 3.5 *Suppose that the* DCD-BUNDLE *is not terminated before the hth iteration. Then, there exist numbers* $\lambda^{h,j} \geq 0$ *for* $j = 1, \ldots, h$ *and* $\tilde{\sigma}_h \geq 0$ *such that*

$$(\tilde{\boldsymbol{\xi}}_h, \tilde{\sigma}_h) = \sum_{j=1}^{h} \lambda^{h,j}(\boldsymbol{\xi}_j, \|\boldsymbol{y}_j - \boldsymbol{x}_h\|), \quad \sum_{j=1}^{h} \lambda^{h,j} = 1 \quad \textit{and} \quad \tilde{\beta}_h \geq \gamma\tilde{\sigma}_h^2.$$

Proposition 3.4 *If the* DCD-BUNDLE *terminates at the hth iteration, then the point* $\boldsymbol{x}_h$ *is Clarke stationary for* f.

Proof If the DCD-BUNDLE terminates, then the condition $\varepsilon = 0$ implies that $w_h = 0$. Therefore, $\tilde{\boldsymbol{\xi}}_h = \boldsymbol{0}$ and by (3.10), (3.12), (3.19) we get $\tilde{\beta}_h = 0$. Since we set $\tilde{\beta}_h = 0$ after serious steps by Lemma 3.5 we have $\tilde{\sigma}_h = 0$. Applying Lemma 3.2 with

$$\bar{\boldsymbol{x}} = \boldsymbol{x}_h, \quad l = h, \quad \bar{\boldsymbol{g}} = \tilde{\boldsymbol{\xi}}_h, \quad \text{and}$$
$$\bar{\boldsymbol{\xi}}_i = \boldsymbol{\xi}_i, \quad \bar{\boldsymbol{y}}_i = \boldsymbol{y}_i, \quad \bar{\lambda}_i = \lambda^{h,i} \quad \text{for} \quad i \leq h,$$

and using Lemma 3.5 we have $\boldsymbol{0} = \tilde{\boldsymbol{\xi}}_h \in \partial f(\boldsymbol{x}_h)$, meaning that $\boldsymbol{x}_h$ is a Clarke stationary point for f.

If the DCD-BUNDLE does not terminate, then $w_h > 0$ for all h. The next lemma, corollary, and proposition are similar, respectively, to Lemma 3.4, Corollary 3.1, and Proposition 3.3 given in the proof of the LMBM. Nevertheless, due to the use of the diagonal updates the limit with respect to h in Lemma 3.6 differs from that in Lemma 3.4. In addition, since the DCD-BUNDLE does not use the line search at serious steps the proof of Proposition 3.5 is simpler than that of Proposition 3.3.

Lemma 3.6 *Suppose that the number of serious steps is finite and the last serious step occurs at the iteration* $m - 1$. *Then* $w_{h+1} \leq w_h$ *for all* $h > m$. *In addition,* $w_h \to 0$ *as* $h \to \infty$.

Corollary 3.2 *Suppose that the number of serious steps is finite and the last serious step occurs at the iteration* $m - 1$. *Then, the point* $\boldsymbol{x}_m$ *is Clarke stationary of the function* f.

Proposition 3.5 *Every accumulation point of the sequence* $\{\boldsymbol{x}_h\}$ *is Clarke stationary for* f.

Proof Let $\bar{x}$ be an accumulation point of the sequence $\{x_h\}$ and $\mathcal{H} \subset \{1, 2, \ldots\}$ be an infinite set such that $\{x_h\}_{h\in\mathcal{H}} \to \bar{x}$. In view of Corollary 3.2, we can restrict our consideration to the case where the number of serious steps is infinite. Denote by

$$\mathcal{H}' = \left\{h : x_{h+1} = x_h + d_h \text{ and there exists} i \in \mathcal{H},\ i \leq h \text{ such that} x_i = x_h\right\}.$$

It is clear that $\mathcal{H}'$ is infinite and $\{x_h\}_{h\in\mathcal{H}'} \to \bar{x}$. The continuity of f implies that $\{f(x_h)\}_{h\in\mathcal{H}'} \to f(\bar{x})$, and, therefore, $f(x_h) \downarrow f(\bar{x})$ due to the monotonicity of the sequence $\{f(x_h)\}$ according to the descent step condition (3.17). Using the condition (3.17) and the fact that $x_{h+1} = x_h$ in null steps, we obtain

$$0 \leq \varepsilon_L w_h \leq f(x_h) - f(x_{h+1}) \to 0 \quad \text{for} \quad h \geq 1. \tag{3.20}$$

Thus, we have

$$\{w_h\}_{h\in\mathcal{H}'} \to 0 \quad \text{and} \quad \{x_h\}_{h\in\mathcal{H}'} \to \bar{x}.$$

Then it follows from Lemma 3.3 that $\mathbf{0} \in \partial f(\bar{x})$. □

Remark 3.4 If $\varepsilon > 0$, then the DCD-BUNDLE terminates after the finite number of iterations. In addition, the proofs remain valid if f_1 is nonsmooth convex and f_2 is smooth convex, or if both f_1 and f_2 are nonsmooth convex functions with a condition that we can compute the subgradient $\boldsymbol{\xi} \in \partial f(x)$ at any x.

Now, we prove that the algorithm for finding inf-stationary points (see Fig. 3.7) terminates after a finite number of iterations. In addition to Assumptions 3.4–3.6, we need the next assumption.

Assumption 3.7 *The gradient $\nabla f_1 : \mathbb{R}^n \to \mathbb{R}^n$ of the function f_1 satisfies the Lipschitz condition.*

Proposition 3.6 *Assume that the subdifferential $\partial f_2(x)$ is not a singleton at x and the subgradients $\boldsymbol{\xi}_2^1, \boldsymbol{\xi}_2^2 \in \partial f_2(x)$ are such that $\boldsymbol{\xi}_2^1 \neq \boldsymbol{\xi}_2^2$. Consider the direction $u = -v/\|v\|$ where*

$$v = \underset{i=1,2}{\operatorname{argmax}}\ \|\nabla f_1(x^*) - \boldsymbol{\xi}_2^i\|.$$

Then

$$f'(x; u) \leq -\|v\|.$$

Proof Since the subdifferential $\partial f_2(x)$ is not a singleton and $\boldsymbol{\xi}_2^1 \neq \boldsymbol{\xi}_2^2$ it follows that $v \neq 0$. For simplicity, assume $v = \nabla f_1(x) - \boldsymbol{\xi}_2^2$. Then the convexity of functions f_1 and f_2 implies that

$$f'(x; u) = f_1'(x; u) - f_2'(x; u)$$

$$
\begin{aligned}
&=\big(\nabla f_1(\boldsymbol{x})\big)^T\boldsymbol{u} - \max_{\boldsymbol{\xi}_2\in\partial f_2(\boldsymbol{x})} \boldsymbol{\xi}_2^T\boldsymbol{u}\\
&\le\big(\nabla f_1(\boldsymbol{x}) - \boldsymbol{\xi}_2^2\big)^T\boldsymbol{u}\\
&=-\ \|\nabla f_1(\boldsymbol{x}) - \boldsymbol{\xi}_2^2\| < 0.
\end{aligned}
$$

Thus, the direction $\boldsymbol{u}$ is a descent direction at the point $\boldsymbol{x}$. □

Proposition 3.7 *If at a Clarke stationary point $\boldsymbol{x} \in \mathbb{R}^n$ of the problem* (2.23) *the subdifferential $\partial f_2(\boldsymbol{x})$ is a singleton, then $\boldsymbol{x}$ is also an inf-stationary point.*

Proof The proof follows from (3.22) and the definition of inf-stationary points. □

Corollary 3.3 *If at a Clarke stationary point $\boldsymbol{x} \in \mathbb{R}^n$ of the problem* (2.23) *the subdifferential $\partial f_2(\boldsymbol{x})$ is not a singleton, then $\boldsymbol{x}$ is not an inf-stationary point.*

Proposition 3.8 *Let $\varepsilon > 0$ be any given number. Then the algorithm for finding* inf-*stationary points terminates after a finite number of iterations at an approximate* inf-*stationary point $\boldsymbol{x}^*$ of the problem* (2.23) *satisfying the condition*

$$
\partial f_2(\boldsymbol{x}^*) \subset \big\{\nabla f_1(\boldsymbol{x}^*)\big\} + B(\boldsymbol{0};\varepsilon). \tag{3.21}
$$

Proof Assume that at the jth iteration we get the Clarke stationary point $\boldsymbol{x}_j$ which does not satisfy the condition (3.21). Then the subdifferential $\partial f_2(\boldsymbol{x}_j)$ is not a singleton and, therefore, there exist subgradients $\boldsymbol{\xi}_2^1,\ \boldsymbol{\xi}_2^2 \in \partial f_2(\boldsymbol{x}_j)$ such that $\boldsymbol{\xi}_2^1 \neq \boldsymbol{\xi}_2^2$. For simplicity, assume that $\boldsymbol{u}_j = -\boldsymbol{v}/\|\boldsymbol{v}\|$ and $\boldsymbol{v} = \nabla f_1(\boldsymbol{x}_j) - \boldsymbol{\xi}_2^1$, where $\|\boldsymbol{v}\| \ge \varepsilon$. Take any $t > 0$. The mean-value theorem (Theorem 2.9) implies that for some $\sigma_j \in (0, 1)$ we have

$$
\begin{aligned}
f(\boldsymbol{x}_j + t\boldsymbol{u}_j) - f(\boldsymbol{x}_j) &= \big(f_1(\boldsymbol{x}_j + t\boldsymbol{u}_j) - f_1(\boldsymbol{x}_j)\big) - \big(f_2(\boldsymbol{x}_j + t\boldsymbol{u}_j) - f_2(\boldsymbol{x}_j)\big)\\
&\le t\big(\nabla f_1(\boldsymbol{x}_j + t\sigma_j\boldsymbol{u}_j)\big)^T\boldsymbol{u}_j - t\big(\boldsymbol{\xi}_2^1\big)^T\boldsymbol{u}_j\\
&\le t\big(\nabla f_1(\boldsymbol{x}_j) - \boldsymbol{\xi}_2^1\big)^T\boldsymbol{u}_j\\
&\quad + t\big(\nabla f_1(\boldsymbol{x}_j + t\sigma_j\boldsymbol{u}_j) - \nabla f_1(\boldsymbol{x}_j)\big)^T\boldsymbol{u}_j.
\end{aligned}
$$

Let $L > 0$ be a Lipschitz constant of the gradient ∇f_1. Then

$$
\|\nabla f_1(\boldsymbol{x}_j + t\sigma_j\boldsymbol{u}_j) - \nabla f_1(\boldsymbol{x}_j)\| \le Lt\sigma_j\|\boldsymbol{u}_j\| = Lt\sigma_j.
$$

This means that

$$
\begin{aligned}
f(\boldsymbol{x}_j + t\boldsymbol{u}_j) - f(\boldsymbol{x}_j) \le& t\big(\nabla f_1(\boldsymbol{x}_j) - \boldsymbol{\xi}_2^1\big)^T\boldsymbol{u}_j + Lt^2\sigma_j\\
=&-t\|\nabla f_1(\boldsymbol{x}_j) - \boldsymbol{\xi}_2^1\| + Lt^2\sigma_j\\
<&\ t(-r + Lt),
\end{aligned}
$$

where $r = \max_{i=1,2} \|\nabla f_1(\boldsymbol{x}_j) - \boldsymbol{\xi}_2^i\|$. Since $\|\boldsymbol{v}\| \geq \varepsilon$ it follows that $r \geq \varepsilon$. Then for $\bar{t} = r/2L$ and for any $\varepsilon_T \in (0, 1/2]$ we have

$$f(\boldsymbol{x}_j + \bar{t}\bar{\boldsymbol{u}}_j) - f(\boldsymbol{x}_j) < -\frac{r^2}{4L} \leq -\varepsilon_T \bar{t} r \leq -\varepsilon_T \bar{t} \varepsilon.$$

This means that at each iteration we have $t_j \geq \bar{t} \geq \varepsilon/2L$ and the function f decreases by at least $\varepsilon_T \varepsilon^2/2L > 0$ at each iteration. By Assumption 3.6 the function f is bounded from below. Thus, the algorithm for finding inf-stationary points must stop after a finite number of iterations. □

3.6 Nonsmooth DC Method

In this section, we introduce another algorithm, the *nonsmooth DC method* (NDCM), for solving unconstrained NSO problems with the DC objective function (2.23), where f_1 is a smooth convex function and f_2 is, in general, a nonsmooth convex function. Similar to the DCD-BUNDLE given in the previous section, the NDCM is originally developed as a part of the clustering algorithm [36].

The method uses the bundling idea in case of null steps, but it only bundles gradients of the first component function f_1. The subgradient of the second component function f_2 is kept fixed during the null steps. In what follows, we assume that at every point $\boldsymbol{x}$ we can evaluate the values of the DC components f_1 and f_2 of the objective f, the gradient $\nabla f_1(\boldsymbol{x})$ of the first component, and one arbitrary subgradient $\boldsymbol{\xi}_2$ from the subdifferential $\partial f_2(\boldsymbol{x})$ of the second component. Due to the smoothness of f_1 we have

$$\partial f(\boldsymbol{x}) = \operatorname{conv}\left\{\nabla f_1(\boldsymbol{x}) - \boldsymbol{\xi}_2 : \boldsymbol{\xi}_2 \in \partial f_2(\boldsymbol{x})\right\}. \tag{3.22}$$

Let us take any $\lambda > 0$ and define the following sets at a point $\boldsymbol{x} \in \mathbb{R}^n$:

$$Q_1(\boldsymbol{x}, \lambda) = \operatorname{conv}\left\{\nabla f_1(\boldsymbol{x} + \lambda \boldsymbol{g}) : \boldsymbol{g} \in S_1\right\} \quad \text{and}$$
$$\tilde{Q}(\boldsymbol{x}, \lambda, \boldsymbol{\xi}_2) = Q_1(\boldsymbol{x}, \lambda) - \boldsymbol{\xi}_2,$$

where $\boldsymbol{\xi}_2 \in \partial f_2(\boldsymbol{x})$ and S_1 is the sphere of the unit ball.

Definition 3.1 A point $\boldsymbol{x}^* \in \mathbb{R}^n$ is called a (λ, δ)-inf-*stationary* of the problem (2.23) (with smooth f_1) if and only if

$$\partial f_2(\boldsymbol{x}^*) \subset Q_1(\boldsymbol{x}^*, \lambda) + B(\boldsymbol{0}; \delta). \tag{3.23}$$

Definition 3.2 A point $x^* \in \mathbb{R}^n$ is called a (λ, δ)-*stationary* of the problem (2.23) if there exists $\xi_2 \in \partial f_2(x^*)$ such that

$$\xi_2 \in Q_1(x^*, \lambda) + B(\mathbf{0}; \delta). \tag{3.24}$$

If a point $x \in \mathbb{R}^n$ is not a (λ, δ)-stationary point, then the set $\tilde{Q}(x, \lambda, \xi_2)$ can be used to find a descent direction for the function f at x. However, the computation of the entire set $\tilde{Q}(x, \lambda, \xi_2)$ is usually not possible. In the NDCM, we only use a finite number of elements from this set to compute search directions.

The NDCM has both inner and outer iterations. In its turn, the inner iteration consists of serious and null steps. We select sequences $\{\delta_h\}$ and $\{\lambda_h\}$ such that $\delta_h, \lambda_h \downarrow 0$ as $h \to \infty$. In principle, any such sequences can be chosen in the NDCM.

The outer iteration depends on the index h and in this iteration parameters δ_h and λ_h are updated. The inner iteration depends on the index s. In the inner iteration, we compute the search direction and we either update the solution or add a new element to the set $\bar{Q}_h^s \subset \tilde{Q}(x_{h_s}, \lambda_h, \xi_{2,h_s})$. In other words, we either take a serious step or a null step occurs. Figure 3.8 illustrates the NDCM.

At the beginning of each inner iteration (i.e., $s = 1$), we first compute the gradient $\nabla f_1(x_{h_1} + \lambda_h g)$ with respect to any initial direction $g \in S_1$ and an arbitrary subgradient $\xi_{2,h_1} \in \partial f_2(x_{h_1})$. We set $z_{h_1} = \nabla f_1(x_{h_1} + \lambda_h g) - \xi_{2,h_1}$, define the initial bundle $\bar{Q}_h^1 = \{z_{h_1}\}$, and compute the search direction d_{h_1} by

$$d_{h_1} = -\frac{z_{h_1}}{\|z_{h_1}\|}.$$

In the following inner iterations (i.e., $s > 1$), we compute the vector

$$z_{h_s} = \underset{z \in \bar{Q}_h^s}{\operatorname{argmin}} \|z\|^2,$$

and the search direction

$$d_{h_s} = -\frac{z_{h_s}}{\|z_{h_s}\|}.$$

Next, we check whether the direction d_{h_s} ($s \geq 1$) is descent or not. If it is, we have

$$f(x_{h_s} + \lambda_h d_{h_s}) - f(x) \leq -\varepsilon_L \lambda_h \|z_{h_s}\|, \tag{3.25}$$

with the given numbers $\varepsilon_L \in (0, 1)$ and $\lambda_h > 0$. In this case, we compute the next (inner) iteration point

$$x_{h_{s+1}} = x_{h_s} + t_{h_s} d_{h_s},$$

where the step size t_{h_s} is defined as

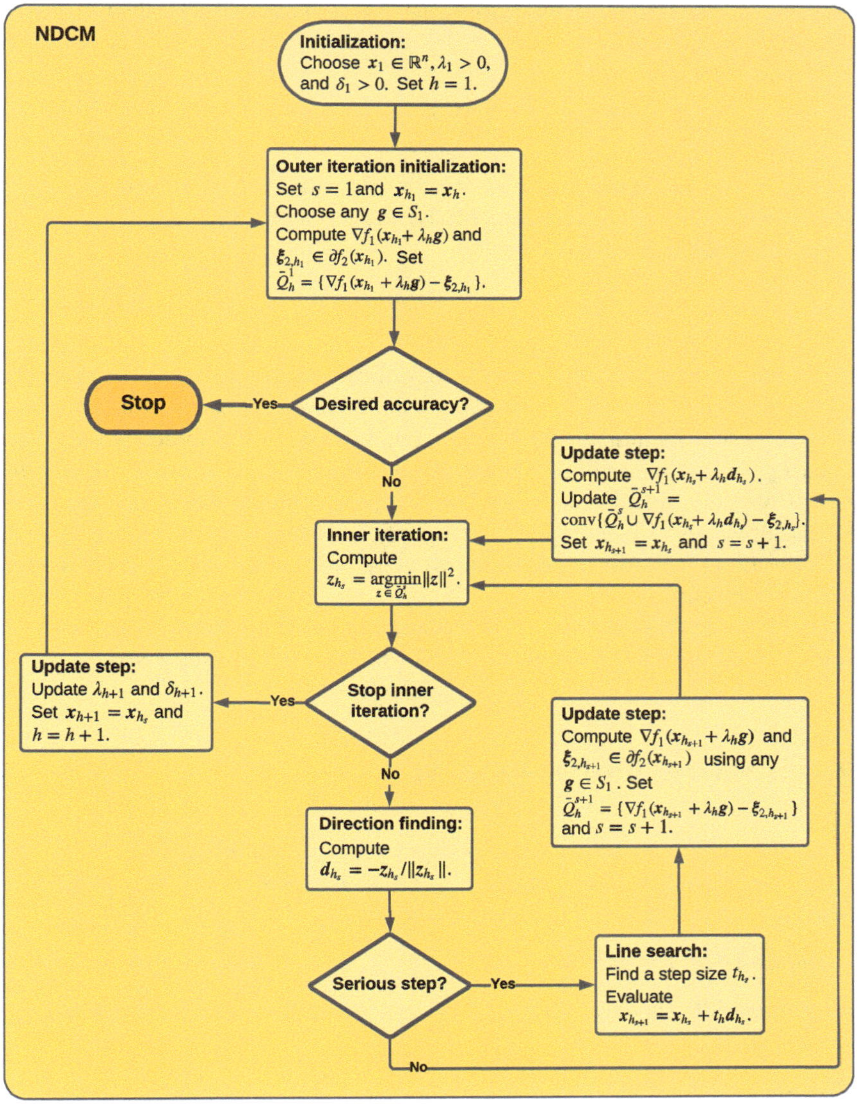

Fig. 3.8 Nonsmooth DC method

$$t_{h_s} = \text{argmax}\left\{t \geq 0 : f(\boldsymbol{x}_{h_s} + t\boldsymbol{d}_{h_s}) - f(\boldsymbol{x}_{h_s}) \leq -\varepsilon_R t \|z_{h_s}\|\right\},$$

with a given $\varepsilon_R \in (0, \varepsilon_L]$. In the case of a serious step, we set

$$\bar{Q}_h^{s+1} = \left\{\nabla f_1(\boldsymbol{x}_{h_{s+1}} + \lambda_h \boldsymbol{g}) - \boldsymbol{\xi}_{2,h_{s+1}}\right\},$$

using any direction $\boldsymbol{g} \in S_1$ and continue to the next inner iteration with $s = s + 1$.

Otherwise, a null step occurs. In this case, we compute a new gradient $\nabla f_1(\boldsymbol{x}_{h_s} + \lambda_h \boldsymbol{d}_{h_s})$, update the set

$$\bar{Q}_h^{s+1} = \text{conv}\left\{\bar{Q}_h^s \cup \{\nabla f_1(\boldsymbol{x}_{h_s} + \lambda_h \boldsymbol{d}_{h_s}) - \boldsymbol{\xi}_{2,h_s}\}\right\},$$

set $\boldsymbol{x}_{h_{s+1}} = \boldsymbol{x}_{h_s}$, and continue the inner iterations with $s = s + 1$. Note that, at null steps the subgradient $\boldsymbol{\xi}_{2,h_s}$ remains unchanged (i.e., we set $\boldsymbol{\xi}_{2,h_{s+1}} = \boldsymbol{\xi}_{2,h_s}$).

The inner iteration stops if

$$\|\boldsymbol{z}_{h_s}\| \leq \delta_h.$$

This condition means that for given values of δ_h and λ_h the last iteration $\boldsymbol{x}_{h_s}$ is a (λ_h, δ_h)-stationary point. Therefore, this point is accepted as a new iteration $\boldsymbol{x}_{h+1}$ and the algorithm returns to the outer iteration to update the values of parameters δ_h and λ_h. In its turn, the outer iteration stops if both $\delta_h < \varepsilon$ and $\lambda_h < \varepsilon$ with a given termination tolerance $\varepsilon > 0$. This stopping criterion means that the further decrease of values of δ_h and λ_h will not significantly improve the solution $\boldsymbol{x}_h$. It has been proved in [36]—and it will be shown in the next subsection—that this solution is Clarke stationary.

Similar to the DCD-BUNDLE, we can use the special procedure given in Fig. 3.7 to escape from the Clarke stationary points which are not inf-stationary.

3.6.1 Convergence of the NDCM

In this subsection, we study the convergence properties of the NDCM. A more detailed description with the proofs of Lemmas is given in [36]. The assumptions needed are the same as with the DCD-BUNDLE:

Assumption 3.8 *The objective function $f = f_1 - f_2$ is DC and f_1 is smooth.*

Assumption 3.9 *The level set* $\text{lev}_{f(\boldsymbol{x}_1)} f$ *is bounded for every starting point $\boldsymbol{x}_1 \in \mathbb{R}^n$.*

We start by recalling that if a point $\boldsymbol{x}$ is not a (λ, δ)-stationary, then the set $\tilde{Q}(\boldsymbol{x}, \lambda, \boldsymbol{\xi}_2)$ can be used to find a descent direction. After that, we show that the number of null steps in the NDCM is finite and, eventually, also the number of inner iterations is finite. We conclude by proving that the NDCM converges to a Clarke stationary point of the problem (2.23).

Proposition 3.9 *Assume that the point $\boldsymbol{x}$ is not (λ, δ)-stationary. Then the direction*

$$\bar{\boldsymbol{d}} = -\frac{\bar{\boldsymbol{z}}}{\|\bar{\boldsymbol{z}}\|},$$

where $\bar{z} = \operatorname{argmin}\{\|z\|^2 : z \in \tilde{Q}(x, \lambda, \xi_2)\} \neq \mathbf{0}$, *is a descent direction of the function* f *at* x, *and*

$$f(x + \lambda\bar{d}) - f(x) \leq -\lambda\|\bar{z}\|.$$

Proof Since the point x is not (λ, δ)-stationary we have $\|\xi_2 - z\| \geq \delta$ for all $\xi_2 \in \partial f_2(x)$ and $z \in Q_1(x, \lambda)$. Therefore, $\|\bar{z}\| \geq \delta$ and

$$f(x + \lambda d) - f(x) \leq \lambda \max_{z \in \tilde{Q}(x,\lambda,\xi_2)} z^T d$$

for all $d \in \mathbb{R}^n$. From the necessary condition for a minimum we have

$$\bar{z}^T(z - \bar{z}) \geq 0 \quad \text{for all} \quad z \in \tilde{Q}(x, \lambda, \xi_2),$$

or equivalently

$$\bar{z}^T z \geq \|\bar{z}\|^2 \quad \text{for all} \quad z \in \tilde{Q}(x, \lambda, \xi_2).$$

Dividing both sides by $-\|\bar{z}\|$ we have $z^T\bar{d} \leq -\|\bar{z}\|$ for all $z \in \tilde{Q}(x, \lambda, \xi_2)$, and the proof follows. □

Lemma 3.7 *Let* $M \in (0, \infty)$ *be such that*

$$\max\Big\{\max\big\{\|\nabla f_1(x + \lambda g)\| : g \in S_1\big\}, \max\big\{\|\xi_2\| : \xi_2 \in \partial f_2(x)\big\}\Big\} \leq M.$$

Then there are at most s_0 *null steps in an inner iteration of the* NDCM, *where*

$$s_0 = \left\lceil \frac{4}{(1 - \varepsilon_L)^2}\left(\frac{M}{\delta}\right)^4 \right\rceil.$$

Here, $\lceil\cdot\rceil$ *is a ceiling of a number and* $\varepsilon_L \in (0, 1)$.

Proposition 3.10 *Assume that* $f^* = \inf\big\{f(x),\ x \in \mathbb{R}^n\big\} > -\infty$. *For any* $\lambda_h, \delta_h > 0$, *the* NDCM *finds* (λ_h, δ_h)*-stationary points using at most* $s_{\max}$ *serious steps where*

$$s_{\max} = \left\lceil \frac{f(x_{h_1}) - f^*}{\varepsilon_R \lambda_h \delta_h} \right\rceil. \tag{3.26}$$

Proof Assume the contrary, that is the sequence $\{x_{h_s}\}$ is infinite and points x_{h_s} are not (λ_h, δ_h)-stationary for any $s = 1, 2, \ldots$. This means that $\|z_{h_s}\| > \delta_h$ for all $s = 1, 2, \ldots$. Since $\varepsilon_R \leq \varepsilon_L$ it follows from (3.25) that the step size $t_{h_s} \geq \lambda_h$ for any $s > 0$. Then we have

$$f(x_{h_{s+1}}) - f(x_{h_s}) \leq -\varepsilon_R \lambda_h \delta_h,$$

and, therefore, we get

$$f(\boldsymbol{x}_{h_{s+1}}) - f(\boldsymbol{x}_{h_1}) \le -\varepsilon_R s \lambda_h \delta_h.$$

This means that $f(\boldsymbol{x}_{h_s}) \to -\infty$ as $s \to \infty$, which contradicts Assumption 3.9. Since $f^* \le f(\boldsymbol{x}_{h_{s+1}})$ it is obvious that the maximum number of iterations $s_{\max}$ is given by (3.26). □

Proposition 3.11 *Assume that* $\varepsilon = 0$. *Then all accumulation points of the sequence* $\{\boldsymbol{x}_h\}$ *generated by the* NDCM *are Clarke stationary points of the problem* (2.23).

Proof Since the NDCM is a descent algorithm the sequence $\{\boldsymbol{x}_h\}$ belongs to the level set $\text{lev}_{f(\boldsymbol{x}_1)} f$. In addition, since $\text{lev}_{f(\boldsymbol{x}_1)} f$ is compact this sequence has at least one accumulation point. Assume that $\bar{\boldsymbol{x}}$ is an accumulation point of $\{\boldsymbol{x}_h\}$ and there exists the subsequence $\{\boldsymbol{x}_{h_j}\}$ such that $\boldsymbol{x}_{h_j} \to \bar{\boldsymbol{x}}$ as $j \to \infty$. After each outer iteration h_j, we obtain a $(\lambda_{h_j}, \delta_{h_j})$-stationary point $\boldsymbol{x}_{h_j+1}$ which means that there exists $\boldsymbol{\xi}_{2,h_j+1} \in \partial f_2(\boldsymbol{x}_{h_j+1})$ such that

$$\boldsymbol{\xi}_{2,h_j+1} \in Q_1(\boldsymbol{x}_{h_j+1}, \lambda_{h_j}) + B(\mathbf{0}; \delta_{h_j}).$$

Replacing h_j by $h_j - 1$ we have

$$\boldsymbol{\xi}_{2,h_j} \in Q_1(\boldsymbol{x}_{h_j}, \lambda_{h_j-1}) + B(\mathbf{0}; \delta_{h_j-1}). \tag{3.27}$$

By continuity of the gradient $\nabla f_1(\boldsymbol{x})$ we have that for any $\gamma > 0$ there exists an index $j_0 > 0$ such that

$$\|\nabla f_1(\boldsymbol{x}_{h_j} + \lambda_{h_j-1}\boldsymbol{g}) - \nabla f_1(\bar{\boldsymbol{x}})\| < \gamma$$

for all $j > j_0$ and $\boldsymbol{g} \in S_1$. This means that for all $j > j_0$ we have

$$Q_1(\boldsymbol{x}_{h_j}, \lambda_{h_j-1}) \subset \left\{\nabla f_1(\bar{\boldsymbol{x}})\right\} + B(\mathbf{0}; \gamma), \tag{3.28}$$

and from (3.27) and (3.28) we get

$$\boldsymbol{\xi}_{2,h_j} \in \nabla f_1(\bar{\boldsymbol{x}}) + B(\mathbf{0}; \gamma + \delta_{h_j-1}). \tag{3.29}$$

The mapping $\boldsymbol{x} \mapsto \partial f_2(\boldsymbol{x})$ is upper semicontinuous. Therefore, for any $\theta > 0$ there exists $\bar{j} > 0$ such that for all $j > \bar{j}$ we have

$$\boldsymbol{\xi}_{2,h_j} \in \partial f_2(\bar{\boldsymbol{x}}) + B(\mathbf{0}; \theta).$$

Without loss of generality, assume that there exists $\bar{\boldsymbol{\xi}}_2 \in \partial f_2(\bar{\boldsymbol{x}})$ such that $\|\boldsymbol{\xi}_{2,h_j} - \bar{\boldsymbol{\xi}}_2\| < \theta$ for all $j > \bar{j}$. Then it follows from (3.29) that

$$\|\nabla f(\bar{x}) - \bar{\xi}_2\| < \theta + \gamma + \delta_{h_j - 1} \quad \text{for all} \quad j > \hat{j} = \max\{j_0, \bar{j}\}.$$

Since γ and θ are arbitrary and $\delta_h \downarrow 0$ as $h \to \infty$ we have

$$\bar{\xi}_2 - \nabla f_1(\bar{x}) = 0.$$

Thus, $\mathbf{0} \in \partial f(\bar{x})$ and $\bar{x}$ is a Clarke stationary point. □

3.7 DC Algorithm

In this section, we describe the well-known *DC algorithm* (DCA) for solving the unconstrained DC programming problem (2.23). Note that unlike the previous sections we now do not require differentiability of the first (or the second) DC component. Nevertheless, if we assume that f_2 is differentiable, then the DCA reduces to the *concave-convex procedure* (CCCP, see, e.g., [274, 309]) frequently used in machine learning applications.

In what follows, we denote by f_i^* the conjugate of the function $f_i, i = 1, 2$ as

$$f_i^*(\xi) = \sup\{\xi^T x - f_i(x) : x \in \mathbb{R}^n\}.$$

We describe the so-called *simplified DCA*. This method consists in the construction of two sequences: $\{x_h\}$ and $\{\xi_h\}$, candidates to primal and dual solutions, respectively. We take any starting point $x_1 \in \mathbb{R}^n$, set $h = 1$ and define

$$\xi_h \in \partial f_2(x_h) \quad \text{and} \quad x_{h+1} \in \partial f_1^*(\xi_h).$$

It is obvious that the point x_{h+1} is a solution to the problem

$$\begin{cases} \text{minimize} & f_1(x) - \xi_{2,h}^T x \\ \text{subject to} & x \in \mathbb{R}^n, \end{cases} \tag{3.30}$$

which is equivalent to

$$\begin{cases} \text{minimize} & f_1(x) - \big(f_2(x_h) + \xi_{2,h}^T (x - x_h)\big) \\ \text{subject to} & x \in \mathbb{R}^n. \end{cases}$$

This is a convex optimization problem whose objective function is obtained from the DC objective by replacing the second DC component f_2 with its affine underestimation. Similarly, $\xi_h \in \partial f_2(x_h)$ means that ξ_h is a solution to the convex problem

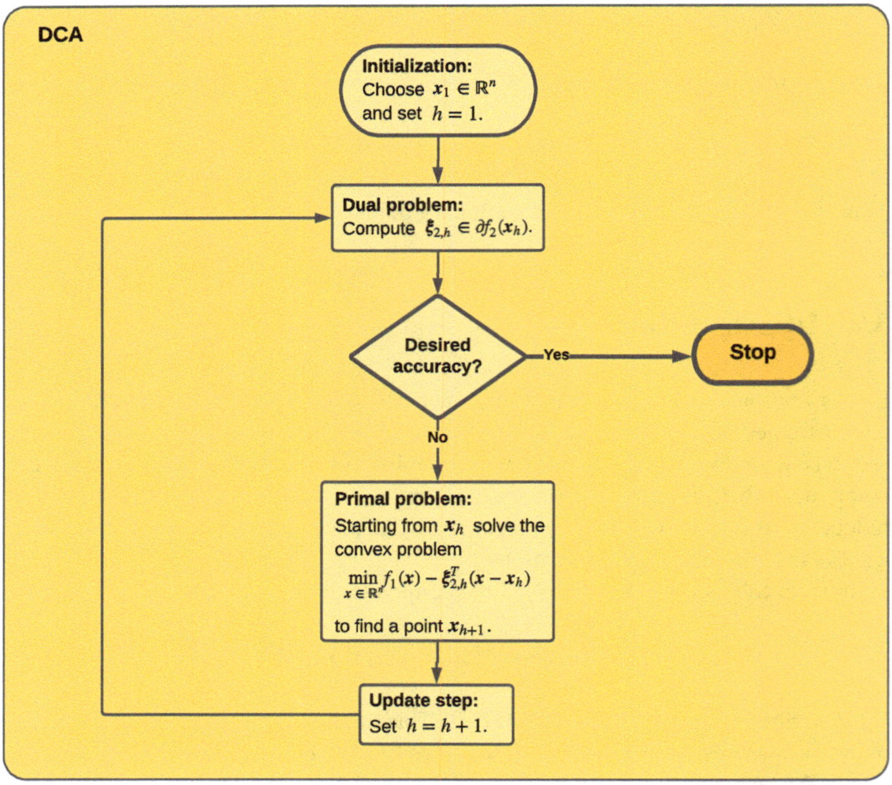

Fig. 3.9 DC algorithm

$$\begin{cases} \text{minimize} & f_2^*(\xi) - \left(f_1^*(\xi_{h-1}) + (\xi - \xi_{h-1})\right)^T x_h \\ \text{subject to} & \xi \in \mathbb{R}^n. \end{cases}$$

The sequences $\{x_h\}$ and $\{\xi_h\}$ are well defined (see Lemma 1 in [183]).

Thus, the basic—and simplified—idea of the DCA is to linearize the concave part $-f_2$ around the current iterate x_h by using the subgradient $\xi_{2,h} \in \partial f_2(x_h)$. Finding the subgradient $\xi_{2,h} \in \partial f_2(x_h)$ is considered as a *dual problem*, while finding the next iterate x_h is considered as a *primal problem*. A detailed study of the DCA can be found in [194, 195].

There are various implementations of the DCA, since the algorithm itself does not specify which optimization method is used to solve the convex primal problem (3.30). The simplified DCA scheme for solving the problem (2.23) is given in Fig. 3.9.

If the first component function f_1 is smooth, then one can apply powerful smooth optimization methods to solve (3.30). Moreover, for smooth f_1 we can easily check whether the criticality is achieved by testing if $\xi_{2,h} = \nabla f_1(x_h)$. Otherwise, the

DCA can be terminated, for instance, if $f(\boldsymbol{x}_{h+1}) = f(\boldsymbol{x}_h)$ which means that $\boldsymbol{\xi}_{2,h} \in \partial f_1(\boldsymbol{x}_h) \cap \partial f_2(\boldsymbol{x}_h)$ and the point $\boldsymbol{x}_h$ is critical for the problem (2.23).

It is well known that accumulation points of the sequence $\{\boldsymbol{x}_h\}$ generated by the DCA are critical points of the problem (2.23). In the case of clustering problems with the squared Euclidean distance, these points are also Clarke stationary since in such problems the function f_1 is smooth and the sets of critical points and Clarke stationary points of the problem (2.23) coincide (see Fig. 2.9).

3.7.1 Convergence of the DCA

We recall the convergence properties of the simplified DCA for general unconstrained DC problems. Since the proof is very specialized we omit it by referring to [194, 278] and give here only the main result without the proof.

Let

$$\Delta \boldsymbol{x}_h = \boldsymbol{x}_{h+1} - \boldsymbol{x}_h \quad \text{and} \quad \Delta \boldsymbol{\xi}_{2,h} = \boldsymbol{\xi}_{2,h+1} - \boldsymbol{\xi}_{2,h},$$

and for a convex function f define

$$\rho(f) = \sup\left\{\rho \geq 0 : f - \frac{\rho}{2}\|\cdot\|^2 \text{ is convex}\right\}. \tag{3.31}$$

Assume that $\rho(f_i)$ and $\rho(f_i^*)$ are defined for the functions f_i and f_i^*, $i = 1, 2$ by applying (3.31). Let ρ_i and ρ_i^* be such that

$$0 \leq \rho_i < \rho(f_i) \text{ and } 0 \leq \rho_i^* < \rho(f_i^*), \quad i = 1, 2.$$

Here, $\rho_i = 0$ if $\rho(f_i) = 0$ and $\rho_i^* = 0$ if $\rho(f_i^*) = 0$. Furthermore, ρ_i may take the value $\rho(f_i)$ (respectively, ρ_i^* may take the value $\rho(f_i^*)$) if it is attained ($i = 1, 2$).

The following assumptions are needed to prove the convergence of the simplified DCA.

Assumption 3.10 *The functions f_1 and f_2 are proper lower semicontinuous convex.*

Assumption 3.11 *The level set* $\text{lev}_{f(\boldsymbol{x}_1)} f$ *is bounded for any starting point* $\boldsymbol{x}_1 \in \mathbb{R}^n$.

The next proposition states that the values of the objective function are decreasing at every iteration of the DCA and the criticality of the solution is achieved when $f(\boldsymbol{x}_{h+1}) = f(\boldsymbol{x}_h)$.

Proposition 3.12 *Assume that Assumptions 3.10 and 3.11 hold true and suppose that the sequences $\{\boldsymbol{x}_h\}$ and $\{\boldsymbol{\xi}_{2,h}\}$ are generated by the simplified DCA. Then we have*

$$
\begin{aligned}
f_1(\boldsymbol{x}_{h+1}) - f_2(\boldsymbol{x}_{h+1}) &\le f_2^*(\boldsymbol{\xi}_{2,h}) - f_1^*(\boldsymbol{\xi}_{2,h}) - \max\left\{\frac{\rho_2}{2}\|\Delta\boldsymbol{x}_h\|^2, \frac{\rho_2^*}{2}\|\Delta\boldsymbol{\xi}_{2,h}\|^2\right\} \\
&\le f_1(\boldsymbol{x}_h) - f_2(\boldsymbol{x}_h) - \max\left\{\frac{\rho_1+\rho_2}{2}\|\Delta\boldsymbol{x}_h\|^2,\right. \\
&\qquad \frac{\rho_1^*}{2}\|\Delta\boldsymbol{\xi}_{2,h-1}\|^2 + \frac{\rho_2}{2}\|\Delta\boldsymbol{x}_h\|^2, \\
&\qquad \left.\frac{\rho_1^*}{2}\|\Delta\boldsymbol{\xi}_{2,h-1}\|^2 + \frac{\rho_2^*}{2}\|\Delta\boldsymbol{\xi}_{2,h}\|^2\right\}
\end{aligned}
$$

for all $h \ge 1$. The equality

$$
f_1(\boldsymbol{x}_{h+1}) - f_2(\boldsymbol{x}_{h+1}) = f_1(\boldsymbol{x}_h) - f_2(\boldsymbol{x}_h)
$$

holds if and only if

$$
\begin{aligned}
&\boldsymbol{x}_h \in \partial f_1^*(\boldsymbol{\xi}_{2,h}), \ \boldsymbol{\xi}_{2,h} \in \partial f_2(\boldsymbol{x}_{h+1}), \quad \textit{and} \\
&(\rho_1+\rho_2)\Delta\boldsymbol{x}_h = \rho_1^*\Delta\boldsymbol{\xi}_{2,h-1} = \rho_2^*\Delta\boldsymbol{\xi}_{2,h} = 0.
\end{aligned}
$$

In this case, we have

$$
f_1(\boldsymbol{x}_{h+1}) - f_2(\boldsymbol{x}_{h+1}) = f_2^*(\boldsymbol{\xi}_{2,h}) - f_1^*(\boldsymbol{\xi}_{2,h}),
$$

and $\boldsymbol{x}_{h+1}$, $\boldsymbol{x}_h$ are critical points of the problem (2.23).

3.8 Discrete Gradient Method

In most NSO algorithms it is assumed that the value of the objective function and its one subgradient can be computed at any point. However, in some practical applications it is not possible—or it may be very time-consuming—to compute subgradients. In principle, derivative free methods can be applied to solve NSO problems. For example, the generalized pattern search methods are well suited for NSO [15, 281]. However, their convergence is proved under some restrictive differentiability assumptions, which are not satisfied in many important practical problems. Particularly, the objective functions are often not strictly differentiable at their local minimizers.

In this section, we describe the *discrete gradient method* (DGM) [28] which can be considered as a semi-derivative free method for solving nonsmooth and, in general, nonconvex optimization problems. The DGM does not use subgradients—even not approximation of them—except final iterations of the solution process (i.e., near the optimal point).

The idea of the DGM is to hybridize derivative free methods with bundle methods. In contrast with bundle methods, which require the computation of a single subgradient of the objective function at each auxiliary point, the DGM computes discrete gradients (see Definition 2.12) using function values only. As noted in Sect. 2.5, discrete gradients can be used to approximate subdifferentials of a broad class of nonsmooth functions. In addition, in the case of a piecewise partially separable objective function, discrete gradients can be computed very efficiently as shown in Sect. 2.6.3. On the other hand, similar to bundle methods the previous values of discrete gradients are gathered into a bundle and the null step is used if the current search direction is not good enough.

The DGM has both inner and outer iterations. In turn, the inner iteration consists of serious and null steps. We select sequences $\{\delta_h\}$ and $\{\lambda_h\}$ such that $\delta_h,\ \lambda_h \downarrow 0$ as $h \to \infty$, any starting point $\boldsymbol{x}_1 \in \mathbb{R}^n$ and set $\boldsymbol{x}_{1_1} = \boldsymbol{x}_1$. The outer iteration depends on the index h and parameters δ_h and λ_h are updated in this iteration. The inner iteration depends on the index s. In the inner iteration, we compute the search direction. We either update the solution or add an element into the set of discrete gradients. In other words, we either take a serious step or a null step occurs. The flowchart of the DGM is given in Fig. 3.10.

At the beginning of each inner iteration (i.e., $s = 1$), we first compute the discrete gradient $\boldsymbol{v}_{h_1} = \boldsymbol{\Gamma}^i(\boldsymbol{x}_{h_1}, \boldsymbol{g}, \boldsymbol{w}, \lambda_h, \alpha)$ (see Definition 2.12) with respect to any initial direction $\boldsymbol{g} \in S_1,\ i \in \mathcal{I}(\boldsymbol{g})$, any fixed $\boldsymbol{w} \in G$ and $\alpha \in (0, 1]$. We set the initial bundle of discrete gradients $\bar{D}(\boldsymbol{x}_{h_1}) = \{\boldsymbol{v}_{h_1}\}$, $\bar{\boldsymbol{v}}_{h_1} = \boldsymbol{v}_{h_1}$ and compute the search direction $\boldsymbol{d}_{h_1}$ by

$$\boldsymbol{d}_{h_1} = -\frac{\bar{\boldsymbol{v}}_{h_1}}{\|\bar{\boldsymbol{v}}_{h_1}\|}.$$

In the following inner iterations (i.e., $s > 1$), we compute the vector

$$\bar{\boldsymbol{v}}_{h_s} = \operatorname*{argmin}_{\boldsymbol{v} \in \bar{D}(\boldsymbol{x}_{h_s})} \|\boldsymbol{v}\|^2,$$

that is the distance between the convex hull of all computed discrete gradients and the origin, and the search direction

$$\boldsymbol{d}_{h_s} = -\frac{\bar{\boldsymbol{v}}_{h_s}}{\|\bar{\boldsymbol{v}}_{h_s}\|}.$$

Next, we check whether the direction $\boldsymbol{d}_{h_s}$ $(s \geq 1)$ is descent or not. If it is, we have

$$f(\boldsymbol{x}_{h_s} + \lambda_h \boldsymbol{d}_{h_s}) - f(\boldsymbol{x}_{h_s}) \leq -\varepsilon_L \lambda_h \|\bar{\boldsymbol{v}}_{h_s}\|, \tag{3.32}$$

with the given numbers $\varepsilon_L \in (0, 1)$ and $\lambda_h > 0$. In this case, we compute the next (inner) iteration point

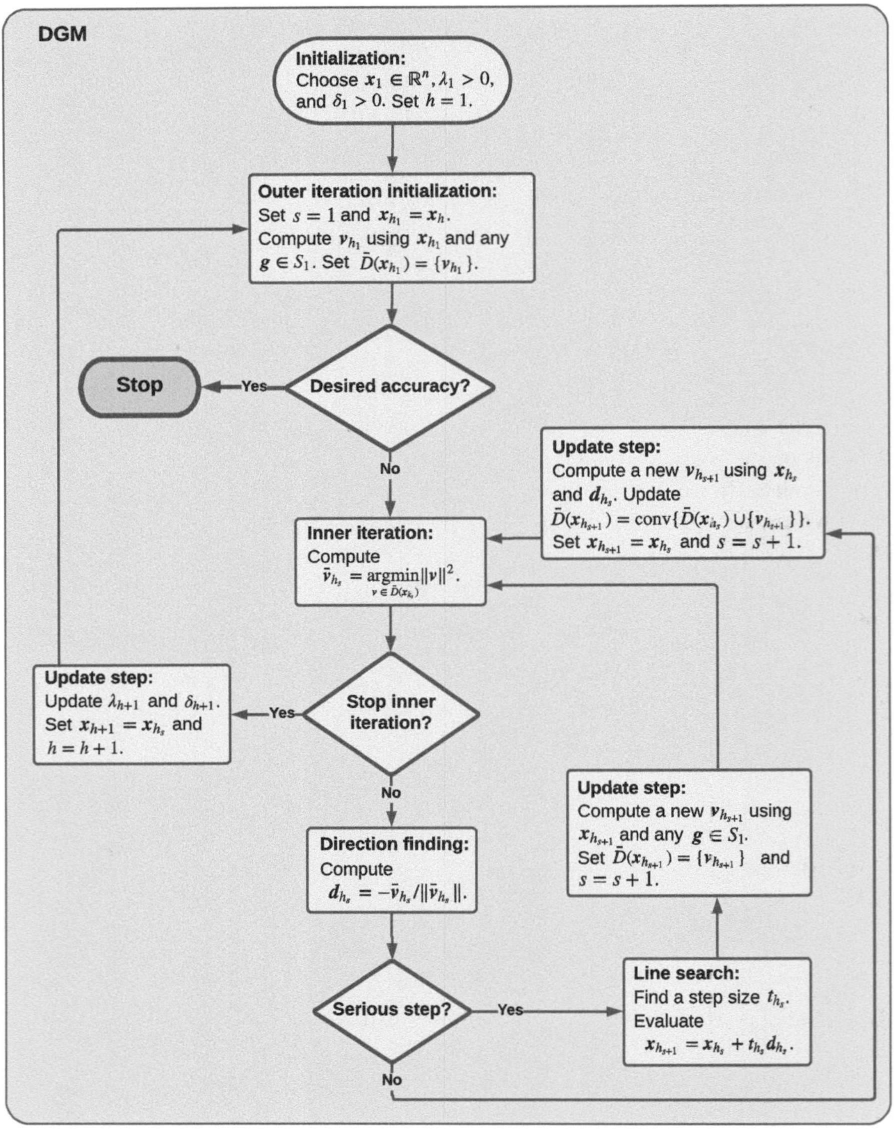

Fig. 3.10 Discrete gradient method

$$x_{h_{s+1}} = x_{h_s} + t_{h_s} d_{h_s},$$

where the step size t_{h_s} is defined as

$$t_{h_s} = \text{argmax} \left\{ t \geq 0 : f(x_{h_s} + t d_{h_s}) - f(x_{h_s}) \leq -\varepsilon_R t \|\bar{v}_{h_s}\| \right\},$$

with a given $\varepsilon_R \in (0, \varepsilon_L]$. In the case of a serious step, we compute the new discrete gradient $\boldsymbol{v}_{h_{s+1}} = \boldsymbol{\Gamma}^i(\boldsymbol{x}_{h_{s+1}}, \boldsymbol{g}, \boldsymbol{w}, \lambda_h, \alpha)$ with respect to any direction $\boldsymbol{g} \in S_1$, set $\bar{D}(\boldsymbol{x}_{h_{s+1}}) = \{\boldsymbol{v}_{h_{s+1}}\}$, and continue to the next inner iteration with $s = s + 1$.

If the direction $\boldsymbol{d}_{h_s}$ does not satisfy the condition (3.32), then a null step occurs. In this case, we compute another discrete gradient $\boldsymbol{v}_{h_{s+1}} = \boldsymbol{\Gamma}^i(\boldsymbol{x}_{h_s}, \boldsymbol{d}_{h_s}, \boldsymbol{w}, \lambda_h, \alpha)$ with respect to the direction $\boldsymbol{d}_{h_s}$, update the bundle of discrete gradients

$$\bar{D}(\boldsymbol{x}_{h_{s+1}}) = \text{conv}\left\{\bar{D}(\boldsymbol{x}_{h_s}) \cup \{\boldsymbol{v}_{h_{s+1}}\}\right\},$$

set $\boldsymbol{x}_{h_{s+1}} = \boldsymbol{x}_{h_s}$, and continue the inner iterations with $s = s + 1$. Note that, at each null step the approximation of the subdifferential $\partial f(\boldsymbol{x})$ is improved.

The inner iteration stops if

$$\|\bar{\boldsymbol{v}}_{h_s}\| \le \delta_h.$$

This condition means that for given values of δ_h and λ_h the last iteration $\boldsymbol{x}_{h_s}$ can be considered as an approximate solution for the objective, and this solution cannot be significantly improved using the same values of parameters. Therefore, this point is accepted as a new iteration $\boldsymbol{x}_{h+1}$ and the algorithm returns to the outer iteration to update the values of parameters δ_h and λ_h. In its turn, the outer iteration stops if both $\delta_h < \varepsilon$ and $\lambda_h < \varepsilon$ with a given termination tolerance $\varepsilon > 0$. This stopping criterion means that the further decrease of values of δ_h and λ_h will not improve both the approximation of the subdifferential and the solution $\boldsymbol{x}_h$.

Since in the DGM the descent direction can be computed for any values of $\lambda > 0$ one can take $\lambda_1 \in (0, 1)$, some $\beta \in (0, 1)$ and update λ_h, for instance, by the formula $\lambda_h = \beta^h \lambda_1$, $h > 1$. Thus, the approximations to subgradients are used only at the final stage, which guarantees the convergence of the method. In most of the iterations such approximations are not used and, thus, the DGM can be considered as a semi-derivative free method.

In the paper [28], it is proved that the DGM is globally convergent for LLC functions under the assumption that the set of discrete gradients uniformly approximates the subdifferential. In addition, improved convergence results for the DGM are given in [181]. In the next subsection, we present the basic scheme for the proof of convergence of the DGM.

3.8.1 Convergence of the DGM

In this subsection, we recall some technical details of the convergence properties of the DGM. A more detailed results are given in [28]. The following assumptions are used in the proofs.

Assumption 3.12 *The objective function $f : \mathbb{R}^n \to \mathbb{R}$ is semismooth.*

Assumption 3.13 *The objective function $f : \mathbb{R}^n \to \mathbb{R}$ is a quasidifferentiable function (see Definition 2.7) whose subdifferential and superdifferential are polytopes at any $\boldsymbol{x} \in \mathbb{R}^n$.*

Assumption 3.14 *The level set $\operatorname{lev}_{f(\boldsymbol{x}_1)} f$ is bounded for any starting point $\boldsymbol{x}_1 \in \mathbb{R}^n$.*

Assumptions 3.12 and 3.13 are satisfied, for instance, for functions represented as a maximum, minimum, or max–min of a finite number of smooth functions. Assumption 3.14 is used in most continuous optimization methods to prove their convergence.

We start by showing that for given δ_h and λ_h the number of serious and null steps in inner iterations of the DGM is finite.

Lemma 3.8 *Let L be a Lipschitz constant of the objective function f and*

$$\bar{C} = C(n)L, \quad \text{where} \quad C(n) = (4n^2 - 3n)^{1/2}.$$

Then for any $\delta_h \in (0, \bar{C})$, the number s_0 of null steps in the inner loop of the DGM *is finite. Here,*

$$s_0 \le 1 + 2\left\lceil \frac{\log_2(\delta_h/\bar{C})}{\log_2 r} \right\rceil,$$

with

$$r = 1 - \left\lceil \frac{1-\varepsilon_L}{2\bar{C}}\delta_h \right\rceil^2.$$

Proposition 3.13 *For any $\delta_h, \lambda_h > 0$, the number $s_{\max}$ of serious steps in the inner loop of the* DGM *is finite with*

$$s_{\max} \le \left\lceil \frac{f(\boldsymbol{x}_h) - f^*}{\varepsilon_L \lambda_h \delta_h} \right\rceil,$$

where $f^* = \inf\{f(\boldsymbol{x}),\ \boldsymbol{x} \in \mathbb{R}^n\} > -\infty$.

Proof At the beginning of the sth inner iteration in the DGM, the value of the objective function f is $f(\boldsymbol{x}_{h_s})$. Furthermore, at each serious step the condition (3.32) is satisfied and $\|\bar{\boldsymbol{v}}_{h_s}\| > \delta_h$. Therefore, we have

$$\begin{aligned} f(\boldsymbol{x}_{h_{s+1}}) - f(\boldsymbol{x}_{h_s}) &\le f(\boldsymbol{x}_{h_s} + \lambda_h \boldsymbol{d}_{h_s}) - f(\boldsymbol{x}_{h_s}) \\ &\le -\varepsilon_L \lambda_h \|\bar{\boldsymbol{v}}_{h_s}\| \\ &< -\varepsilon_L \lambda_h \delta_h. \end{aligned}$$

This means that at each inner iteration the value of the objective function is reduced by at least $u_h = \varepsilon_L \lambda_h \delta_h$ and this number does not depend on the serious steps. By

Assumption 3.14, we have $f^* > -\infty$ and at the initialization of inner iterations we set $\boldsymbol{x}_h = \boldsymbol{x}_{h_1}$. Thus, the number of the serious steps cannot be more than $\lceil (f(\boldsymbol{x}_h) - f^*)/u_h \rceil$. This completes the proof. □

Corollary 3.4 *Assume that conditions of Lemma 3.8 and Proposition 3.13 are satisfied. Then for any h, the number of steps M_h in the hth outer iteration of the* DGM *is finite and $M_h = s_0 s_{\max}$.*

In Theorem 2.22, we showed that for semismooth functions $f : \mathbb{R}^n \to \mathbb{R}$ at a given point $\boldsymbol{x} \in \mathbb{R}^n$ under a mild condition, the closed convex hull of the set of discrete gradients $D(\boldsymbol{x}, \lambda, \alpha)$ is an approximation of the subdifferential $\partial f(\boldsymbol{x})$ for the sufficiently small $\lambda > 0$. However, this is true only at a point $\boldsymbol{x} \in \mathbb{R}^n$. In order to get convergence results for the DGM, we need a relationship between the set $D(\boldsymbol{x}, \lambda, \alpha)$ and $\partial f(\boldsymbol{x})$ also in some neighborhood of $\boldsymbol{x}$. Therefore, we need an additional assumption:

Assumption 3.15 *Let $\boldsymbol{x} \in \mathbb{R}^n$ be a given point. For any $\eta > 0$, there exist $\gamma > 0$, $\lambda_0 > 0$ and $\alpha_0 \in (0, 1]$ such that*

$$D(\boldsymbol{y}, \lambda, \alpha) \subset \partial f\big(\boldsymbol{x} + \bar{B}(\boldsymbol{0}; \eta)\big) + B(\boldsymbol{0}; \eta)$$

for all $\boldsymbol{y} \in B(\boldsymbol{x}; \gamma)$, $\lambda \in (0, \lambda_0)$, and $\alpha \in (0, \alpha_0)$. Here,

$$\partial f\big(\boldsymbol{x} + \bar{B}(\boldsymbol{0}; \eta)\big) = \operatorname{cl}\operatorname{conv} \bigcup_{\boldsymbol{y} \in \bar{B}(\boldsymbol{x}; \eta)} \partial f(\boldsymbol{y}).$$

Proposition 3.14 *Suppose that Assumptions 3.12–3.15 hold for the objective function f in the problem* (3.1)*. Then every accumulation point of the sequence $\{\boldsymbol{x}_h\}$ generated by the* DGM *belongs to the set*

$$X^0 = \big\{\boldsymbol{x} \in \mathbb{R}^n : \boldsymbol{0} \in \partial f(\boldsymbol{x})\big\}.$$

Proof Since the function f is continuous and the set $\operatorname{lev}_{f(\boldsymbol{x}_1)} f$ is bounded we get $f^* > -\infty$. Therefore, by Corollary 3.4 the inner loop of the DGM terminates after finite number of steps and generates a point $\boldsymbol{x}_{h+1}$ such that $\boldsymbol{x}_{h+1} = \boldsymbol{x}_{h_s}$ for some $s > 0$. In addition, we have

$$\min_{\boldsymbol{v} \in \bar{D}(\boldsymbol{x}_{h+1})} \|\boldsymbol{v}\| \le \delta_h. \tag{3.33}$$

It is clear that

$$\bar{D}(\boldsymbol{x}_{h+1}) \subseteq D(\boldsymbol{x}_{h+1}, \lambda_h, \alpha). \tag{3.34}$$

Since $\{f(\boldsymbol{x}_h)\}$ is a decreasing sequence the point $\boldsymbol{x}_h \in \operatorname{lev}_{f(\boldsymbol{x}_1)} f$ for all $h \ge 1$. Thus, the sequence $\{\boldsymbol{x}_h\}$ is bounded and it has at least one accumulation point.

Assume that $\bar{x}$ is an accumulation point of $\{x_h\}$ and there exists the subsequence $\{x_{h_j}\}$ such that $x_{h_j} \to \bar{x}$ as $j \to \infty$. It follows from (3.33) and (3.34) that

$$\min\left\{\|v\| : v \in D(x_{h_j}, \lambda_{h_j-1}, \alpha)\right\} \le \delta_{h_j-1}. \tag{3.35}$$

According to Assumption 3.15 at the point $\bar{x}$ for any $\eta > 0$, there exist $\gamma > 0$, $\alpha_0 > 0$, and $\lambda_0 > 0$ such that

$$D(y, \lambda, \alpha) \subset \partial f\big(\bar{x} + \bar{B}(\mathbf{0}; \eta)\big) + B(\mathbf{0}; \eta) \tag{3.36}$$

for all $y \in B(\bar{x}; \gamma)$, $\alpha \in (0, \alpha_0)$, and $\lambda \in (0, \lambda_0)$. Since the sequence $\{x_{h_j}\}$ converges to $\bar{x}$ for $\gamma > 0$ there exists $j_0 > 0$ such that $x_{h_j} \in B(\bar{x}; \gamma)$ for all $j \ge j_0$. On the other hand, since $\delta_h, \lambda_h \to 0$ as $h \to \infty$ there exists $h_0 > 0$ such that $\delta_h < \eta$ and $\lambda_h < \lambda_0$ for all $h > h_0$. Then there exists $j_1 \ge j_0$ such that $h_j \ge h_0 + 1$ for all $j \ge j_1$. Thus, it follows from (3.35) and (3.36) that

$$\min\left\{\|v\| : v \in \partial f(\bar{x} + \bar{B}(\mathbf{0}; \eta)\right\} \le 2\eta.$$

Since $\eta > 0$ is arbitrary and the mapping $\partial f(x)$ is upper semicontinuous we have $\mathbf{0} \in \partial f(\bar{x})$. This completes the proof. □

3.9 Smoothing Method

In this section, we consider numerical methods for solving NSO problems which are based on *smoothing techniques*. Such an approach allows us to apply powerful smooth optimization algorithms for solving nonsmooth problems. Nevertheless, the application of smoothing techniques requires some special structure of the problem. We consider the finite minimax problem (2.28).

This type of problems are frequently encountered in practical applications including the cluster analysis. Since the functions f_i are smooth the objective function

$$f(x) = \max_{i \in \mathcal{I}} f_i(x), \quad \mathcal{I} = \{1, \ldots, m\}$$

in the problem (2.28) is subdifferentially regular and its subdifferential at a point $x \in \mathbb{R}^n$ can be expressed as (see Sect. 2.8):

$$\partial f(x) = \operatorname{conv}\left\{\nabla f_i(x) : i \in \mathcal{I}(x)\right\},$$

where $\mathcal{I}(x) = \left\{i \in \mathcal{I} : f_i(x) = f(x)\right\}$.

We consider the hyperbolic smoothing technique [30] which allows one to smooth the function f globally. The smoothing is controlled by the precision parameter τ. Given a sequence $\{\tau_h\}$ such that $\tau_h > 0$ and $\tau_h \to 0$ as $h \to \infty$, the finite minimax problem (2.28) can be replaced by the sequence of the following smooth problems:

$$\begin{cases} \text{minimize} & \Phi_{\tau_h}(\boldsymbol{x}, f(\boldsymbol{x})) \\ \text{subject to} & \boldsymbol{x} \in \mathbb{R}^n, \end{cases} \tag{3.37}$$

where

$$\Phi_{\tau_h}(\boldsymbol{x}, t) = t + \sum_{i \in \mathcal{I}} \frac{f_i(\boldsymbol{x}) - t + \sqrt{(f_i(\boldsymbol{x}) - t)^2 + \tau^2}}{2}.$$

Results from Sect. 2.8 demonstrate that algorithms from smooth optimization can be applied to solve the problem (3.37).

The *hyperbolic smoothing method* (HSM) for solving the problem (2.28) is given in Fig. 3.11. Let $\{\varepsilon_h\}$ be a given sequence such that $\varepsilon_h > 0$ and $\varepsilon_h \to 0$ as $h \to \infty$. At every iteration h of the smoothing method, we apply a smooth optimization solver to the problem (3.37) with a starting point $\boldsymbol{x}_h$ to find a point $\bar{\boldsymbol{x}}$ such that

$$\|\nabla \Phi_{\tau_h}(\bar{\boldsymbol{x}}, f(\bar{\boldsymbol{x}}))\| < \varepsilon_h. \tag{3.38}$$

The choice of sequences $\{\tau_h\}$ and $\{\varepsilon_h\}$ might be crucial for some problems. In principle, we can choose the smoothing (precision) parameter $\tau > 0$ sufficiently small to solve the problem (3.37) only once. However, such an approach may make the problem ill-conditioned which will significantly increase computational efforts when solving it. The usage of the sequence $\{\tau_h\}$ may help to prevent such a situation. Moreover, if $\{\tau_h\}$ converges too quickly to 0, then the ill-conditioned behavior of the problem may gradually increase. In this case, a large number of iterations are required to achieve the condition (3.38). In order to avoid this, one should ensure that the sequence $\{\tau_h\}$ converges to 0 slower than the sequence $\{\varepsilon_h\}$.

3.9.1 Convergence of the HSM

Next, we prove that any accumulation point of the sequence $\{\boldsymbol{x}_h\}$ generated by the HSM given in Fig. 3.11 is a stationary point of the finite minimax problem (2.28) [30].

Assumption 3.16 *The level set* $\operatorname{lev}_{f(\boldsymbol{x}_1)} f$ *is bounded for any starting point* $\boldsymbol{x}_1 \in \mathbb{R}^n$.

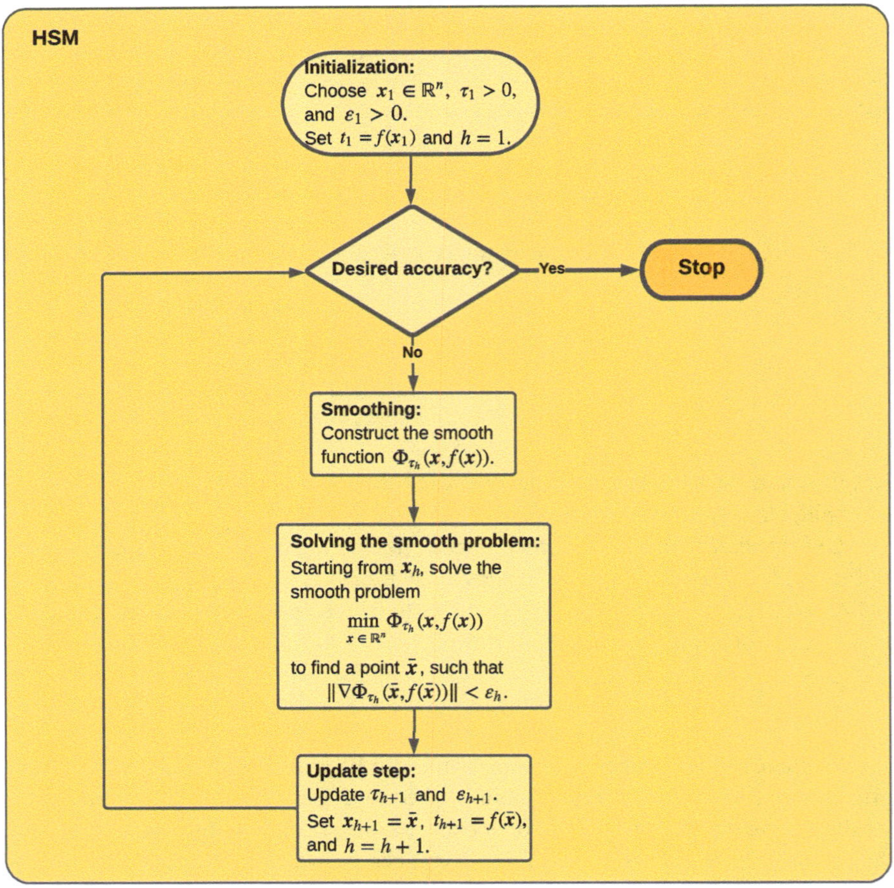

Fig. 3.11 Hyperbolic smoothing method for minimax problems

Proposition 3.15 *Any accumulation point of the sequence* $\{x_h\}$ *generated by the* HSM *is a stationary point of the problem (2.28).*

Proof It is clear that $x_h \in \text{lev}_{f(x_1)} f$ for all $h \geq 1$. Since the set $\text{lev}_{f(x_1)} f$ is bounded the sequence $\{x_h\}$ has at least one accumulation point. Let $\bar{x}$ be an accumulation point of the sequence $\{x_h\}$ and, for simplicity, assume that $x_h \to \bar{x}$ as $h \to \infty$. It follows from Proposition 2.4 that $\mathbf{0}_{n+1} \in \partial F\big(\bar{x}, f(\bar{x})\big)$ which means $\big(\bar{x}, f(\bar{x})\big)$ is a stationary point of the function F (see Sect. 2.8). Then applying Proposition 2.1, we get that $\bar{x}$ is a stationary point of the problem (2.28). □

Part II
Clustering Algorithms

Chapter 4
Optimization Models in Cluster Analysis

4.1 Introduction

In this chapter, we consider unconstrained hard clustering problems. Let A be a set of finite number of points in the n-dimensional space $\mathbb{R}^n$ as defined in Chap. 1:

$$A = \{\boldsymbol{a}_1, \ldots, \boldsymbol{a}_m\}, \ \boldsymbol{a}_i \in \mathbb{R}^n, \quad i = 1, \ldots, m.$$

The hard unconstrained clustering problem is the distribution of points of the set A into a given number k of disjoint clusters A^j, $j = 1, \ldots, k$ with respect to the predefined criteria (1.1).

The notion of the similarity measure is essential to formulate clustering problems. In particular, the similarity measure is defined using different distance functions, in other words using different Minkowski norms. In general, the distance function, defined by (1.2) is

$$d_p(\boldsymbol{b}, \boldsymbol{c}) = \Big(\sum_{i=1}^{n}(b_i - c_i)^p\Big)^{1/p}, \quad \boldsymbol{b}, \boldsymbol{c} \in \mathbb{R}^n. \tag{4.1}$$

Here, $p \in [1, \infty)$. For $p = 1$, we get the similarity measure which is based on the L_1-norm and for $p = \infty$, the similarity measure is defined using the L_∞-norm. In both cases, the similarity measures are also distance functions. For $p = 2$, the squared Euclidean distance (the squared L_2-norm) is used to define the similarity measure:

$$d_2(\boldsymbol{b}, \boldsymbol{c}) = \sum_{i=1}^{n}(b_i - c_i)^2, \quad \boldsymbol{b}, \boldsymbol{c} \in \mathbb{R}^n.$$

A. M. Bagirov et al., *Partitional Clustering via Nonsmooth Optimization*,
Unsupervised and Semi-Supervised Learning,
https://doi.org/10.1007/978-3-030-37826-4_4

In this case, the similarity measure is no longer a distance function. One can also use values $p \in (0, 1)$ in (4.1) to define the similarity measures; however, these measures are not distance functions. Clustering problems defined using the similarity measure based on the L_1-norm are known as the *minimum sum-of-absolutes clustering* (MSAC) problems and those defined using the squared L_2-norm are called the *minimum sum-of-squares clustering* (MSSC) problems.

Next, we describe different optimization models of the clustering problem: the mixed integer programming, the nonconvex NSO, and the nonsmooth DC models. We introduce also the so-called *auxiliary clustering problem* and study smoothing of both the clustering and the auxiliary clustering problems.

4.2 Mixed Integer Programming Model

In this section, we present the mixed integer programming formulation of the clustering problem. Let the association weight w_{ij} of the point $\boldsymbol{a}_i \in A$ with the cluster A^j is given by

$$w_{ij} = \begin{cases} 1, & \text{if the point } \boldsymbol{a}_i \text{ is allocated to the cluster } A^j, \\ 0, & \text{otherwise.} \end{cases}$$

Then the *mixed integer nonlinear programming formulation* of the clustering problem is [26]

$$\begin{cases} \text{minimize} & \zeta_k(\boldsymbol{x}, \boldsymbol{w}) \\ \text{subject to} & w_{ij} \in \{0, 1\}, \quad i = 1, \dots, m, \ j = 1, \dots, k, \\ & \sum_{j=1}^{k} w_{ij} = 1, \quad i = 1, \dots, m, \\ & \boldsymbol{x}_j \in \mathbb{R}^n, \quad j = 1, \dots, k, \end{cases} \tag{4.2}$$

where

$$\zeta_k(\boldsymbol{x}, \boldsymbol{w}) = \frac{1}{m} \sum_{i=1}^{m} \sum_{j=1}^{k} w_{ij} \, d_p(\boldsymbol{x}_j, \boldsymbol{a}_i)$$

is called the *kth cluster function*, $\boldsymbol{x} = (\boldsymbol{x}_1, \dots, \boldsymbol{x}_k) \in \mathbb{R}^{nk}$ and $\boldsymbol{w} = (\boldsymbol{w}_1, \dots, \boldsymbol{w}_k)$, $\boldsymbol{w}_i \in \mathbb{R}^m$.

The problem (4.2) contains mk integer (binary) variables w_{ij}, $i = 1, \dots, m$, $j = 1, \dots, k$ and nk continuous variables $\boldsymbol{x}_j \in \mathbb{R}^n$, $j = 1, \dots, k$. The objective function ζ_k is convex for $k = 1$ and nonconvex for $k > 1$. This function is nonsmooth if similarity measures d_1 or d_∞ are used.

For the similarity measure d_2, the objective function ζ_k is smooth and the cluster centers $\boldsymbol{x}_j$ (called also *centroids*) are computed as

$$\boldsymbol{x}_j = \frac{\sum_{i=1}^m w_{ij}\boldsymbol{a}_i}{\sum_{i=1}^m w_{ij}}, \quad j = 1, \dots, k.$$

In this case, the problem (4.2) becomes an integer programming problem as the centers $\boldsymbol{x}_j,\ j = 1, \dots, k$ are no longer decision variables.

Note that in the case of the similarity measure d_1, the centers $\boldsymbol{x}_j,\ j = 1, \dots, k$ are computed as the medians of clusters. Therefore, similar to the case d_2, the problem (4.2) becomes an integer programming problem.

4.3 Nonsmooth Optimization Model

The NSO model of the clustering problem is formulated as follows [19, 20, 27, 52, 171]:

$$\begin{cases} \text{minimize} & f_k(\boldsymbol{x}) \\ \text{subject to} & \boldsymbol{x} = (\boldsymbol{x}_1, \dots, \boldsymbol{x}_k) \in \mathbb{R}^{nk}, \end{cases} \tag{4.3}$$

where

$$f_k(\boldsymbol{x}) = \frac{1}{m} \sum_{i=1}^m \min_{j=1,\dots,k} d_p(\boldsymbol{x}_j, \boldsymbol{a}_i). \tag{4.4}$$

The problem (4.3) is called the *NSO clustering problem.* It contains nk continuous variables $\boldsymbol{x}_j \in \mathbb{R}^n,\ j = 1, \dots, k$, and this number does not depend on the number of instances. The objective function f_k in the problem (4.3) is called the *kth cluster function.* It is convex for $k = 1$, and nonconvex for $k > 1$. If the similarity measure is defined using the squared Euclidean distance, then the function f_k for $k = 1$ is smooth. However, for $k > 1$ this function is nonsmooth due to the minimum operation. For the similarity measures based on the L_1- and L_∞-norms, the function f_k is nonsmooth for all $k \geq 1$. This is due to the minimum operation and the fact that both functions d_1 and d_∞ are nonsmooth.

Example 4.1 Next, we illustrate the cluster function $f_k(\boldsymbol{x}),\ \boldsymbol{x} \in \mathbb{R}^2$ for $k = 2$. Consider the following set in $\mathbb{R}$:

$$A = \{-3, -2.6, -2.2, -1.8, -1.4, -1, 1, 1.4, 1.8, 2.2, 2.6, 3, 4, 4.4, 4.8, 5.2, 5.6, 6\}. \tag{4.5}$$

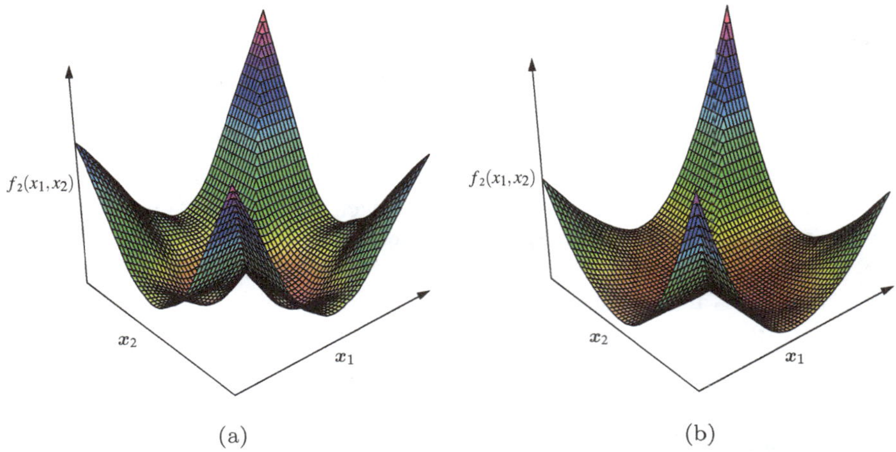

Fig. 4.1 Graphs of cluster function f_2 with similarity measures d_1 and d_2. (**a**) Cluster function f_2 with d_1. (**b**) Cluster function f_2 with d_2

Graphs of the function f_2 on the data set A with the similarity measures d_1 and d_2 are illustrated in Fig. 4.1. Here, the function f_2 with d_1 is as follows:

$$\begin{aligned}
f_2(x_1, x_2) = \frac{1}{18}\Big\{ & \min\{|x_1 + 3|, |x_2 + 3|\}, \min\{|x_1 + 2.6|, |x_2 + 2.6|\}, \\
& \min\{|x_1 + 2.2|, |x_2 + 2.2|\}, \min\{|x_1 + 1.8|, |x_2 + 1.8|\}, \\
& \min\{|x_1 + 1.4|, |x_2 + 1.4|\}, \min\{|x_1 + 1|, |x_2 + 1|\}, \\
& \min\{|x_1 - 1|, |x_2 - 1|\}, \min\{|x_1 - 1.4|, |x_2 - 1.4|\}, \\
& \min\{|x_1 - 1.8|, |x_2 - 1.8|\}, \min\{|x_1 - 2.2|, |x_2 - 2.2|\}, \\
& \min\{|x_1 - 2.6|, |x_2 - 2.6|\}, \min\{|x_1 - 3|, |x_2 - 3|\}, \\
& \min\{|x_1 - 4|, |x_2 - 4|\}, \min\{|x_1 - 4.4|, |x_2 - 4.4|\}, \\
& \min\{|x_1 - 4.8|, |x_2 - 4.8|\}, \min\{|x_1 - 5.2|, |x_2 - 5.2|\}, \\
& \min\{|x_1 - 5.6|, |x_2 - 5.6|\}, \min\{|x_1 - 6|, |x_2 - 6|\}\Big\},
\end{aligned}$$

and with the similarity measure d_2 this function is

$$\begin{aligned}
f_2(x_1, x_2) = \frac{1}{18}\Big\{ & \min\{(x_1 + 3)^2, (x_2 + 3)^2\}, \min\{(x_1 + 2.6)^2, (x_2 + 2.6)^2\}, \\
& \min\{(x_1 + 2.2)^2, (x_2 + 2.2)^2\}, \min\{(x_1 + 1.8)^2, (x_2 + 1.8)^2\}, \\
& \min\{(x_1 + 1.4)^2, (x_2 + 1.4)^2\}, \min\{(x_1 + 1)^2, (x_2 + 1)^2\},
\end{aligned}$$

$$\min\{(x_1-1)^2, (x_2-1)^2\}, \min\{(x_1-1.4)^2, (x_2-1.4)^2\},$$
$$\min\{(x_1-1.8)^2, (x_2-1.8)^2\}, \min\{(x_1-2.2)^2, (x_2-2.2)^2\},$$
$$\min\{(x_1-2.6)^2, (x_2-2.6)^2\}, \min\{(x_1-3)^2, (x_2-3)^2\},$$
$$\min\{(x_1-4)^2, (x_2-4)^2\}, \min\{(x_1-4.4)^2, (x_2-4.4)^2\},$$
$$\min\{(x_1-4.8)^2, (x_2-4.8)^2\}, \min\{(x_1-5.2)^2, (x_2-5.2)^2\},$$
$$\left.\min\{(x_1-5.6)^2, (x_2-5.6)^2\}, \min\{(x_1-6)^2, (x_2-6)^2\}\right\}.$$

We now present some important properties of the cluster function f_k. We start with the description of the subdifferentials of the similarity functions d_1, d_2, and d_∞. Take any $\boldsymbol{a} \in A$. The function $d_1(\cdot, \boldsymbol{a})$ is, in general, nonsmooth convex function and it can be expressed as

$$d_1(\boldsymbol{x}, \boldsymbol{a}) = \sum_{q=1}^{n} \max\{x_q - a_q, a_q - x_q\}, \ \boldsymbol{x} \in \mathbb{R}^n.$$

It follows from Theorem 2.14 that the subdifferential of the function d_1 at a point $\boldsymbol{x} \in \mathbb{R}^n$ is

$$\partial d_1(\boldsymbol{x}, \boldsymbol{a}) = \operatorname{conv}\Big\{\boldsymbol{\xi} = (\xi_1, \ldots, \xi_n) \in \mathbb{R}^n : \xi_q = \operatorname{sign}(x_q - a_q), \text{ if } x_q \neq a_q,$$
$$\xi_q \in [-1, 1], \qquad \text{otherwise}\Big\}, \tag{4.6}$$

where $q = 1, \ldots, n$ and

$$\operatorname{sign}(z) = \begin{cases} -1, & \text{if } z < 0, \\ 0, & \text{if } z = 0, \\ 1, & \text{if } z > 0. \end{cases}$$

It follows from (4.6) that for all $\boldsymbol{x} \in \mathbb{R}^n$ and $\boldsymbol{\xi} \in \partial d_1(\boldsymbol{x}, \boldsymbol{a})$, we have

$$\|\boldsymbol{\xi}\|_1 \leq n \ \text{ and } \ \|\boldsymbol{\xi}\| \leq \sqrt{n}. \tag{4.7}$$

Note that we have $\|.\| = \|.\|_2$.

For a given $\boldsymbol{a} \in A$, the function $d_2(\cdot, \boldsymbol{a})$ is continuously differentiable and its gradient at a point $\boldsymbol{x} \in \mathbb{R}^n$ is

$$\nabla d_2(\boldsymbol{x}, \boldsymbol{a}) = 2(\boldsymbol{x} - \boldsymbol{a}). \tag{4.8}$$

Finally, consider the distance function d_∞. In this case, we have

$$d_\infty(\boldsymbol{x}, \boldsymbol{a}) = \max_{q=1,\ldots,n} |x_q - a_q| = \max_{q=1,\ldots,n} \max \{x_q - a_q, a_q - x_q\}.$$

Define the set

$$\bar{\mathcal{R}}_a(\boldsymbol{x}) = \{q \in \{1, \ldots, n\} : |x_q - a_q| = d_\infty(\boldsymbol{x}, \boldsymbol{a})\}, \tag{4.9}$$

and introduce the sets

$$\begin{aligned} \mathcal{I}_-(\boldsymbol{a}, \boldsymbol{x}) &= \{q \in \{1, \ldots, n\} : x_q < a_q\}, \\ \mathcal{I}_+(\boldsymbol{a}, \boldsymbol{x}) &= \{q \in \{1, \ldots, n\} : x_q > a_q\}, \quad \text{and} \\ \mathcal{I}_0(\boldsymbol{a}, \boldsymbol{x}) &= \{q \in \{1, \ldots, n\} : x_q = a_q\}. \end{aligned}$$

The condition $\mathcal{I}_0(\boldsymbol{a}, \boldsymbol{x}) \cap \bar{\mathcal{R}}_a(\boldsymbol{x}) \neq \emptyset$ implies that $\bar{\mathcal{R}}_a(\boldsymbol{x}) = \mathcal{I}_0(\boldsymbol{a}, \boldsymbol{x}) = \{1, \ldots, n\}$, and therefore $d_\infty(\boldsymbol{x}, \boldsymbol{a}) = 0$. Then

$$\partial d_\infty(\boldsymbol{x}, \boldsymbol{a}) = \operatorname{conv}\{-\boldsymbol{e}_q, \boldsymbol{e}_q, \ q \in \{1, \ldots, n\}\}. \tag{4.10}$$

Here, $\boldsymbol{e}_q \in \mathbb{R}^n$ is the qth unit vector. If $\mathcal{I}_0(\boldsymbol{a}, \boldsymbol{x}) \cap \bar{\mathcal{R}}_a(\boldsymbol{x}) = \emptyset$, then

$$\bar{\mathcal{R}}_a(\boldsymbol{x}) \subset \mathcal{I}_-(\boldsymbol{a}, \boldsymbol{x}) \cup \mathcal{I}_+(\boldsymbol{a}, \boldsymbol{x}).$$

In this case, we get

$$\partial d_\infty(\boldsymbol{x}, \boldsymbol{a}) = \operatorname{conv}\{(0, \ldots, 0, \operatorname{sign}(x_q - a_q), 0 \ldots, 0) : q \in \bar{\mathcal{R}}_a(\boldsymbol{x})\}. \tag{4.11}$$

It follows from (4.10) and (4.11) that for all $\boldsymbol{x} \in \mathbb{R}^n$ and $\boldsymbol{\xi} \in \partial d_\infty(\boldsymbol{x}, \boldsymbol{a})$ we have

$$\|\boldsymbol{\xi}\| \leq 1 \quad \text{and} \quad \|\boldsymbol{\xi}\|_\infty \leq 1. \tag{4.12}$$

Proposition 4.1 *The functions d_1, d_2, and d_∞ are LLC on $\mathbb{R}^n$.*

Proof Take any bounded subset $X \subset \mathbb{R}^n$, and assume that it is convex (otherwise, one can take its convex hull). Select any two points $\boldsymbol{x}, \boldsymbol{y} \in X$. Applying the mean-value Theorem 2.9, we get

$$d_p(\boldsymbol{y}, \boldsymbol{a}) - d_p(\boldsymbol{x}, \boldsymbol{a}) \in \partial d_p(\boldsymbol{z}, \boldsymbol{a})^T(\boldsymbol{y} - \boldsymbol{x}), \quad p = 1, 2, \infty$$

for some $\boldsymbol{z} \in (\boldsymbol{x}, \boldsymbol{y})$. Since X is convex $\boldsymbol{z} \in X$. Then applying Cauchy–Schwarz inequality we have

$$|d_p(\boldsymbol{y}, \boldsymbol{a}) - d_p(\boldsymbol{x}, \boldsymbol{a})| \leq \max_{\boldsymbol{\xi} \in \partial d_p(\boldsymbol{x}, \boldsymbol{a})} \|\boldsymbol{\xi}\| \|\boldsymbol{x} - \boldsymbol{y}\|.$$

It follows from (4.7) and (4.12) that the distance functions d_1 and d_∞ satisfy the Lipschitz condition on $\mathbb{R}^n$ with the Lipschitz constants $L = \sqrt{n}$ and $L = 1$, respectively. For the gradient $\nabla d_2(\boldsymbol{x}, \boldsymbol{a})$ given in (4.8) we get

$$\|\nabla d_2(\boldsymbol{x}, \boldsymbol{a})\| \leq 2\big(\|\boldsymbol{x}\| + \|\boldsymbol{a}\|\big).$$

Since the set X is bounded there exists $M_1 > 0$ such that $\|\boldsymbol{x}\| \leq M_1$ for all $\boldsymbol{x} \in X$. Let

$$M_2 = \max_{\boldsymbol{a} \in A} \|\boldsymbol{a}\| < +\infty.$$

Then, the function d_2 satisfies the Lipschitz condition on the set X with the Lipschitz constant $L = 2(M_1 + M_2)$. □

Remark 4.1 Notice that we consider Lipschitz condition for all similarity functions using the L_2-norm.

For each $\boldsymbol{a} \in A$, consider the function

$$\psi_{\boldsymbol{a}}(\boldsymbol{x}_1, \ldots, \boldsymbol{x}_k) = \min_{j=1,\ldots,k} d_p(\boldsymbol{x}_j, \boldsymbol{a}), \quad \boldsymbol{x}_j \in \mathbb{R}^n. \tag{4.13}$$

Proposition 4.2 *The function $\psi_{\boldsymbol{a}}$ defined in* (4.13) *is LLC on* $\mathbb{R}^{nk}$.

Proof Let $X \subset \mathbb{R}^{nk}$ be any bounded subset. Take any $\boldsymbol{x}, \boldsymbol{y} \in X$ and consider the sets

$$\mathcal{R}_{\boldsymbol{a}}(\boldsymbol{x}) = \big\{j \in \{1, \ldots, k\} : d_p(\boldsymbol{x}_j, \boldsymbol{a}) = \psi_{\boldsymbol{a}}(\boldsymbol{x})\big\}, \quad \text{and} \tag{4.14}$$
$$\mathcal{R}_{\boldsymbol{a}}(\boldsymbol{y}) = \big\{j \in \{1, \ldots, k\} : d_p(\boldsymbol{y}_j, \boldsymbol{a}) = \psi_{\boldsymbol{a}}(\boldsymbol{y})\big\}.$$

Select any $j_1 \in \mathcal{R}_{\boldsymbol{a}}(\boldsymbol{x})$ and $j_2 \in \mathcal{R}_{\boldsymbol{a}}(\boldsymbol{y})$. Then it follows from Proposition 4.1 that there exists $L > 0$, not depending on $\boldsymbol{x}$ and $\boldsymbol{y}$, such that

$$\psi_{\boldsymbol{a}}(\boldsymbol{y}) - \psi_{\boldsymbol{a}}(\boldsymbol{x}) \geq d_p(\boldsymbol{y}_{j_2}, \boldsymbol{a}) - d_p(\boldsymbol{x}_{j_2}, \boldsymbol{a}) \geq -L\|\boldsymbol{y}_{j_2} - \boldsymbol{x}_{j_2}\| \geq -L\|\boldsymbol{y} - \boldsymbol{x}\|,$$
$$\psi_{\boldsymbol{a}}(\boldsymbol{y}) - \psi_{\boldsymbol{a}}(\boldsymbol{x}) \leq d_p(\boldsymbol{y}_{j_1}, \boldsymbol{a}) - d_p(\boldsymbol{x}_{j_1}, \boldsymbol{a}) \leq L\|\boldsymbol{y}_{j_1} - \boldsymbol{x}_{j_1}\| \leq L\|\boldsymbol{y} - \boldsymbol{x}\|.$$

Thus, we have

$$\big|\psi_{\boldsymbol{a}}(\boldsymbol{y}) - \psi_{\boldsymbol{a}}(\boldsymbol{x})\big| \leq L\|\boldsymbol{y} - \boldsymbol{x}\|,$$

where $L = \sqrt{n}$ for $p = 1$, $L = 1$ for $p = \infty$, and $L = 2(M_1 + M_2)$ for $p = 2$. □

Proposition 4.3 *The function f_k defined in* (4.4) *is LLC on $\mathbb{R}^{nk}$ for the similarity functions d_1, d_2, and d_∞.*

Proof The Lipschitz continuity of the function f_k follows from the facts that the functions $\psi_{\boldsymbol{a}}$, defined in (4.13), are LLC for any $\boldsymbol{a} \in A$ and the sum of LLC functions is also LLC, where the Lipschitz constant of the function f_k is $L = \sqrt{n}$ for $p = 1$, $L = 1$ for $p = \infty$, and $L = 2(M_1 + M_2)$ for $p = 2$. □

The functions d_1, d_2, and d_∞ are finite valued convex functions on $\mathbb{R}^n$ and therefore, they are directionally differentiable. The directional derivatives of these functions can be obtained using their subdifferentials and the formula (2.4). Therefore, the function $\psi_{\boldsymbol{a}}$, defined in (4.13), is also directionally differentiable and

$$\psi'_{\boldsymbol{a}}(\boldsymbol{x};\boldsymbol{u}) = \min_{j\in\mathcal{R}_{\boldsymbol{a}}(\boldsymbol{x})} d'_p\big((\boldsymbol{x}_j,\boldsymbol{a});\boldsymbol{u}_j\big), \quad \boldsymbol{u} = (\boldsymbol{u}_1,\ldots,\boldsymbol{u}_k) \in \mathbb{R}^{nk}. \tag{4.15}$$

Here, $d'_p\big((\boldsymbol{x}_j,\boldsymbol{a});\boldsymbol{u}_j\big)$ is the directional derivative of the function d_p at the point $\boldsymbol{x}_j \in \mathbb{R}^n$ in the direction $\boldsymbol{u}_j \in \mathbb{R}^n$, $j \in \mathcal{R}_{\boldsymbol{a}}(\boldsymbol{x})$.

Proposition 4.4 *The function f_k, given by (4.4) with the similarity functions d_1, d_2, or d_∞, is directionally differentiable at any $\boldsymbol{x} = (\boldsymbol{x}_1,\ldots,\boldsymbol{x}_k) \in \mathbb{R}^{nk}$ and*

$$f'_k(\boldsymbol{x};\boldsymbol{u}) = \frac{1}{m}\sum_{\boldsymbol{a}\in A} \min_{j\in\mathcal{R}_{\boldsymbol{a}}(\boldsymbol{x})} d'_p\big((\boldsymbol{x}_j,\boldsymbol{a});\boldsymbol{u}_j\big), \quad \boldsymbol{u} = (\boldsymbol{u}_1,\ldots,\boldsymbol{u}_k) \in \mathbb{R}^{nk}. \tag{4.16}$$

Proof The directional differentiability of the function f_k follows from its representation as a sum of minimum functions $\psi_{\boldsymbol{a}}$, $\boldsymbol{a} \in A$, and the expression (4.16) follows from (4.15) for the directional derivative of the function $\psi_{\boldsymbol{a}}$. □

Corollary 4.1 *Assume that at a point $\boldsymbol{x} = (\boldsymbol{x}_1,\ldots,\boldsymbol{x}_k) \in \mathbb{R}^{nk}$ there exists $\boldsymbol{a} \in A$ such that the cardinality $|\mathcal{R}_{\boldsymbol{a}}(\boldsymbol{x})| \geq 2$ and $d'_p\big((\boldsymbol{x}_{j_1},\boldsymbol{a});\boldsymbol{u}_{j_1}\big) \neq d'_p\big((\boldsymbol{x}_{j_2},\boldsymbol{a});\boldsymbol{u}_{j_2}\big)$ for some $j_1, j_2 \in \mathcal{R}_{\boldsymbol{a}}(\boldsymbol{x})$ and $\boldsymbol{u}_j \in \mathbb{R}^n$ where $\boldsymbol{u} = (\boldsymbol{u}_1,\ldots,\boldsymbol{u}_k) \in \mathbb{R}^{nk}$. Then the function f_k is not regular at $\boldsymbol{x}$.*

Proof The generalized directional derivative $f_k^\circ(\boldsymbol{x};\boldsymbol{u})$ of the function f_k at the point $\boldsymbol{x} \in \mathbb{R}^{nk}$ in the direction $\boldsymbol{u} = (\boldsymbol{u}_1,\ldots,\boldsymbol{u}_k) \in \mathbb{R}^{nk}$ is given by

$$f_k^\circ(\boldsymbol{x};\boldsymbol{u}) = \frac{1}{m}\sum_{\boldsymbol{a}\in A} \max_{j\in\mathcal{R}_{\boldsymbol{a}}(\boldsymbol{x})} d'_p\big((\boldsymbol{x}_j,\boldsymbol{a});\boldsymbol{u}_j\big).$$

This obviously implies that if the conditions of the corollary are satisfied then $f'_k(\boldsymbol{x};\boldsymbol{u}) < f_k^\circ(\boldsymbol{x};\boldsymbol{u})$ for some $\boldsymbol{u} \in \mathbb{R}^{nk}$. This completes the proof. □

If all conditions of Corollary 4.1 are satisfied for some $\boldsymbol{a} \in A$, then it follows from the proof of this corollary that at the point $\boldsymbol{x} \in \mathbb{R}^{nk}$ there exists $\boldsymbol{u} \in \mathbb{R}^{nk}$ such that $\psi'_{\boldsymbol{a}}(\boldsymbol{x};\boldsymbol{u}) < \psi^\circ_{\boldsymbol{a}}(\boldsymbol{x};\boldsymbol{u})$. This together with Corollary 4.1 implies that, in general, functions $\psi_{\boldsymbol{a}}$, $\boldsymbol{a} \in A$ and f_k are not regular. Thus, in general, at the point $\boldsymbol{x} = (\boldsymbol{x}_1,\ldots,\boldsymbol{x}_k) \in \mathbb{R}^{nk}$ we have

$$\begin{aligned} \partial\psi_{\boldsymbol{a}}(\boldsymbol{x}) &\subseteq \operatorname{conv}\big\{\partial d_p(\boldsymbol{x}_j,\boldsymbol{a}) :\ j \in \mathcal{R}_{\boldsymbol{a}}(\boldsymbol{x})\big\}, \quad \text{and} \\ \partial f_k(\boldsymbol{x}) &\subseteq \frac{1}{m}\sum_{\boldsymbol{a}\in A} \partial\psi_{\boldsymbol{a}}(\boldsymbol{x}). \end{aligned} \tag{4.17}$$

Next, we show that the function f_k is piecewise separable.

Proposition 4.5 *The function f_k, defined by (4.4), is piecewise separable for the similarity functions d_1, d_2, and d_∞.*

Proof It is clear that functions d_1 and d_2 are separable and the function d_∞ is piecewise separable. Since the function

$$\psi_{\boldsymbol{a}}(\boldsymbol{x}) = \min_{j=1,\ldots,k} d_p(\boldsymbol{x}_j, \boldsymbol{a}), \quad \boldsymbol{x} = (\boldsymbol{x}_1, \ldots, \boldsymbol{x}_k) \in \mathbb{R}^{nk}, \ \boldsymbol{a} \in A$$

is represented as a minimum of the finite number of functions each depending on a subset of variables it is piecewise separable according to Theorem 2.25. The function f_k—as a sum of piecewise separable functions $\psi_{\boldsymbol{a}}$, $\boldsymbol{a} \in A$—is also piecewise separable by this theorem. □

4.4 Nonsmooth DC Optimization Model

The objective function f_k in the clustering problem (4.3) can be represented as a difference of two convex functions [36, 80, 170]

$$f_k(\boldsymbol{x}) = f_{k1}(\boldsymbol{x}) - f_{k2}(\boldsymbol{x}), \quad \boldsymbol{x} = (\boldsymbol{x}_1, \ldots, \boldsymbol{x}_k) \in \mathbb{R}^{nk}, \tag{4.18}$$

where

$$f_{k1}(\boldsymbol{x}) = \frac{1}{m}\sum_{i=1}^{m}\sum_{j=1}^{k} d_p(\boldsymbol{x}_j, \boldsymbol{a}_i), \quad \text{and} \tag{4.19}$$

$$f_{k2}(\boldsymbol{x}) = \frac{1}{m}\sum_{i=1}^{m} \max_{j=1,\ldots,k} \sum_{s=1, s\neq j}^{k} d_p(\boldsymbol{x}_s, \boldsymbol{a}_i).$$

Therefore, the problem (4.3) can be reformulated as a DC optimization clustering problem

$$\begin{cases} \text{minimize} & f_k(\boldsymbol{x}) = f_{k1}(\boldsymbol{x}) - f_{k2}(\boldsymbol{x}) \\ \text{subject to} & \boldsymbol{x} = (\boldsymbol{x}_1, \ldots, \boldsymbol{x}_k) \in \mathbb{R}^{nk}. \end{cases} \tag{4.20}$$

As mentioned before, the function d_p is convex at $\boldsymbol{x}$ for $p = 1$, $p = 2$, and $p = \infty$. Therefore, the function f_{k1} as a sum of convex functions is convex. Furthermore, since the maximum of a finite number of convex functions is convex, the function f_{k2} is also convex as a sum of maxima of sum of convex functions. For the similarity measure d_2 the function f_{k1} is smooth and the function f_{k2} is, in general, nonsmooth. If $p = 1$ or $p = \infty$, then both f_{k1} and f_{k2} are nonsmooth functions.

Proposition 4.6 *The functions f_{k1} and f_{k2} are LLC.*

Proof The local Lipschitz continuity of DC components f_{k1} and f_{k2} follows from their convexity. For $p = 1$ the Lipschitz constant $L = \sqrt{n}$, for $p = 2$ this constant is $2(M_1 + M_2)$ and for $p = \infty$ one has $L = 1$ (see the proof of Proposition 4.1). □

It is obvious that the functions f_{k1} and f_{k2} are directionally differentiable as they are finite valued convex functions on $\mathbb{R}^{nk}$ and

$$f_k'(\boldsymbol{x}; \boldsymbol{u}) = f_{k1}'(\boldsymbol{x}; \boldsymbol{u}) - f_{k2}'(\boldsymbol{x}; \boldsymbol{u}), \quad \boldsymbol{x}, \boldsymbol{u} \in \mathbb{R}^{nk}. \tag{4.21}$$

Next, we describe subdifferentials of the functions f_{k1} and f_{k2} for each similarity measure d_p. For a given $\boldsymbol{a}_i \in A,\ i = 1, \ldots, m$, consider the following function:

$$\varphi_i(\boldsymbol{x}) = \max_{j=1,\ldots,k} \sum_{s=1, s \neq j}^{k} d_p(\boldsymbol{x}_s, \boldsymbol{a}_i). \tag{4.22}$$

Then the function f_{k2} can be represented as

$$f_{k2}(\boldsymbol{x}) = \frac{1}{m} \sum_{i=1}^{m} \varphi_i(\boldsymbol{x}).$$

For each $\boldsymbol{a}_i \in A$, introduce the set

$$\tilde{\mathcal{R}}_i(\boldsymbol{x}) = \Big\{ j \in \{1, \ldots, k\} : \sum_{s=1, s \neq j}^{k} d_p(\boldsymbol{x}_s, \boldsymbol{a}_i) = \varphi_i(\boldsymbol{x}) \Big\}. \tag{4.23}$$

Then the subdifferential $\partial \varphi_i(\boldsymbol{x})$ at the point $\boldsymbol{x} \in \mathbb{R}^{nk}$ can be expressed as

$$\partial \varphi_i(\boldsymbol{x}) = \operatorname{conv} \Big\{ \big(\boldsymbol{\xi}_1, \ldots, \boldsymbol{\xi}_{j-1}, \boldsymbol{0}, \boldsymbol{\xi}_{j+1}, \ldots, \boldsymbol{\xi}_k\big),\ \boldsymbol{\xi}_t \in \partial d_p(\boldsymbol{x}_t, \boldsymbol{a}_i),$$
$$t = 1, \ldots, k,\ t \neq j,\ j \in \tilde{\mathcal{R}}_i(\boldsymbol{x}) \Big\}.$$

For $p = 1$ the subdifferential $\partial d_1(\boldsymbol{x}_t, \boldsymbol{a}_i)$ is defined by (4.6), for $p = 2$ the subdifferential $\partial d_2(\boldsymbol{x}_t, \boldsymbol{a}_i)$ is a singleton and given in (4.8) and for $p = \infty$ the subdifferential $\partial d_\infty(\boldsymbol{x}_t, \boldsymbol{a}_i)$ is given by (4.10) and (4.11), $t = 1, \ldots, k$. For $p = 2$ we can get more simple representation of the subdifferential $\partial \varphi_i(\boldsymbol{x})$ as

$$\partial \varphi_i(\boldsymbol{x}) = \operatorname{conv} \Big\{ 2\big(\boldsymbol{x}_1 - \boldsymbol{a}_i, \ldots, \boldsymbol{x}_{j-1} - \boldsymbol{a}_i, \boldsymbol{0}, \boldsymbol{x}_{j+1} - \boldsymbol{a}_i, \ldots, \boldsymbol{x}_k - \boldsymbol{a}_i\big),$$
$$j \in \tilde{\mathcal{R}}_i(\boldsymbol{x}) \Big\}.$$

For each $\boldsymbol{a}_i \in A,\ i \in \{1, \ldots, m\}$, consider the set

$$\Pi_i(\boldsymbol{x}) = \Big\{\boldsymbol{\xi} = (\boldsymbol{\xi}_1, \ldots, \boldsymbol{\xi}_k) : \boldsymbol{\xi}_j \in \partial d_p(\boldsymbol{x}_j, \boldsymbol{a}_i),\ j = 1, \ldots, k\Big\}.$$

Thus, taking into account that the functions f_{k1} and f_{k2} are convex we have

$$\partial f_{k1}(\boldsymbol{x}) = \frac{1}{m}\sum_{i=1}^{m} \Pi_i(\boldsymbol{x}), \quad \text{and} \tag{4.24}$$

$$\partial f_{k2}(\boldsymbol{x}) = \frac{1}{m}\sum_{i=1}^{m} \partial\varphi_i(\boldsymbol{x}). \tag{4.25}$$

For the similarity measure d_2, the set $\Pi_i(\boldsymbol{x})$ is a singleton and we have

$$\Pi_i(\boldsymbol{x}) = \Big\{2\big(\boldsymbol{x}_1 - \boldsymbol{a}_i, \ldots, \boldsymbol{x}_k - \boldsymbol{a}_i\big)\Big\}.$$

In addition, it follows from (4.8) that the function f_{k1} is continuously differentiable in this case and we have

$$\nabla f_{k1}(\boldsymbol{x}) = 2(\boldsymbol{x} - \tilde{\boldsymbol{a}}), \tag{4.26}$$

where $\tilde{\boldsymbol{a}} = (\bar{\boldsymbol{a}}, \ldots, \bar{\boldsymbol{a}})$ and $\bar{\boldsymbol{a}}$ is the center of the set A, that is

$$\bar{\boldsymbol{a}} = \frac{1}{m}\sum_{i=1}^{m} \boldsymbol{a}_i.$$

Proposition 4.7 *For $p = 2$, the gradient ∇f_{k1} of the function f_{k1} satisfies the Lipschitz condition with the Lipschitz constant $L = 2$:*

$$\|\nabla f_{k1}(\boldsymbol{x}) - \nabla f_{k1}(\boldsymbol{y})\| \le 2\|\boldsymbol{x} - \boldsymbol{y}\| \quad \text{for all} \quad \boldsymbol{x}, \boldsymbol{y} \in \mathbb{R}^{nk}.$$

Proof The proof follows from the expression (4.26) of the gradient. □

Since, in general, DC functions are not regular we have

$$\partial f_k(\boldsymbol{x}) \subseteq \partial f_{k1}(\boldsymbol{x}) - \partial f_{k2}(\boldsymbol{x}), \quad \boldsymbol{x} \in \mathbb{R}^{nk}.$$

The equality holds for the similarity measure d_2.

Proposition 4.8 *For $p = 2$, the Clarke subdifferential $\partial f_k(\boldsymbol{x})$ of the cluster function f_k at $\boldsymbol{x} \in \mathbb{R}^{nk}$ is*

$$\partial f_k(\boldsymbol{x}) = \nabla f_{k1}(\boldsymbol{x}) - \partial f_{k2}(\boldsymbol{x}).$$

Proof Denote by $D(\boldsymbol{x}) = \nabla f_{k1}(\boldsymbol{x}) - \partial f_{k2}(\boldsymbol{x})$. For the general nonsmooth DC functions we have $\partial f_k(\boldsymbol{x}) \subseteq D(\boldsymbol{x})$, therefore, we only prove the opposite inclusion. Since the finite valued convex functions f_{k1} and f_{k2} are directionally differentiable the function f_k is also directionally differentiable and we have

$$f_k'(\boldsymbol{x}; \boldsymbol{u}) = f_{k1}'(\boldsymbol{x}; \boldsymbol{u}) - f_{k2}'(\boldsymbol{x}; \boldsymbol{u}), \quad \boldsymbol{u} \in \mathbb{R}^{nk}.$$

The function f_k is DC, and therefore, the function $-f_k$ is DC and we get

$$(-f_k)'(\boldsymbol{x}; \boldsymbol{u}) = f_{k2}'(\boldsymbol{x}; \boldsymbol{u}) - f_{k1}'(\boldsymbol{x}; \boldsymbol{u}), \quad \boldsymbol{u} \in \mathbb{R}^{nk}.$$

In addition, we have (see, e.g. (2.4))

$$\begin{aligned} f_{k1}'(\boldsymbol{x}; \boldsymbol{u}) &= (\nabla f_{k1}(\boldsymbol{x}))^T \boldsymbol{u}, \quad \text{and} \\ f_{k2}'(\boldsymbol{x}; \boldsymbol{u}) &= \max_{\boldsymbol{\xi}_2 \in \partial f_{k2}(\boldsymbol{x})} \boldsymbol{\xi}_2^T \boldsymbol{u}. \end{aligned}$$

Then

$$\begin{aligned} (-f_k)^\circ(\boldsymbol{x}; \boldsymbol{u}) &\geq (-f_k)'(\boldsymbol{x}; \boldsymbol{u}) \\ &= f_{k2}'(\boldsymbol{x}; \boldsymbol{u}) - f_{k1}'(\boldsymbol{x}; \boldsymbol{u}) \\ &= \max_{\boldsymbol{\xi}_2 \in \partial f_{k2}(\boldsymbol{x})} \big(\boldsymbol{\xi}_2 - \nabla f_{k1}(\boldsymbol{x})\big)^T \boldsymbol{u} \end{aligned}$$

for all $\boldsymbol{u} \in \mathbb{R}^{nk}$, and therefore, we have $(-f_k)^\circ(\boldsymbol{x}; \boldsymbol{u}) \geq \boldsymbol{v}^T \boldsymbol{u}$ for all $\boldsymbol{u} \in \mathbb{R}^{nk}$ and $\boldsymbol{v} \in -D(\boldsymbol{x})$. This means that $-D(\boldsymbol{x}) \subseteq \partial(-f_k)(\boldsymbol{x}) = -\partial f_k(\boldsymbol{x})$ considering the convexity of the set $D(\boldsymbol{x})$. Thus, we have $D(\boldsymbol{x}) \subseteq \partial f_k(\boldsymbol{x})$ and this completes the proof. □

Proposition 4.9 *For the functions f_{k1} and f_{k2} the following holds:*

(i) *the function f_{k1} is separable for the functions d_1 and d_2 and piecewise separable for the function d_∞; and*
(ii) *the function f_{k2} is piecewise separable for functions d_1, d_2, and d_∞.*

Proof For the case (i), since both d_1 and d_2 are separable according to Theorem 2.25 in these two cases the function f_{k1} as a sum of separable functions is also separable. The function d_∞ is piecewise separable and therefore, the function f_{k1} is also piecewise separable as a sum of piecewise separable functions.

For the case (ii), it follows from Theorem 2.25 that functions represented as a maximum of separable or piecewise separable functions are piecewise separable and therefore, the function φ_i defined in (4.22) is piecewise separable for similarity measures d_1, d_2, and d_∞. Then again by applying Theorem 2.25 we get the proof. □

4.5 Auxiliary Clustering Problem

The objective function in the clustering problem (4.3) is nonconvex and the problem itself is a global optimization problem. This function has many local minimizers and its global or nearly global minimizers are of interest. Conventional global optimization techniques may become extremely time-consuming for solving the clustering problem in large data sets with relatively large number of clusters. Local search optimization algorithms are much faster than their global search counterparts; however, starting from a given point they may end up at the closest local minimizer which can be significantly different from the global minimizer.

The success of local search methods in finding global or nearly global solutions to the clustering problems highly depends on the choice of starting points. Different approaches can be used to generate such points (see [63], for some of these approaches). We introduce the auxiliary clustering problem to design algorithms for finding a set of "good/promising" starting points.

Assume that the solution $\boldsymbol{x}_1, \ldots, \boldsymbol{x}_{k-1},\ k \geq 2$ to the $(k-1)$-clustering problem is known. Denote by r^i_{k-1} the distance (the squared distance for $p = 2$) between the data point $\boldsymbol{a}_i,\ i = 1, \ldots, m$ and its cluster center

$$r^i_{k-1} = \min_{j=1,\ldots,k-1} d_p(\boldsymbol{x}_j, \boldsymbol{a}_i). \tag{4.27}$$

We will also use the notation $r^{\boldsymbol{a}}_{k-1}$ for the data point $\boldsymbol{a} \in A$. The *kth auxiliary cluster function* is defined as [19]

$$\bar{f}_k(\boldsymbol{y}) = \frac{1}{m} \sum_{i=1}^{m} \min\left\{r^i_{k-1}, d_p(\boldsymbol{y}, \boldsymbol{a}_i)\right\}, \quad \boldsymbol{y} \in \mathbb{R}^n. \tag{4.28}$$

This function is nonsmooth and as a sum of minima of convex functions it is, in general, nonconvex. The kth auxiliary cluster function has n variables and the number of variables does not depend on the number of clusters k.

The *kth auxiliary clustering problem* is formulated as [19]

$$\begin{cases} \text{minimize} & \bar{f}_k(\boldsymbol{y}) \\ \text{subject to} & \boldsymbol{y} \in \mathbb{R}^n. \end{cases} \tag{4.29}$$

It is obvious that

$$\bar{f}_k(\boldsymbol{y}) = f_k(\boldsymbol{x}_1, \ldots, \boldsymbol{x}_{k-1}, \boldsymbol{y}) \quad \text{for all} \quad \boldsymbol{y} \in \mathbb{R}^n.$$

Example 4.2 Next, we illustrate the auxiliary cluster function $\bar{f}_k$. Consider the set (4.5) and assume that $k = 3$. This means that the solution to the 2-clustering problem is known which is $(-2, 3.5)$. In order to construct the function $\bar{f}_k$, first we compute the numbers $r^i_2,\ i = 1, \ldots, 18$. For the similarity measure d_1 we have

$$r_2^1 = 1,\ r_2^2 = 0.6,\ r_2^3 = 0.2,\ r_2^4 = 0.2,\ r_2^5 = 0.6,\ r_2^6 = 1,\ r_2^7 = 2.5,\ r_2^8 = 2.1,$$
$$r_2^9 = 1.7,\ r_2^{10} = 1.3,\ r_2^{11} = 0.9,\ r_2^{12} = 0.5,\ r_2^{13} = 0.5,\ r_2^{14} = 0.9,\ r_2^{15} = 1.3,$$
$$r_2^{16} = 1.7,\ r_2^{17} = 2.1,\ r_2^{18} = 2.5.$$

Then the function $\bar{f}_3$ with the similarity measure d_1 is

$$\begin{aligned}\bar{f}_3(y) = \frac{1}{18}\Big\{ & \min\{1, |y+3|\}, \min\{0.6, |y+2.6|\}, \min\{0.2, |y+2.2|\},\\ & \min\{0.2, |y+1.8|\}, \min\{0.6, |y+1.4|\}, \min\{1, |y+1|\},\\ & \min\{2.5, |y-1|\}, \min\{2.1, |y-1.4|\}, \min\{1.7, |y-1.8|\},\\ & \min\{1.3, |y-2.2|\}, \min\{0.9, |y-2.6|\}, \min\{0.5, |y-3|\},\\ & \min\{0.5, |y-4|\}, \min\{0.9, |y-4.4|\}, \min\{1.3, |y-4.8|\},\\ & \min\{1.7, |y-5.2|\}, \min\{2.1, |y-5.6|\}, \min\{2.5, |y-6|\}\Big\}.\end{aligned}$$

Now, consider the similarity measure d_2. For this measure, we have

$$r_2^1 = 1,\ r_2^2 = 0.36,\ r_2^3 = 0.04,\ r_2^4 = 0.04,\ r_2^5 = 0.36,\ r_2^6 = 1,\ r_2^7 = 6.25,$$
$$r_2^8 = 4.41,\ r_2^9 = 2.89,\ r_2^{10} = 1.69,\ r_2^{11} = 0.81,\ r_2^{12} = 0.25,\ r_2^{13} = 0.25,$$
$$r_2^{14} = 0.81,\ r_2^{15} = 1.69,\ r_2^{16} = 2.89,\ r_2^{17} = 4.41,\ r_2^{18} = 6.25.$$

Then we get

$$\begin{aligned}\bar{f}_3(y) = \frac{1}{18}\Big\{ & \min\{1, (y+3)^2\}, \min\{0.36, (y+2.6)^2\}, \min\{0.04, (y+2.2)^2\},\\ & \min\{0.04, (y+1.8)^2\}, \min\{0.36, (y+1.4)^2\}, \min\{1, (y+1)^2\},\\ & \min\{6.25, (y-1)^2\}, \min\{4.41, (y-1.4)^2\}, \min\{2.89, (y-1.8)^2\},\\ & \min\{1.69, (y-2.2)^2\}, \min\{0.81, (y-2.6)^2\}, \min\{0.25, (y-3)^2\},\\ & \min\{0.25, (y-4)^2\}, \min\{0.81, (y-4.4)^2\}, \min\{1.69, (y-4.8)^2\},\\ & \min\{2.89, (y-5.2)^2\}, \min\{4.41, (y-5.6)^2\}, \min\{6.25, (y-6)^2\}\Big\}.\end{aligned}$$

Graphs of the function $\bar{f}_3$ with the similarity measures d_1 and d_2 are depicted in Fig. 4.2. It is easy to see that in both cases, there are some regions where the functions are constant.

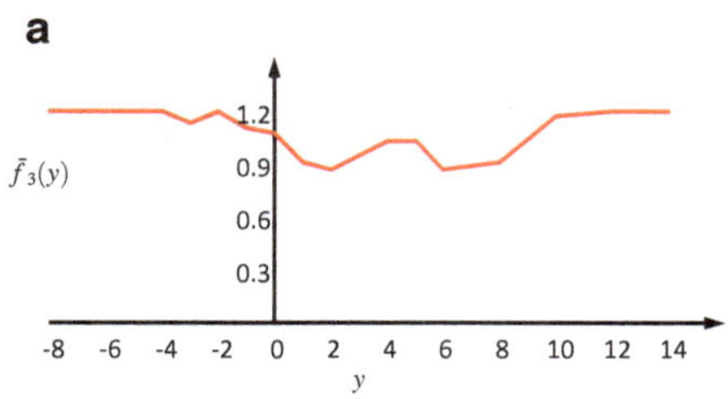

Fig. 4.2 Graphs of auxiliary cluster function $\bar{f}_3$ with similarity measures d_1 and d_2. (**a**) Auxiliary cluster function $\bar{f}_3$ with d_1. (**b**) Auxiliary cluster function $\bar{f}_3$ with d_2

Next, we study some important properties of the auxiliary cluster function $\bar{f}_k$. For a given $\boldsymbol{a} \in A$, consider the function

$$\theta_{\boldsymbol{a}}(\boldsymbol{y}) = \min \left\{ r_{k-1}^{\boldsymbol{a}}, d_p(\boldsymbol{y}, \boldsymbol{a}) \right\}.$$

Proposition 4.10 *The function $\theta_{\boldsymbol{a}}$ is LLC for any $\boldsymbol{a} \in A$.*

Proof Take any bounded subset $X \subset \mathbb{R}^n$ and $\boldsymbol{y}, \boldsymbol{z} \in X$. Then the following cases are possible:

(i) $r_{k-1}^{\boldsymbol{a}} \le d_p(\boldsymbol{y}, \boldsymbol{a})$ and $r_{k-1}^{\boldsymbol{a}} \le d_p(\boldsymbol{z}, \boldsymbol{a})$: then we have

$$\theta_{\boldsymbol{a}}(\boldsymbol{z}) - \theta_{\boldsymbol{a}}(\boldsymbol{y}) = 0;$$

(ii) $r_{k-1}^{\boldsymbol{a}} > d_p(\boldsymbol{y}, \boldsymbol{a})$ and $r_{k-1}^{\boldsymbol{a}} \le d_p(\boldsymbol{z}, \boldsymbol{a})$: then we get

$$0 \le \theta_{\boldsymbol{a}}(\boldsymbol{z}) - \theta_{\boldsymbol{a}}(\boldsymbol{y}) \le d_p(\boldsymbol{z}, \boldsymbol{a}) - d_p(\boldsymbol{y}, \boldsymbol{a});$$

(iii) $r_{k-1}^{\boldsymbol{a}} \le d_p(\boldsymbol{y}, \boldsymbol{a})$ and $r_{k-1}^{\boldsymbol{a}} > d_p(\boldsymbol{z}, \boldsymbol{a})$: then we have

$$d_p(\boldsymbol{z}, \boldsymbol{a}) - d_p(\boldsymbol{y}, \boldsymbol{a}) \le \theta_{\boldsymbol{a}}(\boldsymbol{z}) - \theta_{\boldsymbol{a}}(\boldsymbol{y}) \le 0;$$

(iv) $r_{k-1}^{\boldsymbol{a}} > d_p(\boldsymbol{y}, \boldsymbol{a})$ and $r_{k-1}^{\boldsymbol{a}} > d_p(\boldsymbol{z}, \boldsymbol{a})$: this means that

$$\theta_{\boldsymbol{a}}(\boldsymbol{z}) - \theta_{\boldsymbol{a}}(\boldsymbol{y}) = d_p(\boldsymbol{z}, \boldsymbol{a}) - d_p(\boldsymbol{y}, \boldsymbol{a}).$$

In all cases, we have

$$|\theta_{\boldsymbol{a}}(\boldsymbol{z}) - \theta_{\boldsymbol{a}}(\boldsymbol{y})| \le |d_p(\boldsymbol{z}, \boldsymbol{a}) - d_p(\boldsymbol{y}, \boldsymbol{a})|.$$

Since the function d_p is LLC for $p = 1$, $p = 2$, and $p = \infty$ it follows that the function $\theta_{\boldsymbol{a}}$ is also LLC. Its Lipschitz constant is the same as constants of the corresponding similarity functions d_p. □

The function $\theta_{\boldsymbol{a}}$ is directionally differentiable and at a point $\boldsymbol{y} \in \mathbb{R}^n$ for the direction $\boldsymbol{u} \in \mathbb{R}^n$ we have

$$\theta'_{\boldsymbol{a}}(\boldsymbol{y}; \boldsymbol{u}) = \begin{cases} 0, & \text{if } r^{\boldsymbol{a}}_{k-1} < d_p(\boldsymbol{y}, \boldsymbol{a}), \\ \min\big\{0, d'_p\big((\boldsymbol{y}, \boldsymbol{a}); \boldsymbol{u}\big)\big\}, & \text{if } r^{\boldsymbol{a}}_{k-1} = d_p(\boldsymbol{y}, \boldsymbol{a}), \\ d'_p\big((\boldsymbol{y}, \boldsymbol{a}); \boldsymbol{u}\big), & \text{if } r^{\boldsymbol{a}}_{k-1} > d_p(\boldsymbol{y}, \boldsymbol{a}). \end{cases}$$

Since the function $\theta_{\boldsymbol{a}}$ is LLC it has generalized directional derivatives and at a point $\boldsymbol{y} \in \mathbb{R}^n$ with respect to a direction $\boldsymbol{u} \in \mathbb{R}^n$ one has

$$\theta^\circ_{\boldsymbol{a}}(\boldsymbol{y}; \boldsymbol{u}) = \begin{cases} 0, & \text{if } r^{\boldsymbol{a}}_{k-1} < d_p(\boldsymbol{y}, \boldsymbol{a}), \\ \max\big\{0, d'_p\big((\boldsymbol{y}, \boldsymbol{a}); \boldsymbol{u}\big)\big\}, & \text{if } r^{\boldsymbol{a}}_{k-1} = d_p(\boldsymbol{y}, \boldsymbol{a}), \\ d'_p\big((\boldsymbol{y}, \boldsymbol{a}); \boldsymbol{u}\big), & \text{if } r^{\boldsymbol{a}}_{k-1} > d_p(\boldsymbol{y}, \boldsymbol{a}). \end{cases}$$

It follows from representations of the directional and generalized directional derivatives of the function $\theta_{\boldsymbol{a}}$ that if at a point $\boldsymbol{y} \in \mathbb{R}^n$ one has $r^{\boldsymbol{a}}_{k-1} = d_p(\boldsymbol{y}, \boldsymbol{a})$ and there exists $\boldsymbol{u} \in \mathbb{R}^n$ such that $d'_p\big((\boldsymbol{y}, \boldsymbol{a}); \boldsymbol{u}\big) > 0$, then $\theta'_{\boldsymbol{a}}(\boldsymbol{y}; \boldsymbol{u}) < \theta^\circ_{\boldsymbol{a}}(\boldsymbol{y}; \boldsymbol{u})$ and the function $\theta_{\boldsymbol{a}}$ is not regular at $\boldsymbol{y}$. Since the function $\theta_{\boldsymbol{a}}$ is LLC it is subdifferentiable and, in general, we have

$$\partial\theta_{\boldsymbol{a}}(\boldsymbol{y}; \boldsymbol{u}) \begin{cases} = \{\boldsymbol{0}\}, & \text{if } r^{\boldsymbol{a}}_{k-1} < d_p(\boldsymbol{y}, \boldsymbol{a}), \\ \subseteq \big\{\boldsymbol{0}, \partial d_p(\boldsymbol{y}, \boldsymbol{a})\big\}, & \text{if } r^{\boldsymbol{a}}_{k-1} = d_p(\boldsymbol{y}, \boldsymbol{a}), \\ = \partial d_p(\boldsymbol{y}, \boldsymbol{a}), & \text{if } r^{\boldsymbol{a}}_{k-1} > d_p(\boldsymbol{y}, \boldsymbol{a}). \end{cases}$$

Proposition 4.11 *The auxiliary cluster function $\bar{f}_k$, defined in* (4.28)*, is LLC on* $\mathbb{R}^n$.

Proof The proof follows from Proposition 4.10 that the functions $\theta_{\boldsymbol{a}}$, $\boldsymbol{a} \in A$ are LLC. Moreover, the Lipschitz constant of the function $\bar{f}_k$ is the same as that of the functions $\theta_{\boldsymbol{a}}$, $\boldsymbol{a} \in A$ for $p = 1$, $p = 2$, and $p = \infty$. □

The function $\bar{f}_k$ is directionally differentiable and we have

$$\bar{f}'_k(\boldsymbol{y}; \boldsymbol{u}) = \frac{1}{|A|} \sum_{\boldsymbol{a} \in A} \theta'_{\boldsymbol{a}}(\boldsymbol{y}; \boldsymbol{u}), \quad \boldsymbol{y},\ \boldsymbol{u} \in \mathbb{R}^n.$$

Furthermore, since the function $\bar{f}_k$ is LLC it has the generalized directional derivatives and it is also subdifferentiable. However, in general, this function is not regular as a sum of nonregular functions $\theta_{\boldsymbol{a}}$, $\boldsymbol{a} \in A$. Then at a point $\boldsymbol{y} \in \mathbb{R}^n$, one only has

$$\partial \bar{f}_k(\boldsymbol{y}) \subseteq \frac{1}{|A|} \sum_{\boldsymbol{a} \in A} \partial\theta_{\boldsymbol{a}}(\boldsymbol{y}).$$

In order to get more constructive formula for the subdifferential $\partial \bar{f}_k(\mathbf{y})$ at the point $\mathbf{y} \in \mathbb{R}^n$, consider the sets

$$\begin{aligned} B_1(\mathbf{y}) &= \{\mathbf{a} \in A : r_{k-1}^{\mathbf{a}} < d_p(\mathbf{y}, \mathbf{a})\}, \\ B_2(\mathbf{y}) &= \{\mathbf{a} \in A : r_{k-1}^{\mathbf{a}} = d_p(\mathbf{y}, \mathbf{a})\}, \quad \text{and} \\ B_3(\mathbf{y}) &= \{\mathbf{a} \in A : r_{k-1}^{\mathbf{a}} > d_p(\mathbf{y}, \mathbf{a})\}. \end{aligned} \tag{4.30}$$

The set $B_1(\mathbf{y})$ contains all data points $\mathbf{a} \in A$ which are closer to their cluster centers than to the point $\mathbf{y}$, the set $B_2(\mathbf{y})$ contains all data points which are the same distance from their cluster centers and the point $\mathbf{y}$, and finally, the set $B_3(\mathbf{y})$ contains all data points which are closer to the point $\mathbf{y}$ than to their cluster centers. We assume that $B_3(\mathbf{y}) \neq A$ for any $\mathbf{y} \in \mathbb{R}^n$. This condition means that there is no point $\mathbf{y} \in \mathbb{R}^n$ which can attract all data points than their cluster centers. The case $B_3(\mathbf{y}) = A$ for some $\mathbf{y} \in \mathbb{R}^n$ is rather pathological.

Now the function $\bar{f}_k$ can be represented as

$$\bar{f}_k(\mathbf{y}) = \frac{1}{m}\Bigg(\sum_{\mathbf{a} \in B_1(\mathbf{y})} r_{k-1}^{\mathbf{a}} + \sum_{\mathbf{a} \in B_2(\mathbf{y})} \min\big\{r_{k-1}^{\mathbf{a}}, d_p(\mathbf{y}, \mathbf{a})\big\} + \sum_{\mathbf{a} \in B_3(\mathbf{y})} d_p(\mathbf{y}, \mathbf{a}) \Bigg),$$

and thus the subdifferential of $\bar{f}_k$ at the point $\mathbf{y} \in \mathbb{R}^n$ when $B_1(\mathbf{y}) = A$ is

$$\partial \bar{f}_k(\mathbf{y}) = \{\mathbf{0}\},$$

otherwise

$$\partial \bar{f}_k(\mathbf{y}) \subseteq \frac{1}{m}\Bigg(\sum_{\mathbf{a} \in B_2(\mathbf{y})} \operatorname{conv}\big\{\mathbf{0}, \partial d_p(\mathbf{y}, \mathbf{a})\big\} + \sum_{\mathbf{a} \in B_3(\mathbf{y})} \partial d_p(\mathbf{y}, \mathbf{a}) \Bigg). \tag{4.31}$$

Proposition 4.12 *The auxiliary cluster function $\bar{f}_k$, defined in* (4.28), *is piecewise separable for $p = 1,\ p = 2$, and $p = \infty$.*

Proof According to Theorem 2.25 the function $\theta_{\mathbf{a}},\ \mathbf{a} \in A$ is piecewise separable for $p = 1,\ p = 2$ and $p = \infty$. Then due to Theorem 2.25 the function $\bar{f}_k$ is also piecewise separable as a sum of such functions. □

4.5.1 DC Representation of Auxiliary Cluster Function

Similar to the cluster function f_k, the auxiliary cluster function $\bar{f}_k$ is also DC [36, 170] and

$$\bar{f}_k(\mathbf{y}) = \bar{f}_{k1}(\mathbf{y}) - \bar{f}_{k2}(\mathbf{y}), \tag{4.32}$$

where

$$\bar{f}_{k1}(\mathbf{y}) = \frac{1}{m}\sum_{i=1}^{m}\left(r_{k-1}^{i} + d_p(\mathbf{y}, \mathbf{a}_i)\right), \quad \text{and} \tag{4.33}$$

$$\bar{f}_{k2}(\mathbf{y}) = \frac{1}{m}\sum_{i=1}^{m}\max\left\{r_{k-1}^{i}, d_p(\mathbf{y}, \mathbf{a}_i)\right\}.$$

It is clear that both $\bar{f}_{k1}$ and $\bar{f}_{k2}$ are convex functions. Then the auxiliary clustering problem (4.29) can be reformulated as

$$\begin{cases} \text{minimize} & \bar{f}_k(\mathbf{y}) = \bar{f}_{k1}(\mathbf{y}) - \bar{f}_{k2}(\mathbf{y}) \\ \text{subject to} & \mathbf{y} \in \mathbb{R}^n. \end{cases} \tag{4.34}$$

Proposition 4.13 *The functions $\bar{f}_{k1}$ and $\bar{f}_{k2}$ are LLC on $\mathbb{R}^n$.*

Proof Let $X \subset \mathbb{R}^n$ be any bounded subset. It is obvious that the function $\bar{f}_{k1}$ is LLC, and on the set X its Lipschitz constant is the same as constants of the corresponding similarity functions (see Proposition 4.1). To prove the locally Lipschitz continuity of the function $\bar{f}_{k2}$, consider the function

$$\bar{\theta}_{\mathbf{a}}(\mathbf{y}) = \max\left\{r_{k-1}^{\mathbf{a}}, d_p(\mathbf{y}, \mathbf{a})\right\}, \quad \mathbf{a} \in A.$$

Take any $\mathbf{y}, \mathbf{z} \in X$. Similar to Proposition 4.10, we have the following cases:

(i) $r_{k-1}^{\mathbf{a}} < d_p(\mathbf{y}, \mathbf{a})$ and $r_{k-1}^{\mathbf{a}} < d_p(\mathbf{z}, \mathbf{a})$: then we have

$$\bar{\theta}_{\mathbf{a}}(\mathbf{z}) - \bar{\theta}_{\mathbf{a}}(\mathbf{y}) = d_p(\mathbf{z}, \mathbf{a}) - d_p(\mathbf{y}, \mathbf{a});$$

(ii) $r_{k-1}^{\mathbf{a}} \geq d_p(\mathbf{y}, \mathbf{a})$ and $r_{k-1}^{\mathbf{a}} < d_p(\mathbf{z}, \mathbf{a})$: then we get

$$0 \leq \bar{\theta}_{\mathbf{a}}(\mathbf{z}) - \bar{\theta}_{\mathbf{a}}(\mathbf{y}) \leq d_p(\mathbf{z}, \mathbf{a}) - d_p(\mathbf{y}, \mathbf{a});$$

(iii) $r_{k-1}^{\mathbf{a}} < d_p(\mathbf{y}, \mathbf{a})$ and $r_{k-1}^{\mathbf{a}} \geq d_p(\mathbf{z}, \mathbf{a})$: then we have

$$d_p(\mathbf{z}, \mathbf{a}) - d_p(\mathbf{y}, \mathbf{a}) \leq \bar{\theta}_{\mathbf{a}}(\mathbf{z}) - \bar{\theta}_{\mathbf{a}}(\mathbf{y}) \leq 0;$$

(iv) $r_{k-1}^{\mathbf{a}} \geq d_p(\mathbf{y}, \mathbf{a})$ and $r_{k-1}^{\mathbf{a}} \geq d_p(\mathbf{z}, \mathbf{a})$: this means that

$$\bar{\theta}_{\mathbf{a}}(\mathbf{z}) - \bar{\theta}_{\mathbf{a}}(\mathbf{y}) = 0.$$

In all cases, we have

$$|\bar{\theta}_{\mathbf{a}}(\mathbf{z}) - \bar{\theta}_{\mathbf{a}}(\mathbf{y})| \leq |d_p(\mathbf{z}, \mathbf{a}) - d_p(\mathbf{y}, \mathbf{a})|.$$

Since the function d_p is LLC for $p = 1,\ p = 2$, and $p = \infty$, the function $\bar{\theta}_{\boldsymbol{a}}$ is also LLC. Then we get that the function $\bar{f}_{k2}$ is LLC and its Lipschitz constant is the same as constants of the corresponding similarity functions. □

Using the sets $B_i(\boldsymbol{y}),\ \boldsymbol{y} \in \mathbb{R}^n,\ i = 1, 2, 3$, defined in (4.30), the function $\bar{f}_{k2}$ at the point $\boldsymbol{y}$ can be rewritten as

$$\bar{f}_{k2}(\boldsymbol{y}) = \frac{1}{m}\Bigg(\sum_{\boldsymbol{a}\in B_1(\boldsymbol{y})} d_p(\boldsymbol{y},\boldsymbol{a}) + \sum_{\boldsymbol{a}\in B_2(\boldsymbol{y})} \max\left\{r_{k-1}^{\boldsymbol{a}}, d_p(\boldsymbol{y},\boldsymbol{a})\right\} + \sum_{\boldsymbol{a}\in B_3(\boldsymbol{y})} r_{k-1}^{\boldsymbol{a}}\Bigg). \tag{4.35}$$

As finite valued convex functions defined on $\mathbb{R}^n$ the functions $\bar{f}_{k1}$ and $\bar{f}_{k2}$ are directionally differentiable, and we have

$$\bar{f}'_{k1}(\boldsymbol{y};\boldsymbol{u}) = \frac{1}{m}\sum_{\boldsymbol{a}\in A} d'_p\big((\boldsymbol{y},\boldsymbol{a});\boldsymbol{u}\big), \quad \text{and} \tag{4.36}$$

$$\bar{f}'_{k2}(\boldsymbol{y};\boldsymbol{u}) = \frac{1}{m}\Bigg(\sum_{\boldsymbol{a}\in B_1(\boldsymbol{y})} d'_p\big((\boldsymbol{y},\boldsymbol{a});\boldsymbol{u}\big) + \sum_{\boldsymbol{a}\in B_2(\boldsymbol{y})} \max\left\{0, d'_p\big((\boldsymbol{y},\boldsymbol{a});\boldsymbol{u}\big)\right\}\Bigg). \tag{4.37}$$

Then the subdifferential $\partial \bar{f}_{k1}(\boldsymbol{y})$ of the function $\bar{f}_{k1}$ at the point $\boldsymbol{y} \in \mathbb{R}^n$ is

$$\partial \bar{f}_{k1}(\boldsymbol{y}) = \frac{1}{m}\sum_{\boldsymbol{a}\in A} \partial d_p(\boldsymbol{y},\boldsymbol{a}). \tag{4.38}$$

In the case when $B_3(\boldsymbol{y}) = A$ at the point $\boldsymbol{y} \in \mathbb{R}^n$ we have

$$\partial \bar{f}_{k2}(\boldsymbol{y}) = \{\boldsymbol{0}\} \tag{4.39}$$

and when $B_3(\boldsymbol{y}) \neq A$ the subdifferential $\partial \bar{f}_{k2}(\boldsymbol{y})$ of the function $\bar{f}_{k2}$ at $\boldsymbol{y} \in \mathbb{R}^n$ is

$$\partial \bar{f}_{k2}(\boldsymbol{y}) = \frac{1}{m}\Bigg(\sum_{\boldsymbol{a}\in B_1(\boldsymbol{y})} \partial d_p(\boldsymbol{y},\boldsymbol{a}) + \sum_{\boldsymbol{a}\in B_2(\boldsymbol{y})} \operatorname{conv}\left\{\boldsymbol{0}, \partial d_p(\boldsymbol{y},\boldsymbol{a})\right\}\Bigg). \tag{4.40}$$

For $p = 2$, the function $\bar{f}_{k1}$ is differentiable and

$$\nabla \bar{f}_{k1}(\boldsymbol{y}) = 2(\boldsymbol{y} - \bar{\boldsymbol{a}}), \tag{4.41}$$

where $\bar{a}$ is the center of the set A. In this case, we get a simple formula for the subdifferential $\partial \bar{f}_{k2}(y)$ as

$$\partial \bar{f}_{k2}(y) = \frac{2}{m}\Big(\sum_{a \in B_1(y)} (y - a) + \sum_{a \in B_2(y)} \operatorname{conv}\big\{0, y - a\big\}\Big). \tag{4.42}$$

Proposition 4.14 *Let $p = 2$. Then the Clarke subdifferential $\partial \bar{f}_k(y)$ of the function $\bar{f}_k$ at $y \in \mathbb{R}^n$ can be given as*

$$\partial \bar{f}_k(y) = \nabla \bar{f}_{k1}(y) - \partial \bar{f}_{k2}(y).$$

Proof The proof is similar to that of Proposition 4.8. □

Proposition 4.15 *For the DC components $\bar{f}_{k1}$ and $\bar{f}_{k2}$ the following holds:*

(i) *the function $\bar{f}_{k1}$ is separable for $p = 1$ and $p = 2$ and it is piecewise separable for $p = \infty$; and*
(ii) *the function $\bar{f}_{k2}$ is piecewise separable for $p = 1$, $p = 2$, and $p = \infty$.*

Proof

(i) Since the functions d_1 and d_2 are separable the function $\bar{f}_{k1}$ is also separable as a sum of separable functions. Piecewise separability of the function $\bar{f}_{k1}$ follows from piecewise separability of d_∞ and Theorem 2.25.
(ii) Due to Theorem 2.25, the maximum of finite number of separable or piecewise separable functions is piecewise separable. Then the function $\bar{f}_{k2}$ is piecewise separable as a sum of piecewise separable functions. □

4.6 Optimality Conditions

In this section, we formulate optimality conditions for the clustering and the auxiliary clustering problems (4.3) and (4.29) using, in particular, their respective DC models (4.20) and (4.34).

4.6.1 Optimality Conditions for Clustering Problem

Different optimality conditions can be obtained for the clustering problem using its formulations (4.3) and (4.20). We start with the formulation of the optimality condition using the Clarke subdifferential.

Proposition 4.16 *Let $x^* \in \mathbb{R}^{nk}$ be a local minimizer of the problem* (4.3). *Then*

$$0 \in \partial f_k(x^*). \tag{4.43}$$

Proof According to Proposition 4.3, the function f_k is LLC. Then the proof follows from Theorem 2.17. □

Points satisfying the condition (4.43) are called the Clarke stationary points of the clustering problem (4.3).

Corollary 4.2 *Clarke stationary points* $\boldsymbol{x} = (\boldsymbol{x}_1, \ldots, \boldsymbol{x}_k) \in \mathbb{R}^{nk}$ *of the clustering problem* (4.3) *satisfy the following condition:*

$$\mathbf{0} \in \sum_{\boldsymbol{a} \in A} \operatorname{conv} \Big\{ \partial d_p(\boldsymbol{x}_j, \boldsymbol{a}),\ j \in \mathcal{R}_{\boldsymbol{a}}(\boldsymbol{x}) \Big\},$$

for $p = 1, 2, \infty$. *The set* $\mathcal{R}_{\boldsymbol{a}}(\boldsymbol{x})$ *is defined in* (4.14).

Proof The proof follows from Proposition 4.16 and the inclusion (4.17). □

Next, we consider optimality conditions obtained using the nonsmooth DC model (4.20) of the clustering problem. The following theorem follows from Theorem 2.27.

Theorem 4.1 *Let* $\boldsymbol{x}^* \in \mathbb{R}^{nk}$ *be a local minimizer of the problem* (4.20). *Then*

$$\partial f_{k2}(\boldsymbol{x}^*) \subseteq \partial f_{k1}(\boldsymbol{x}^*), \quad \textit{and} \tag{4.44}$$

$$\partial f_{k1}(\boldsymbol{x}^*) \cap \partial f_{k2}(\boldsymbol{x}^*) \neq \emptyset. \tag{4.45}$$

Proof The condition (4.44) follows from (2.24), and the condition (4.45) results from (2.26) in Theorem 2.27. □

Points satisfying the condition (4.44) are called inf-stationary, and points satisfying the condition (4.45) are called critical points of the clustering problem (4.20).

We formulate optimality conditions for each similarity functions d_1, d_2, and d_∞ separately. We start with the function d_1. The subdifferentials of the functions f_{k1} and f_{k2} are given in (4.24) and (4.25), respectively. These subdifferentials at the point $\boldsymbol{x} \in \mathbb{R}^{nk}$ can be rewritten as

$$\partial f_{k1}(\boldsymbol{x}) = \frac{1}{m} \sum_{\boldsymbol{a} \in A} \Big(\partial d_1(\boldsymbol{x}_1, \boldsymbol{a}), \partial d_1(\boldsymbol{x}_2, \boldsymbol{a}), \ldots, \partial d_1(\boldsymbol{x}_k, \boldsymbol{a}) \Big), \tag{4.46}$$

and the subdifferential of the function $\varphi_{\boldsymbol{a}}$, defined in (4.22), at $\boldsymbol{x}$ is

$$\partial \varphi_{\boldsymbol{a}}(\boldsymbol{x}) = \operatorname{conv} \Big\{ \Big(\partial d_1(\boldsymbol{x}_1, \boldsymbol{a}), \ldots, \partial d_1(\boldsymbol{x}_{j-1}, \boldsymbol{a}), \mathbf{0}, \partial d_1(\boldsymbol{x}_{j+1}, \boldsymbol{a}), \ldots,$$
$$\partial d_1(\boldsymbol{x}_k, \boldsymbol{a}) \Big),\ j \in \tilde{\mathcal{R}}_{\boldsymbol{a}}(\boldsymbol{x}) \Big\}.$$

The set $\tilde{\mathcal{R}}_{\boldsymbol{a}}(\boldsymbol{x})$ is defined in (4.23). Then the subdifferential $\partial f_{k2}(\boldsymbol{x})$ can be expressed as

$$\partial f_{k2}(\boldsymbol{x}) = \frac{1}{m} \sum_{\boldsymbol{a} \in A} \partial \varphi_{\boldsymbol{a}}(\boldsymbol{x}). \tag{4.47}$$

For a given $\boldsymbol{x} \in \mathbb{R}^{nk}$, consider the sets

$$C_j(\boldsymbol{x}) = \big\{\boldsymbol{a} \in A \ : \ j \in \tilde{\mathcal{R}}_{\boldsymbol{a}}(\boldsymbol{x})\big\}, \quad j = 1, \ldots, k. \tag{4.48}$$

Proposition 4.17 *Assume that a point* $\boldsymbol{x}^* = (\boldsymbol{x}_1^*, \ldots, \boldsymbol{x}_k^*) \in \mathbb{R}^{nk}$ *is a local minimizer of the problem* (4.20) *with the similarity measure* d_1 *and the sets* $C_j(\boldsymbol{x}^*)$ *are not empty for any* $j = 1, \ldots, k$. *Then*

$$\mathbf{0} \in \Big(\sum_{\boldsymbol{a} \in C_1(\boldsymbol{x}^*)} \partial d_1(\boldsymbol{x}_1^*, \boldsymbol{a}), \ldots, \sum_{\boldsymbol{a} \in C_k(\boldsymbol{x}^*)} \partial d_1(\boldsymbol{x}_k^*, \boldsymbol{a}) \Big). \tag{4.49}$$

Proof The point $\boldsymbol{x}^*$ is a local minimizer of the problem (4.20), and therefore, this point is also an inf-stationary point. Then applying (4.44) we get $\partial f_{k2}(\boldsymbol{x}^*) \subseteq \partial f_{k1}(\boldsymbol{x}^*)$. Using the expressions of subdifferentials of functions f_{k1} and f_{k2}, given in (4.46) and (4.47), we have

$$\sum_{\boldsymbol{a} \in A} \partial \varphi_{\boldsymbol{a}}(\boldsymbol{x}^*) \subseteq \sum_{\boldsymbol{a} \in A} \Big(\partial d_1(\boldsymbol{x}_1^*, \boldsymbol{a}), \partial d_1(\boldsymbol{x}_2^*, \boldsymbol{a}), \ldots, \partial d_1(\boldsymbol{x}_k^*, \boldsymbol{a}) \Big). \tag{4.50}$$

Here, the subdifferential of the function $\varphi_{\boldsymbol{a}}$ at $\boldsymbol{x}^*$ is

$$\partial \varphi_{\boldsymbol{a}}(\boldsymbol{x}^*) = \Big(\sum_{\boldsymbol{a} \in A \setminus C_1(\boldsymbol{x}^*)} \partial d_1(\boldsymbol{x}_1^*, \boldsymbol{a}), \ldots, \sum_{\boldsymbol{a} \in A \setminus C_k(\boldsymbol{x}^*)} \partial d_1(\boldsymbol{x}_k^*, \boldsymbol{a}) \Big).$$

Then the proof follows from this expression and the inclusion (4.50). □

Now, consider the function d_2. In this case, the DC component f_{k1} is continuously differentiable and its gradient is given by (4.26). The subdifferential of the function $\varphi_{\boldsymbol{a}}$ at a point $\boldsymbol{x} \in \mathbb{R}^{nk}$ can be rewritten as

$$\partial \varphi_{\boldsymbol{a}}(\boldsymbol{x}) = \operatorname{conv}\Big\{ 2\big((\boldsymbol{x}_1 - \boldsymbol{a}), \ldots, (\boldsymbol{x}_{j-1} - \boldsymbol{a}), \mathbf{0}, (\boldsymbol{x}_{j+1} - \boldsymbol{a}), \ldots, (\boldsymbol{x}_k - \boldsymbol{a})\big), \\ j \in \tilde{\mathcal{R}}_{\boldsymbol{a}}(\boldsymbol{x}) \Big\}.$$

Proposition 4.18 *Let* $\boldsymbol{x}^* \in \mathbb{R}^{nk}$ *be a local minimizer of the problem (4.20) with the similarity measure* d_2. *Then the objective function* f_k *is continuously differentiable at this point and*

$$\partial f_k(\boldsymbol{x}^*) = \{\nabla f_k(\boldsymbol{x}^*)\} = \{\mathbf{0}\}.$$

Proof It follows from (4.26) that the subdifferential $\partial f_{k1}(\boldsymbol{x}^*)$ is a singleton for any $\boldsymbol{x}^* \in \mathbb{R}^{nk}$. Since the local minimizer $\boldsymbol{x}^*$ of the problem (4.20) is its inf-stationary point the subdifferential ∂f_{k2} is a singleton at the point $\boldsymbol{x}^*$. This means that the subdifferentials $\partial \varphi_{\boldsymbol{a}}(\boldsymbol{x}^*)$, $\boldsymbol{a} \in A$ are also singletons at the point $\boldsymbol{x}^*$ which in turn means that the index sets $\tilde{\mathcal{R}}_{\boldsymbol{a}}(\boldsymbol{x}^*)$ are singletons for all $\boldsymbol{a} \in A$. This implies that for each $\boldsymbol{a} \in A$ there exists a unique $j \in \{1, \ldots, k\}$ such that $\tilde{\mathcal{R}}_{\boldsymbol{a}}(\boldsymbol{x}^*) = \{j\}$. It follows from the DC representation of the function f_k that this j stands for the index of the cluster to which the data point $\boldsymbol{a}$ belongs, and for this point there exists only one cluster center $\boldsymbol{x}_j^*$ such that

$$\min_{s=1,\ldots,k} d_2(\boldsymbol{x}_s^*, \boldsymbol{a}) = d_2(\boldsymbol{x}_j^*, \boldsymbol{a}).$$

Therefore, $d_2(\boldsymbol{x}_s^*, \boldsymbol{a}) > d_2(\boldsymbol{x}_j^*, \boldsymbol{a})$ for any $s = 1, \ldots, k,\ s \neq j$, and we get that the sets $C_j(\boldsymbol{x}^*)$, defined by (4.48), are pairwise disjoint:

$$C_{j_1}(\boldsymbol{x}^*) \cap C_{j_2}(\boldsymbol{x}^*) = \emptyset, \quad j_1, j_2 = 1, \ldots, k,\ j_1 \neq j_2.$$

Using these sets the gradient of the function f_{k2} can be expressed as

$$\nabla f_{k2}(\boldsymbol{x}^*) = \frac{2}{m}\left(\sum_{\boldsymbol{a} \in A \setminus C_1(\boldsymbol{x}^*)} (\boldsymbol{x}_1^* - \boldsymbol{a}), \ldots, \sum_{\boldsymbol{a} \in A \setminus C_k(\boldsymbol{x}^*)} (\boldsymbol{x}_k^* - \boldsymbol{a})\right).$$

This together with (4.26) implies that

$$\nabla f_k(\boldsymbol{x}^*) = \frac{2}{m}\left(\sum_{\boldsymbol{a} \in C_1(\boldsymbol{x}^*)} (\boldsymbol{x}_1^* - \boldsymbol{a}), \ldots, \sum_{\boldsymbol{a} \in C_k(\boldsymbol{x}^*)} (\boldsymbol{x}_k^* - \boldsymbol{a})\right).$$

This means that the cluster function f_k is continuously differentiable at any inf-stationary point $\boldsymbol{x}^*$ of the problem (4.20) and the Clarke subdifferential of this function are singletons at such points, that is

$$\partial f_k(\boldsymbol{x}^*) = \{\nabla f_k(\boldsymbol{x}^*)\}.$$

Then the necessary condition for a minimum implies that $\nabla f_k(\boldsymbol{x}^*) = \boldsymbol{0}$. □

Proposition 4.19 *The sets of Clarke stationary and critical points of the problem (4.20) with the similarity measure d_2 coincide and at these points*

$$\nabla f_{k1}(\boldsymbol{x}) \in \partial f_{k2}(\boldsymbol{x}).$$

Proof It follows from Proposition 4.8 that

$$\partial f_k(\boldsymbol{x}) = \nabla f_{k1}(\boldsymbol{x}) - \partial f_{k2}(\boldsymbol{x}), \quad \boldsymbol{x} \in \mathbb{R}^{nk}.$$

If a point $\boldsymbol{x} \in \mathbb{R}^{nk}$ is Clarke stationary, then $\mathbf{0} \in \partial f_k(\boldsymbol{x})$ and we get $\nabla f_{k1}(\boldsymbol{x}) \in \partial f_{k2}(\boldsymbol{x})$ and therefore, $\boldsymbol{x}$ is a critical point.

If $\boldsymbol{x} \in \mathbb{R}^{nk}$ is a critical point, then $\nabla f_{k1}(\boldsymbol{x}) \in \partial f_{k2}(\boldsymbol{x})$ and therefore, $\mathbf{0} \in \partial f_k(\boldsymbol{x})$ and $\boldsymbol{x}$ is a Clarke stationary point. □

Remark 4.2 If a point $\boldsymbol{x} \in \mathbb{R}^{nk}$ is not inf-stationary, then, in general, the function f_{k2} is not strictly differentiable at this point and the subdifferential $\partial f_{k2}(\boldsymbol{x})$ is not a singleton. Then it is easy to calculate two different subgradients from ∂f_{k2}. Consider the following sets:

$$A_1 = \{\boldsymbol{a} \in A \ : \ |\tilde{\mathcal{R}}_{\boldsymbol{a}}(\boldsymbol{x})| = 1\}, \quad \text{and}$$
$$A_2 = \{\boldsymbol{a} \in A \ : \ |\tilde{\mathcal{R}}_{\boldsymbol{a}}(\boldsymbol{x})| \geq 2\}.$$

If $A_1 = A$, then $\partial f_{k2}(\boldsymbol{x})$ is a singleton. If $|A_2| \geq 1$, then $\partial f_{k2}(\boldsymbol{x})$ is not a singleton. Take any $\boldsymbol{a} \in A_2$. Since $|\tilde{\mathcal{R}}_{\boldsymbol{a}}(\boldsymbol{x})| \geq 2$ this point is attracted by at least two cluster centers, say $j_1, j_2 \in \{1, \ldots, k\}$, $j_1 \neq j_2$. Using these two cluster centers, we compute two subgradients $\boldsymbol{\xi}_1, \boldsymbol{\xi}_2 \in \mathbb{R}^{nk}$ of the function $\varphi_{\boldsymbol{a}}$, defined by (4.22), as follows:

$$\boldsymbol{\xi}_1 = 2\big((\boldsymbol{x}_1 - \boldsymbol{a}), \ldots, (\boldsymbol{x}_{j_1-1} - \boldsymbol{a}), \mathbf{0}, (\boldsymbol{x}_{j_1+1} - \boldsymbol{a}), \ldots, (\boldsymbol{x}_k - \boldsymbol{a})\big), \quad \text{and}$$
$$\boldsymbol{\xi}_2 = 2\big((\boldsymbol{x}_1 - \boldsymbol{a}), \ldots, (\boldsymbol{x}_{j_2-1} - \boldsymbol{a}), \mathbf{0}, (\boldsymbol{x}_{j_2+1} - \boldsymbol{a}), \ldots, (\boldsymbol{x}_k - \boldsymbol{a})\big).$$

Since $j_1 \neq j_2$ and $\boldsymbol{a}$ is not a cluster center it follows that $\boldsymbol{\xi}_1 \neq \boldsymbol{\xi}_2$.

Finally, we discuss optimality conditions for the problem (4.20) when the similarity measure d_∞ is applied in its objective function. For a given $\boldsymbol{x}_t \in \mathbb{R}^n$ and $\boldsymbol{a} \in A$ if $d_\infty(\boldsymbol{x}_t, \boldsymbol{a}) = 0$, then the subdifferential $\partial d_\infty(\boldsymbol{x}, \boldsymbol{a})$ is defined by (4.10), otherwise it is defined by (4.11). Then applying (4.24) we get that the subdifferential $\partial f_{k1}(\boldsymbol{x})$ of the function f_{k1} at $\boldsymbol{x} \in \mathbb{R}^{nk}$ is

$$\partial f_{k1}(\boldsymbol{x}) = \frac{1}{m} \sum_{\boldsymbol{a} \in A} \big(\partial d_\infty(\boldsymbol{x}_1, \boldsymbol{a}), \ldots, \partial d_\infty(\boldsymbol{x}_k, \boldsymbol{a})\big),$$

and applying (4.25) the subdifferential $\partial f_{k2}(\boldsymbol{x})$ of the function f_{k2} at $\boldsymbol{x} \in \mathbb{R}^{nk}$ can be expressed as

$$\partial f_{k2}(\boldsymbol{x}) = \frac{1}{m} \sum_{\boldsymbol{a} \in A} \operatorname{conv} \Big\{\big(\partial d_\infty(\boldsymbol{x}_1, \boldsymbol{a}), \ldots, \partial d_\infty(\boldsymbol{x}_{j-1}, \boldsymbol{a}), \mathbf{0},$$
$$\partial d_\infty(\boldsymbol{x}_{j+1}, \boldsymbol{a}), \ldots, \partial d_\infty(\boldsymbol{x}_k, \boldsymbol{a})\big), \ j \in \tilde{\mathcal{R}}_{\boldsymbol{a}}(\boldsymbol{x})\Big\}.$$

Then for the point $\boldsymbol{x}^* = (\boldsymbol{x}_1^*, \ldots, \boldsymbol{x}_k^*) \in \mathbb{R}^{nk}$ to be a local minimizer of the problem (4.20) it is necessary that

$$\partial f_{k2}(\boldsymbol{x}^*) \subseteq \partial f_{k1}(\boldsymbol{x}^*). \tag{4.51}$$

Using the sets $C_j(\boldsymbol{x}^*),\ j = 1, \dots, k$, defined in (4.48), we can get the necessary condition similar to that given in Proposition 4.17 for the clustering problem based on d_1.

Proposition 4.20 *Assume that a point* $\boldsymbol{x}^* = (\boldsymbol{x}_1^*, \dots, \boldsymbol{x}_k^*) \in \mathbb{R}^{nk}$ *is a local minimizer of the problem* (4.20) *with the similarity measure* d_∞ *and the sets* $C_j(\boldsymbol{x}^*)$ *are not empty for any* $j = 1, \dots, k$. *Then*

$$\mathbf{0} \in \left(\sum_{\boldsymbol{a} \in C_1(\boldsymbol{x}^*)} \partial d_\infty(\boldsymbol{x}_1^*, \boldsymbol{a}), \dots, \sum_{\boldsymbol{a} \in C_k(\boldsymbol{x}^*)} \partial d_\infty(\boldsymbol{x}_k^*, \boldsymbol{a}) \right). \tag{4.52}$$

Proof The proof repeats to that of Proposition 4.17 and therefore, is omitted. □

Proposition 4.21 *Assume* $\boldsymbol{x}^* = (\boldsymbol{x}_1^*, \dots, \boldsymbol{x}_k^*) \in \mathbb{R}^{nk}$ *is the inf-stationary point of the problem* (4.20) *with the similarity measure* d_∞ *and the sets* $\bar{\mathcal{R}}_{\boldsymbol{a}}(\boldsymbol{x}^*)$, *defined by* (4.9), *are singletons for any* $\boldsymbol{a} \in A$, $x_{ji}^* \neq a_i$ *for all* $\boldsymbol{a} \in A$, $j \in \{1, \dots, k\}$ *and* $i \in \{1, \dots, n\}$. *Then the function* f_k *is strictly differentiable at* $\boldsymbol{x}^*$ *and we have* $\nabla f_k(\boldsymbol{x}^*) = \mathbf{0}$.

Proof Under given conditions, the subdifferential $\partial f_{k1}(\boldsymbol{x}^*)$ is a singleton as subdifferentials $d_\infty(\boldsymbol{x}_j^*, \boldsymbol{a})$ are singletons for all $\boldsymbol{a} \in A$ and $j = 1, \dots, k$. Since the point $\boldsymbol{x}^*$ is inf-stationary it follows from (4.51) that the subdifferential $\partial f_{k2}(\boldsymbol{x}^*)$ is also a singleton. This means that $\nabla f_{k2}(\boldsymbol{x}^*) = \nabla f_{k1}(\boldsymbol{x}^*)$, the function f_k is strictly differentiable at $\boldsymbol{x}^*$ and we get $\nabla f_k(\boldsymbol{x}^*) = \mathbf{0}$. □

4.6.2 Optimality Conditions for Auxiliary Clustering Problem

In this subsection, we discuss optimality conditions for the auxiliary clustering problem using its general formulation (4.29) and also the DC model (4.34). We formulate such conditions for similarity measures d_1, d_2, and d_∞. We start with the formulation of conditions by applying the Clarke subdifferential.

Proposition 4.22 *Let* $\boldsymbol{y}^* \in \mathbb{R}^n$ *be a local minimizer of the problem* (4.29). *Then*

$$\mathbf{0} \in \partial \bar{f}_k(\boldsymbol{y}^*). \tag{4.53}$$

Proof It follows from Proposition 4.12 that the function $\bar{f}_k$ is LLC on $\mathbb{R}^n$ and therefore, the proof follows from Theorem 2.17. □

Points satisfying the condition (4.53) are called the Clarke stationary points of the auxiliary clustering problem (4.29).

Corollary 4.3 *Let $\boldsymbol{y}^* \in \mathbb{R}^n$ be a stationary point of the problem* (4.29). *Assume that the set $B_3(\boldsymbol{y}^*)$, defined in* (4.30), *is not empty. Then*

$$\mathbf{0} \in \sum_{\boldsymbol{a} \in B_2(\boldsymbol{y}^*)} \operatorname{conv}\big\{\mathbf{0}, \partial d_p(\boldsymbol{y}^*, \boldsymbol{a})\big\} + \sum_{\boldsymbol{a} \in B_3(\boldsymbol{y}^*)} \partial d_p(\boldsymbol{y}^*, \boldsymbol{a}), \quad p = 1, 2, \infty.$$

Proof The proof follows from (4.31) and Proposition 4.22. □

Next, we formulate optimality conditions obtained using the DC representation (4.32) of the auxiliary cluster function $\bar{f}_k$. The subdifferentials of the functions $\bar{f}_{k1}$ and $\bar{f}_{k2}$, in general form, are given in (4.38) and (4.40), respectively.

Theorem 4.2 *Let $\boldsymbol{y}^* \in \mathbb{R}^n$ be a local minimizer of the problem* (4.34). *Then*

$$\partial \bar{f}_{k2}(\boldsymbol{y}^*) \subseteq \partial \bar{f}_{k1}(\boldsymbol{y}^*), \quad \text{and} \tag{4.54}$$

$$\partial \bar{f}_{k1}(\boldsymbol{y}^*) \cap \partial \bar{f}_{k2}(\boldsymbol{y}^*) \neq \emptyset. \tag{4.55}$$

Proof The proof follows from Theorem 2.27. □

Points satisfying the condition (4.54) are called inf-stationary, and points satisfying the condition (4.55) are called critical points of the auxiliary clustering problem (4.29).

Corollary 4.4 *Let the set $B_2(\boldsymbol{y}^*) = \emptyset$. Then for a point $\boldsymbol{y}^*$ to be a local minimizer of the problem* (4.34) *it is necessary that*

$$\sum_{\boldsymbol{a} \in B_1(\boldsymbol{y}^*)} \partial d_p(\boldsymbol{y}^*, \boldsymbol{a}) \subseteq \sum_{\boldsymbol{a} \in B_1(\boldsymbol{y}^*) \cup B_3(\boldsymbol{y}^*)} \partial d_p(\boldsymbol{y}^*, \boldsymbol{a}). \tag{4.56}$$

Proof The proof follows from expressions for $\partial \bar{f}_{k1}(\boldsymbol{y}^*)$, $\partial \bar{f}_{k2}(\boldsymbol{y}^*)$ and the inclusion (4.54). □

Remark 4.3 All inf-stationary points of the problem (4.34) satisfy the condition (4.56). The condition $B_2(\boldsymbol{y}^*) = \emptyset$ means that any data point is attracted either by its cluster center or by the point $\boldsymbol{y}^*$.

Now, we formulate optimality conditions for the problem (4.29) using the similarity measures d_1, d_2, and d_∞. We start with the measure d_1.

Proposition 4.23 *Assume $\boldsymbol{y}^* \in \mathbb{R}^n$ is the inf-stationary point of the problem* (4.34) *with the similarity measure d_1, the set $B_2(\boldsymbol{y}^*) = \emptyset$ and $\boldsymbol{y}_i^* \neq a_i$, $i = 1, \ldots, n$ for all $\boldsymbol{a} \in A$. Then the point y^* satisfies the condition*

$$\sum_{\boldsymbol{a} \in B_3(\boldsymbol{y}^*)} \Big(\operatorname{sign}(y_1^* - a_1), \ldots, \operatorname{sign}(y_n^* - a_n)\Big) = \mathbf{0}.$$

Proof The condition $y_i^* \neq a_i,\ i = 1, \ldots, n$ for all $\boldsymbol{a} \in A$ and the expression (4.6) imply that the function d_1 is strictly differentiable at $\boldsymbol{y}^*$, the subdifferentials $\partial d_1(\boldsymbol{y}^*)$ are singletons and

$$\nabla d_1(\boldsymbol{y}^*, \boldsymbol{a}) = \Big(\operatorname{sign}(y_1^* - a_1), \ldots, \operatorname{sign}(y_n^* - a_n)\Big).$$

Then we get from Corollary 4.4 that

$$\sum_{\boldsymbol{a} \in B_1(\boldsymbol{y}^*)} \nabla d_1(\boldsymbol{y}^*, \boldsymbol{a}) = \sum_{\boldsymbol{a} \in B_1(\boldsymbol{y}^*)} \nabla d_1(\boldsymbol{y}^*, \boldsymbol{a}) + \sum_{\boldsymbol{a} \in B_3(\boldsymbol{y}^*)} \nabla d_1(\boldsymbol{y}^*, \boldsymbol{a}).$$

This completes the proof. □

Next, we consider the auxiliary clustering problem with the similarity measure d_2. In this case the function $\bar{f}_{k1}$ is differentiable and its gradient is given by (4.41) and the function $\bar{f}_{k2}$ is, in general, nonsmooth and its subdifferential is given in (4.42). According to Proposition 4.14 for all $\boldsymbol{y} \in \mathbb{R}^n$ the Clarke subdifferential of the function $\bar{f}_k$ can be represented as

$$\partial \bar{f}_k(\boldsymbol{y}) = \nabla \bar{f}_{k1}(\boldsymbol{y}) - \partial \bar{f}_{k2}(\boldsymbol{y}).$$

Proposition 4.24 *At any inf-stationary point $\boldsymbol{y}^*$ of the problem (4.34) with the similarity measure d_2, the subdifferential $\partial \bar{f}_{k2}(\boldsymbol{y}^*)$ is a singleton and*

$$\partial \bar{f}_{k2}(\boldsymbol{y}^*) = \frac{2}{m}\left\{\sum_{\boldsymbol{a} \in B_1(\boldsymbol{y}^*)} (\boldsymbol{y}^* - \boldsymbol{a})\right\}. \tag{4.57}$$

Proof Since the subdifferential $\partial \bar{f}_{k1}$ is a singleton at any $\boldsymbol{y} \in \mathbb{R}^n$ it follows from the definition of inf-stationary points that the set $\partial \bar{f}_{k2}(\boldsymbol{y}^*)$ must be a singleton at all such points. The first term in (4.42) is a singleton. If the set $B_2(\boldsymbol{y}^*)$ is empty, then we get the expression (4.57) for the subdifferential $\partial \bar{f}_{k2}(\boldsymbol{y}^*)$. Assume that the set $B_2(\boldsymbol{y}^*)$ is not empty. Since the subdifferential $\partial \bar{f}_{k2}(\boldsymbol{y}^*)$ is a singleton the second term in its expression (4.42) must be a singleton at the inf-stationary point $\boldsymbol{y}^*$. This means that $\boldsymbol{y}^* = \boldsymbol{a}$ for every $\boldsymbol{a} \in B_2(\boldsymbol{y}^*)$, and therefore, the set $B_2(\boldsymbol{y}^*)$ is a singleton and the subdifferential $\partial \bar{f}_{k2}(\boldsymbol{y}^*)$ is given by (4.57). □

Remark 4.4 The condition $\boldsymbol{y}^* = \boldsymbol{a}$ for every $\boldsymbol{a} \in B_2(\boldsymbol{y}^*)$ implies that the set $B_2(\boldsymbol{y}^*)$ cannot have more than one point and therefore, $\boldsymbol{y}^* = \boldsymbol{x}_j^*$ for some $j = 1, \ldots, k-1$, where $\boldsymbol{x}_j^*$ is the center of the cluster A^j. This means that in most inf-stationary points $\boldsymbol{y}^* \in \mathbb{R}^n$ the set $B_2(\boldsymbol{y}^*)$ is likely to be empty.

Proposition 4.25 *The sets of Clarke stationary and critical points of the problem (4.34) with the similarity measure d_2 coincide, and they are given by*

$$Y = \big\{\boldsymbol{y} \in \mathbb{R}^n \ :\ \nabla \bar{f}_{k1}(\boldsymbol{y}) \in \partial \bar{f}_{k2}(\boldsymbol{y})\big\}. \tag{4.58}$$

Proof Assume the point $\bar{y}$ is a critical point of the problem (4.34). Since the subdifferential $\bar{f}_{k1}$ at any $y \in \mathbb{R}^n$ is a singleton we get $\nabla \bar{f}_{k1}(\bar{y}) \in \partial \bar{f}_{k2}(\bar{y})$. Then it follows from Proposition 4.14 that $\mathbf{0} \in \partial \bar{f}_k(\bar{y})$, meaning that $\bar{y}$ is Clarke stationary.

Now, assume that $\bar{y}$ is Clarke stationary. Then Proposition 4.14 implies that $\nabla \bar{f}_{k1}(\bar{y}) \in \partial \bar{f}_{k2}(\bar{y})$ and therefore, $\bar{y}$ is a critical point. □

Remark 4.5 It is obvious that any inf-stationary point of the problem (4.34) is also Clarke stationary and critical points of this problem.

Proposition 4.26 *For a point $y^* \in \mathbb{R}^n$ such that $B_3(y^*) \neq \emptyset$, to be a local minimizer of the problem (4.34) with the similarity measure d_2 it is necessary that*

$$\sum_{a \in B_3(y^*)} (y^* - a) = \mathbf{0}. \tag{4.59}$$

Proof Any local minimizer of the problem (4.34) is an inf-stationary point to this problem. Then the proof follows from the expression of the gradient $\nabla \bar{f}_{k1}(y^*)$, the expression (4.57) in Proposition 4.24, Remark 4.4 and the assumption that $B_3(y^*) \neq \emptyset$. □

Next, we demonstrate how two different subgradients from the subdifferential $\partial \bar{f}_{k2}(y)$, $y \in \mathbb{R}^n$ can be computed if this subdifferential is not a singleton. This result will be used to develop an algorithm that can escape a Clarke stationary point and converge to an inf-stationary point.

Remark 4.6 To compute different subgradients, we can choose $\xi_2^1, \xi_2^2 \in \partial \bar{f}_{k2}(y)$ using (4.42) as follows:

$$\xi_2^1 = \frac{2}{m} \sum_{a \in B_1(y)} (y - a), \quad \text{and}$$

$$\xi_2^2 = \xi_2^1 + \bar{\xi}_2, \quad \bar{\xi}_2 = \operatorname*{argmax}_{a \in B_2(y)} \|y - a\|. \tag{4.60}$$

Note that if $\bar{\xi}_2 = \mathbf{0}$, then $\partial \bar{f}_{k2}(y)$ is a singleton.

Finally, we study optimality conditions for the problem (4.34), where the distance function d_∞ is used in its objective function. Using the subdifferentials ∂d_∞, given in (4.10) and (4.11), we can write the subdifferentials of functions $\bar{f}_{k1}$ and $\bar{f}_{k2}$ at $y \in \mathbb{R}^n$ as

$$\partial \bar{f}_{k1}(y) = \frac{1}{m} \sum_{a \in A} \partial d_\infty(y, a), \quad \text{and}$$

$$\partial \bar{f}_{k2}(y) = \frac{1}{m} \left(\sum_{a \in B_1(y)} \partial d_\infty(y, a) + \sum_{a \in B_2(y)} \operatorname{conv}\left\{\mathbf{0}, \partial d_\infty(y, a)\right\} \right).$$

Proposition 4.27 *Assume that* $\boldsymbol{y}^* \in \mathbb{R}^n$ *is the inf-stationary point of the problem* (4.34) *and the sets* $\bar{\mathcal{R}}_{\boldsymbol{a}}(\boldsymbol{y}^*)$, *defined by* (4.9), *are singletons for any* $\boldsymbol{a} \in A$, $y_i^* \neq a_i$ *for all* $\boldsymbol{a} \in A$ *and* $i \in \{1, \dots, n\}$. *Then the function* $\bar{f}_k$ *with the similarity measure* d_∞ *is strictly differentiable at* $\boldsymbol{y}^*$ *and* $\nabla \bar{f}_k(\boldsymbol{y}^*) = \boldsymbol{0}$.

Proof Under given conditions, the subdifferential $\partial \bar{f}_{k1}(\boldsymbol{y}^*)$ is a singleton as subdifferentials $d_\infty(\boldsymbol{y}^*, \boldsymbol{a})$ are singletons for all $\boldsymbol{a} \in A$. Since the point $\boldsymbol{y}^*$ is inf-stationary it follows from (4.54) that the subdifferential $\partial \bar{f}_{k2}(\boldsymbol{y}^*)$ is also a singleton. This means that $\nabla \bar{f}_{k2}(\boldsymbol{y}^*) = \nabla \bar{f}_{k1}(\boldsymbol{y}^*)$, the function $\bar{f}_k$ is strictly differentiable at $\boldsymbol{y}^*$ and $\nabla \bar{f}_k(\boldsymbol{y}^*) = \boldsymbol{0}$. □

4.7 Smoothing of Cluster Functions

In this section, we describe the hyperbolic smoothing of the cluster function f_k and the auxiliary cluster function $\bar{f}_k$, defined in (4.4) and (4.28), respectively. We start with the discussion on the smoothing of similarity measures. Since the similarity function d_2 is differentiable there is no need for further smoothing when d_2 is used in the cluster and the auxiliary cluster functions. However, the distance functions d_1 and d_∞ are nondifferentiable and need to be smoothed to make possible the use of smooth optimization methods.

4.7.1 *Hyperbolic Smoothing of Functions d_1 and d_∞*

Take any $\boldsymbol{a} \in \mathbb{R}^n$ and fix it. Then for any $\boldsymbol{x} \in \mathbb{R}^n$, we have

$$\begin{aligned} d_1(\boldsymbol{x}, \boldsymbol{a}) &= \|\boldsymbol{x} - \boldsymbol{a}\|_1 \\ &= \sum_{q=1}^{n} |x_q - a_q| \\ &= \sum_{q=1}^{n} \max\left\{x_q - a_q, a_q - x_q\right\} \\ &= \sum_{q=1}^{n} \left(x_q - a_q + 2\max\left\{0, a_q - x_q\right\}\right). \end{aligned}$$

Applying the HSM, described in Sect. 2.8, $d_1(\boldsymbol{x}, \boldsymbol{a})$ can be approximated by the smooth function $\eta_{1,\tau}(\cdot, \boldsymbol{a}) : \mathbb{R}^n \to \mathbb{R}$ as follows:

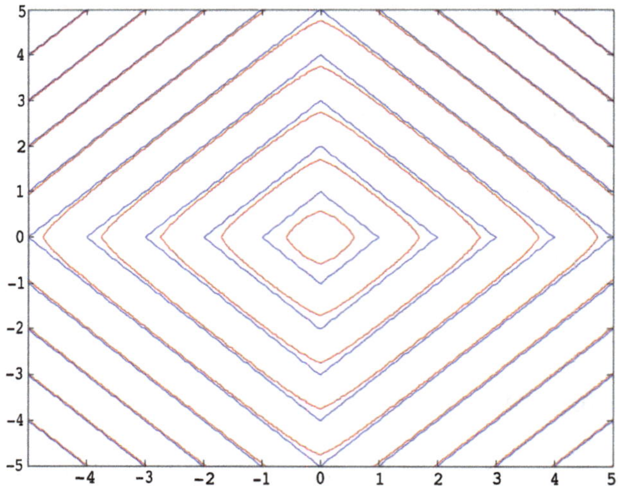

Fig. 4.3 Smooth approximation of the similarity function d_1

$$\eta_{1,\tau}(\boldsymbol{x}, \boldsymbol{a}) = \sum_{q=1}^{n} \sqrt{(x_q - a_q)^2 + \tau^2}, \tag{4.61}$$

where $\tau > 0$ is a smoothing parameter. Figure 4.3 illustrates the level curves of the function d_1 (blue) and its hyperbolic smooth approximation $\eta_{1,\tau}$ (red).

Next, we explain the smoothing of the function d_∞. Take any $\boldsymbol{a} \in \mathbb{R}^n$ and fix it. For any $\boldsymbol{x} \in \mathbb{R}^n$, we get

$$\begin{aligned} d_\infty(\boldsymbol{x}, \boldsymbol{a}) &= \|\boldsymbol{x} - \boldsymbol{a}\|_\infty \\ &= \max_{q=1,\ldots,n} |x_q - a_q| \\ &= \max_{q=1,\ldots,n} \max\{x_q - a_q, a_q - x_q\} \\ &= \max_{q=1,\ldots,2n} \Omega_{q\boldsymbol{a}}(\boldsymbol{x}), \end{aligned}$$

where

$$\Omega_{q\boldsymbol{a}}(\boldsymbol{x}) = x_q - a_q \quad \text{and} \quad \Omega_{(q+n)\boldsymbol{a}}(\boldsymbol{x}) = -x_q + a_q, \quad q = 1, \ldots, n.$$

Consider the function

$$\Omega(\boldsymbol{x}, \theta_{\boldsymbol{a}}) = \theta_{\boldsymbol{a}} + \sum_{q=1}^{2n} \max\left\{0, \Omega_{q\boldsymbol{a}}(\boldsymbol{x}) - \theta_{\boldsymbol{a}}\right\},$$

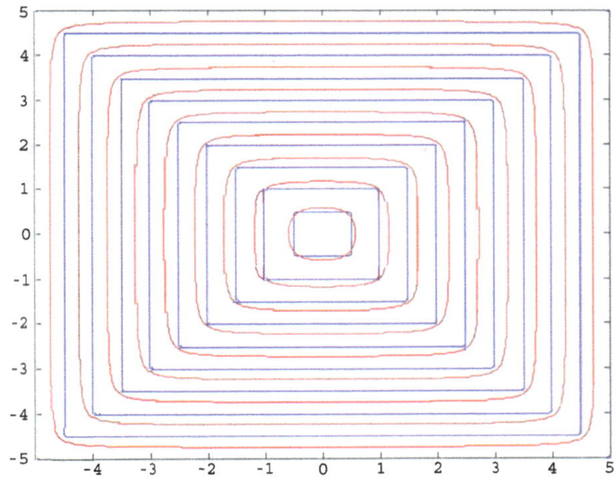

Fig. 4.4 Smooth approximation of the similarity function d_∞

where $\theta_{\boldsymbol{a}} = d_\infty(\boldsymbol{x}, \boldsymbol{a})$. Then the hyperbolic smoothing of the function Ω and, consequently, the function $d_\infty(\cdot, \boldsymbol{a})$ is given by

$$\eta_{\infty,\tau}(\boldsymbol{x}, \theta_{\boldsymbol{a}}) = \theta_{\boldsymbol{a}} + \frac{1}{2}\sum_{q=1}^{2n}\left(\Omega_{q\boldsymbol{a}}(\boldsymbol{x}) - \theta_{\boldsymbol{a}} + \sqrt{(\Omega_{q\boldsymbol{a}}(\boldsymbol{x}) - \theta_{\boldsymbol{a}})^2 + \tau^2}\right). \tag{4.62}$$

Figure 4.4 presents the level curves of the function d_∞ (blue) and its hyperbolic smooth approximation $\eta_{\infty,\tau}$ (red).

4.7.2 Hyperbolic Smoothing of the Cluster Function

In this subsection, we describe the hyperbolic smoothing of the cluster function f_k, defined in (4.4). It is clear that

$$\min_{j=1,\ldots,k} d_p(\boldsymbol{x}_j, \boldsymbol{a}_i) = -\max_{j=1,\ldots,k} -d_p(\boldsymbol{x}_j, \boldsymbol{a}_i), \quad i = 1, \ldots, m.$$

Then the function f_k can be rewritten as

$$f_k(\boldsymbol{x}_1, \ldots, \boldsymbol{x}_k) = -\frac{1}{m}\sum_{i=1}^{m}\max_{j=1,\ldots,k} -d_p(\boldsymbol{x}_j, \boldsymbol{a}_i).$$

Define the following function $F_k : \mathbb{R}^{nk} \to \mathbb{R}$ similar to the function (2.32):

$$F_k(\boldsymbol{x}, \boldsymbol{t}) = -\frac{1}{m} \sum_{i=1}^{m} \left(t_i + \sum_{j=1}^{k} \max\left\{0, -d_p(\boldsymbol{x}_j, \boldsymbol{a}_i) - t_i\right\} \right),$$

where $\boldsymbol{x} = (\boldsymbol{x}_1, \dots, \boldsymbol{x}_k)$, $\boldsymbol{t} = (t_1, \dots, t_m)$, and scalars t_i, $i = 1, \dots, m$ are defined as

$$t_i = \max_{j=1,\dots,k} -d_p(\boldsymbol{x}_j, \boldsymbol{a}_i) = -\min_{j=1,\dots,k} d_p(\boldsymbol{x}_j, \boldsymbol{a}_i) \le 0, \quad i = 1, \dots, m. \tag{4.63}$$

Then applying (2.33) we get the hyperbolic smoothing $\Phi_{k,\tau}$ of the function F_k:

$$\begin{aligned} \Phi_{k,\tau}(\boldsymbol{x}, \boldsymbol{t}) &= -\frac{1}{m} \sum_{i=1}^{m} \left(t_i + \sum_{j=1}^{k} \frac{-d_p(\boldsymbol{x}_j, \boldsymbol{a}_i) - t_i + \sqrt{(d_p(\boldsymbol{x}_j, \boldsymbol{a}_i) + t_i)^2 + \tau^2}}{2} \right) \\ &= \frac{1}{m} \sum_{i=1}^{m} \left(-t_i + \sum_{j=1}^{k} \frac{t_i + d_p(\boldsymbol{x}_j, \boldsymbol{a}_i) - \sqrt{(d_p(\boldsymbol{x}_j, \boldsymbol{a}_i) + t_i)^2 + \tau^2}}{2} \right), \end{aligned} \tag{4.64}$$

where $\tau > 0$ is a smoothing parameter.

For the distance function d_1, we replace it by its approximation $\eta_{1,\tau}$, given in (4.61), and get the following smooth approximation of the cluster function f_k:

$$U_{k,\tau}(\boldsymbol{x}, \boldsymbol{t}) = \frac{1}{m} \sum_{i=1}^{m} \left(-t_i + \sum_{j=1}^{k} \frac{t_i + \eta_{1,\tau}(\boldsymbol{x}_j, \boldsymbol{a}_i) - \sqrt{(\eta_{1,\tau}(\boldsymbol{x}_j, \boldsymbol{a}_i) + t_i)^2 + \tau^2}}{2} \right).$$

Then for a sequence $\{\tau_h\}$ such that $\tau_h \downarrow 0$ as $h \to \infty$, the problem (4.3) is replaced by the sequence of smooth problems

$$\begin{cases} \text{minimize} & U_{k,\tau_h}(\boldsymbol{x}, \boldsymbol{t}) \\ \text{subject to} & \boldsymbol{x} = (\boldsymbol{x}_1, \dots, \boldsymbol{x}_k) \in \mathbb{R}^{nk}, \end{cases} \tag{4.65}$$

where $\boldsymbol{t} = (t_1, \dots, t_m)$, t_i, $i = 1, \dots, m$ are defined using (4.63).

In the case of the similarity measure d_2, we get the following approximation for the cluster function f_k:

$$\begin{aligned} Q_{k,\tau}(\boldsymbol{x}, \boldsymbol{t}) &= -\frac{1}{m} \sum_{i=1}^{m} \left(t_i + \sum_{j=1}^{k} \frac{-d_2(\boldsymbol{x}_j, \boldsymbol{a}_i) - t_i + \sqrt{(d_2(\boldsymbol{x}_j, \boldsymbol{a}_i) + t_i)^2 + \tau^2}}{2} \right) \\ &= \frac{1}{m} \sum_{i=1}^{m} \left(-t_i + \sum_{j=1}^{k} \frac{t_i + d_2(\boldsymbol{x}_j, \boldsymbol{a}_i) - \sqrt{(d_2(\boldsymbol{x}_j, \boldsymbol{a}_i) + t_i)^2 + \tau^2}}{2} \right), \end{aligned} \tag{4.66}$$

where $t = (t_1, \ldots, t_m),\ t_i,\ i = 1, \ldots, m$ are defined using (4.63). Then by taking any sequence $\{\tau_h\}$ such that $\tau_h \downarrow 0$ as $h \to \infty$ the clustering problem (4.3) is replaced by the sequence of the following smooth problems:

$$\begin{cases} \text{minimize} & Q_{k,\tau_h}(x, t) \\ \text{subject to} & x = (x_1, \ldots, x_k) \in \mathbb{R}^{nk}. \end{cases} \tag{4.67}$$

Finally, in the case of the distance function d_∞ replacing this function by its approximation $\eta_{\infty,\tau}$, given in (4.62), we get the following smooth approximation of the cluster function f_k:

$$\begin{aligned} V_{k,\tau}(x, t, \Theta) &= -\frac{1}{m}\sum_{i=1}^{m} t_i \\ &+ \frac{1}{m}\sum_{i=1}^{m}\sum_{j=1}^{k} \frac{t_i + \eta_{\infty,\tau}(x_j, \theta_{a_i}^j) - \sqrt{(\eta_{\infty,\tau}(x_j, \theta_{a_i}^j) + t_i)^2 + \tau^2}}{2}. \end{aligned}$$

Here, $t = (t_1, \ldots, t_m),\ t_i,\ i = 1, \ldots, m$ are defined using (4.63), $\Theta = [\theta_{a_i}^j]_{ij},\ i = 1, \ldots, m,\ j = 1, \ldots, k$, and $\theta_{a_i}^j = d_\infty(x_j, a_i)$.

Take any sequence $\{\tau_h\}$ such that $\tau_h \downarrow 0$ as $h \to \infty$. Then the problem (4.3) is replaced by the sequence of smooth problems

$$\begin{cases} \text{minimize} & V_{k,\tau_h}(x, t, \Theta) \\ \text{subject to} & x = (x_1, \ldots, x_k) \in \mathbb{R}^{nk}. \end{cases} \tag{4.68}$$

4.7.3 Smoothing of Auxiliary Cluster Function

In this subsection, we describe the hyperbolic smoothing of the auxiliary cluster function $\bar{f}_k$, defined in (4.28). The function $\bar{f}_k$ can be rewritten as

$$\begin{aligned} \bar{f}_k(y) &= \frac{1}{m}\sum_{i=1}^{m}\Big(r_{k-1}^i + \min\big\{0, d_p(y, a_i) - r_{k-1}^i\big\}\Big) \\ &= \frac{1}{m}\sum_{i=1}^{m}\Big(r_{k-1}^i - \max\big\{0, r_{k-1}^i - d_p(y, a_i)\big\}\Big) \\ &= -\frac{1}{m}\sum_{i=1}^{m}\Big(-r_{k-1}^i + \max\big\{0, r_{k-1}^i - d_p(y, a_i)\big\}\Big). \end{aligned}$$

Applying (2.31), we obtain the following hyperbolic smoothing function $\bar{\Phi}_{k,\tau} : \mathbb{R}^n \to \mathbb{R}$ to $\bar{f}_k$:

$$\begin{aligned}\bar{\Phi}_{k,\tau}(\boldsymbol{y}) &= \frac{1}{m}\sum_{i=1}^{m} r_{k-1}^i \\ &\quad - \frac{1}{m}\sum_{i=1}^{m} \frac{r_{k-1}^i - d_p(\boldsymbol{y}, \boldsymbol{a}_i) + \sqrt{(r_{k-1}^i - d_p(\boldsymbol{y}, \boldsymbol{a}_i))^2 + \tau^2}}{2} \\ &= \frac{1}{m}\sum_{i=1}^{m} \frac{r_{k-1}^i + d_p(\boldsymbol{y}, \boldsymbol{a}_i) - \sqrt{(r_{k-1}^i - d_p(\boldsymbol{y}, \boldsymbol{a}_i))^2 + \tau^2}}{2}. \end{aligned} \tag{4.69}$$

If we use d_1 as the similarity measure, then the smooth approximation of the function $\bar{f}_k$ is obtained from (4.69) by replacing d_1 with its smooth approximation $\eta_{1,\tau}$, defined in (4.61):

$$\bar{U}_{k,\tau}(\boldsymbol{y}) = \frac{1}{m}\sum_{i=1}^{m} \frac{r_{k-1}^i + \eta_{1,\tau}(\boldsymbol{y}, \boldsymbol{a}_i) - \sqrt{(r_{k-1}^i - \eta_{1,\tau}(\boldsymbol{y}, \boldsymbol{a}_i))^2 + \tau^2}}{2}.$$

Then by selecting any sequence $\{\tau_h\}$ such that $\tau_h \downarrow 0$ as $h \to \infty$, the auxiliary clustering problem (4.29) is replaced by the sequence of the following smooth problems:

$$\begin{cases} \text{minimize} & \bar{U}_{k,\tau_h}(\boldsymbol{y}) \\ \text{subject to} & \boldsymbol{y} \in \mathbb{R}^n. \end{cases} \tag{4.70}$$

In the case of the auxiliary cluster function with the similarity measure d_2, one needs to simply substitute $p = 2$ in the expression of the function $\bar{\Phi}_{k,\tau}$. Then using the sequence $\{\tau_h\}$, we can replace the auxiliary clustering problem (4.29) by the sequence of the following smooth problems:

$$\begin{cases} \text{minimize} & \bar{\Phi}_{k,\tau_h}(\boldsymbol{y}) \\ \text{subject to} & \boldsymbol{y} \in \mathbb{R}^n. \end{cases} \tag{4.71}$$

Finally, in the case of the cluster function with the distance function d_∞ the auxiliary cluster function (4.69) can be approximated by using the function $\eta_{\infty,\tau}$, given in (4.62), that is

$$\bar{V}_{k,\tau}(\boldsymbol{y}, \theta_{\boldsymbol{a}}) = \frac{1}{m}\sum_{i=1}^{m} \frac{r_{k-1}^i + \eta_{\infty,\tau}(\boldsymbol{y}, \theta_{\boldsymbol{a}_i}) - \sqrt{(r_{k-1}^i - \eta_{\infty,\tau}(\boldsymbol{y}, \theta_{\boldsymbol{a}_i}))^2 + \tau^2}}{2}.$$

Here, $\theta_{a_i} = d_\infty(y, a_i),\ i = 1, \ldots, m$. Then we can replace the problem (4.29) by the sequence of the smooth problems

$$\begin{cases} \text{minimize} & \bar{V}_{k,\tau_h}(y, \theta_a) \\ \text{subject to} & y \in \mathbb{R}^n, \end{cases} \tag{4.72}$$

with the sequence $\{\tau_h\},\ \tau_h \downarrow 0$ as $h \to \infty$.

4.7.4 Partial Smoothing of DC Cluster Function

As mentioned above, if the function d_p in (4.20) is defined using L_1- or L_∞-norms, then both functions f_{k1} and f_{k2} are nonsmooth. In these cases, due to the fact that the general nonsmooth DC functions are not subdifferentially regular, the subdifferential calculus exists only in the form of inclusions. One possible approach to get the full subdifferential calculus for these functions is to smooth either the first, second or both DC components of the cluster function (4.18). Note that smoothing of both components leads to more complex functions involving many smoothing parameters. In this book, we smooth the first DC component as the function f_{k1} is quite simple nonsmooth function, whereas the second component f_{k2} is more complex.

Assume that the distance function d_1 is used in the function (4.18). The function $\eta_{1,\tau}$, given in (4.61), is the smooth approximation of the distance function d_1. Then the function

$$\widehat{U}_{k1,\tau}(x) = \frac{1}{m} \sum_{i=1}^{m} \sum_{j=1}^{k} \eta_{1,\tau}(x_j, a_i) \tag{4.73}$$

is the smooth approximation for the first DC component f_{k1}. Using it we get the following approximation for the cluster function f_k:

$$\widehat{U}_{k,\tau}(x) = \widehat{U}_{k1,\tau}(x) - f_{k2}(x).$$

Then the problem (4.20) can be replaced by [22, 23]

$$\begin{cases} \text{minimize} & \widehat{U}_{k,\tau_h}(x) \\ \text{subject to} & x = (x_1, \ldots, x_k) \in \mathbb{R}^{nk}, \end{cases} \tag{4.74}$$

where the sequence $\{\tau_h\}$ is such that $\tau_h > 0$ and $\tau_h \downarrow 0$ as $h \to \infty$.

Next, we describe the partial smoothing of the function (4.18) when the distance function d_∞ is applied. Using the smooth function $\eta_{\infty,\tau}$, given in (4.62), we have the following approximation for the function f_{k1}:

$$\widehat{V}_{k1,\tau}(\boldsymbol{x}) = \frac{1}{m}\sum_{i=1}^{m}\sum_{j=1}^{k} \eta_{\infty,\tau}(\boldsymbol{x}_j, \theta^j_{\boldsymbol{a}_i}). \tag{4.75}$$

Here, $\theta^j_{\boldsymbol{a}_i} = d_\infty(\boldsymbol{x}_j, \boldsymbol{a}_i),\ i = 1, \ldots, m,\ j = 1, \ldots, k$. Then the approximation for the function (4.18) is

$$\widehat{V}_{k,\tau}(\boldsymbol{x}) = \widehat{V}_{k1,\tau}(\boldsymbol{x}) - f_{k2}(\boldsymbol{x}),$$

and for a sequence $\{\tau_h\},\ \tau_h > 0,\ \tau_h \downarrow 0$ as $h \to \infty$, the problem (4.20) is replaced by

$$\begin{cases} \text{minimize} & \widehat{V}_{k,\tau_h}(\boldsymbol{x}) \\ \text{subject to} & \boldsymbol{x} = (\boldsymbol{x}_1, \ldots, \boldsymbol{x}_k) \in \mathbb{R}^{nk}. \end{cases} \tag{4.76}$$

Note that the Clarke subdifferential calculus can be efficiently applied to compute subgradients of the objective functions in the clustering problems (4.74) and (4.76).

4.7.5 Partial Smoothing of DC Auxiliary Cluster Function

Similar to the partial smoothing of the DC cluster function, we can formulate the partial smoothing of the auxiliary cluster function $\bar{f}_k$, defined in (4.32).

For the distance function d_1 using its smooth approximation $\eta_{1,\tau}$, given in (4.61), we have the smooth function

$$\widetilde{U}_{k1,\tau}(\boldsymbol{y}) = \frac{1}{m}\sum_{i=1}^{m} \big(r^i_{k-1} + \eta_{1,\tau}(\boldsymbol{y}, \boldsymbol{a}_i)\big), \tag{4.77}$$

which is the approximation of the function $\bar{f}_{k1}$. In this case an approximation for the function $\bar{f}_k$ can be defined as

$$\widetilde{U}_{k,\tau}(\boldsymbol{y}) = \widetilde{U}_{k1,\tau}(\boldsymbol{y}) - \bar{f}_{k2}(\boldsymbol{y}).$$

By selecting a sequence $\{\tau_h\}$ such that $\tau_h > 0$ and $\tau_h \downarrow 0$ as $h \to \infty$, we can replace the problem (4.34) by [23]

$$\begin{cases} \text{minimize} & \widetilde{U}_{k,\tau_h}(\boldsymbol{y}) \\ \text{subject to} & \boldsymbol{y} \in \mathbb{R}^{n}. \end{cases} \tag{4.78}$$

For the distance function d_∞, we use its smooth approximation $\eta_{\infty,\tau}$, given in (4.62). Then we get the smooth function

$$\widetilde{V}_{k1,\tau}(\mathbf{y}) = \frac{1}{m}\sum_{i=1}^{m}\big(r_{k-1}^{i} + \eta_{\infty,\tau}(\mathbf{y}, \theta_{\mathbf{a}_i})\big). \tag{4.79}$$

Here, $\theta_{\mathbf{a}_i} = d_\infty(\mathbf{y}, \mathbf{a}_i),\ i = 1, \dots, m$. Therefore, an approximation for the function $\bar{f}_k$ is

$$\widetilde{V}_{k,\tau}(\mathbf{y}) = \widetilde{V}_{k1,\tau}(\mathbf{y}) - \bar{f}_{k2}(\mathbf{y}).$$

Then the problem (4.34) is replaced by

$$\begin{cases} \text{minimize} & \widetilde{V}_{k,\tau_h}(\mathbf{y}) \\ \text{subject to} & \mathbf{y} \in \mathbb{R}^n, \end{cases} \tag{4.80}$$

with the sequence $\{\tau_h\},\ \tau_h \downarrow 0$ as $h \to \infty$. Note that the full subdifferential calculus for the objective functions in the problems (4.78) and (4.80) exists. Therefore, the Clarke subdifferential calculus can be efficiently applied to compute subgradients of these functions and to design algorithms for finding Clarke stationary points of the problems.

Chapter 5
Heuristic Clustering Algorithms

5.1 Introduction

The number of ways in which a set of m objects can be partitioned into k non-empty groups is given by the Stirling number [64]:

$$S(m,k) = \frac{1}{k!}\sum_{i=0}^{k}(-1)^{k-i}\binom{k}{i}i^m, \tag{5.1}$$

where

$$\binom{k}{i} = \frac{k!}{i!(k-i)!}$$

is the binomial coefficient. The Stirling number can be approximated by $k^m/k!$. A complete enumeration of all possible clusterings in order to determine the global minimum of the nonconvex clustering problem is computationally prohibitive for large data sets [174]. In fact, it has been proven that the clustering problem is NP-hard even for two clusters [7] or two attributes [205]. Therefore, various heuristics have been developed to solve the clustering problems.

In this chapter, we consider mainly heuristic partitional clustering algorithms for solving hard clustering problems, which do not explicitly use the NSO model. A partitional clustering algorithm produces a single partition of data with no hierarchical structure. This means that the algorithm requires the number of clusters to be specified—as a rule—a priori. A partitional algorithm usually optimizes an objective function defined using the data set. Most of these algorithms need to be run multiple times with different starting states, and the best configuration produced is the one used as the (optimal) clustering.

A. M. Bagirov et al., *Partitional Clustering via Nonsmooth Optimization*,
Unsupervised and Semi-Supervised Learning,
https://doi.org/10.1007/978-3-030-37826-4_5

We start by describing the k-means algorithm that is undoubtedly the most well-known and widely used partitional clustering algorithm. We also briefly describe the main ideas of different versions of k-means. Particularly, we describe the global k-means algorithm that computes clusters incrementally. Algorithms based on the incremental approach start with the calculation of one cluster center and gradually add a new cluster center at each iteration of the algorithm. Such an approach leads to the finding of at least a nearly global minimizer of the clustering problem.

The k-means algorithm and its variants use the squared Euclidean distance function (1.4) as a similarity measure. In Sect. 5.3 we recall the k-medians algorithm that aims to solve clustering problems with the d_1 distance function (1.3). Further, in Sect. 5.4 we give a short description of the k-medoids algorithm, where the final cluster centers are the most centrally located data points in the clusters.

The last three sections of this chapter present clustering algorithms that are not partitional and/or not applicable for hard clustering problems. First, we describe the fuzzy c-means algorithm that allows a data point to belong to more than one cluster, that is, the soft clustering problem is considered. As may be inferred by the name, the fuzzy c-means clustering algorithm is an extension of the k-means algorithm. Second, we recall the basic idea of clustering algorithms based on mixture models. In some sense these algorithms can be considered as fuzzy clustering algorithms with the membership matrix defined as a probability of each data point belonging to a particular cluster.

The artificial neural networks (ANNs) have been used extensively for both supervised data classification and clustering [264]. The most common ANN used for clustering include the Kohonen's learning vector quantization and the self-organizing map [185] and adaptive resonance theory models [61]. These networks have simple architectures with single layers and the weights in the networks are learnt by iteratively changing them until a predefined termination criterion is satisfied. These learning or weight changing procedures are similar to some used in classical clustering approaches. For instance, the procedure used in the learning vector quantization is similar to the k-means algorithm. The ANNs do not use any clustering models considered in this book. Therefore, to give an idea of the ANN in clustering applications we only provide an overview of the self-organizing map in Sect. 5.7.

5.2 k-Means Algorithm and Its Variants

The k-means algorithm is the most commonly used technique in partitional clustering. Early versions of this algorithm were introduced in [40, 108, 200, 204, 277]. The paper [298] places the k-means algorithm among the ten most important data mining algorithms.

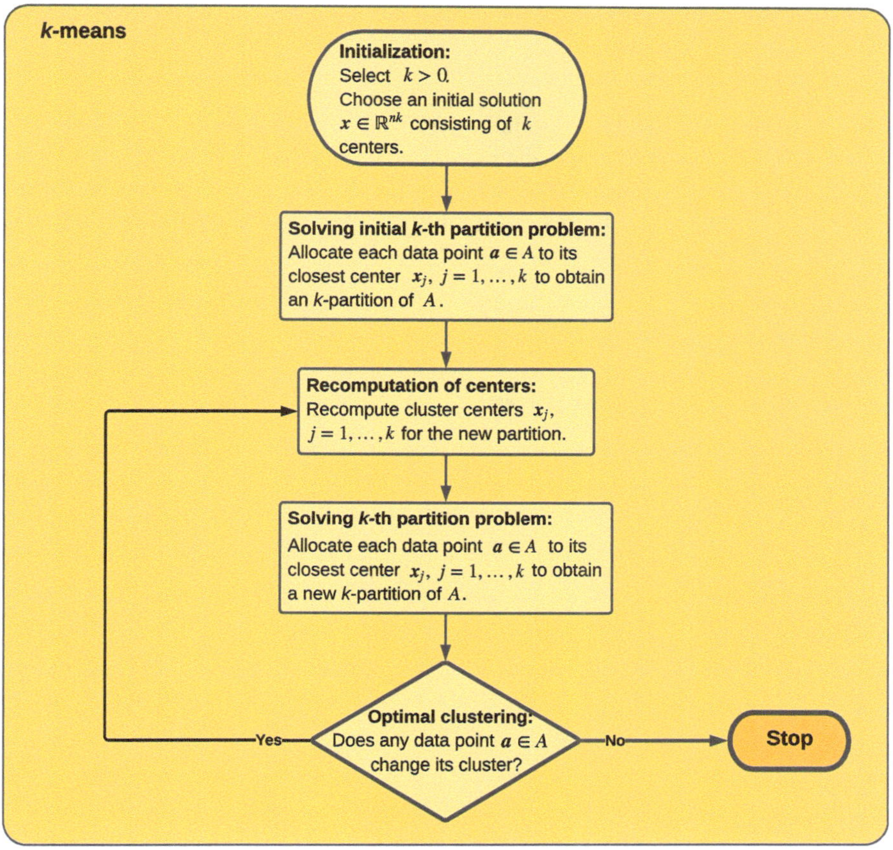

Fig. 5.1 k-means algorithm

5.2.1 k-Means Algorithm

The k-means algorithm aims to solve the MSSC problem that is when the similarity measure d_2 is applied. This algorithm minimizes the objective function (4.2) of the mixed integer programming formulation of the clustering problem.

The k-means algorithm utilizes an iterative scheme which starts with an arbitrary selected initial cluster configuration of the data, then alters the cluster membership in an iterative manner to obtain a better configuration. The popularity of the k-means algorithm is due to the fact that it is very simple and easy to implement. The flowchart of the basic k-means algorithm is given in Fig. 5.1.

At the beginning, the k-means algorithm randomly chooses k cluster centers with a predefined k. Then it alternates between two major steps until a stopping criterion is satisfied. These steps are as follows:

- distribution of data points among clusters utilizing the minimum squared Euclidean distance; and
- recomputing of cluster centers.

In other words, the k-means algorithm iteratively reassigns data points to clusters based on the similarity between the points and the cluster centers until there is no further reassignment or significant decrease in the value of the clustering function.

Next, we present the k-means algorithm in the step by step form. Then we give a more detailed description of the procedures used.

Algorithm 5.1 k-means algorithm

Input: Data set A and the number of clusters k to be computed.
Output: Solution to the k-partition problem.

1: *(Initialization)* Choose a seed solution consisting of k centers (not necessarily belonging to A).
2: Allocate each data point $\boldsymbol{a} \in A$ to its closest center and obtain a k-partition of A.
3: (*Stopping criterion*) If some predefined stopping criterion is met, then **stop**.
4: Recompute centers for the new partition and go to Step 2.

In Step 4, the following problem is solved to find the center $\boldsymbol{x}_j$ of the cluster $A^j,\ j = 1, \ldots, k$:

$$\begin{cases} \text{minimize} & \sum_{\boldsymbol{a} \in A^j} d_2(\boldsymbol{x}_j, \boldsymbol{a}) \\ \text{subject to} & \boldsymbol{x}_j \in \mathbb{R}^n. \end{cases}$$

This problem is convex and its objective function is strongly convex. Therefore, the problem has a unique solution. The necessary and sufficient condition for optimality is

$$\sum_{\boldsymbol{a} \in A^j} (\boldsymbol{x}_j - \boldsymbol{a}) = \boldsymbol{0},$$

which leads to the following formula for the center $\boldsymbol{x}_j,\ j = 1, \ldots, k$:

$$\boldsymbol{x}_j = \frac{1}{m_j} \sum_{\boldsymbol{a} \in A^j} \boldsymbol{a},$$

where m_j is the number of objects in the cluster $A^j,\ j = 1, \ldots, k$. Thus, there is no need to solve any optimization problem to find cluster centers in Step 4 of Algorithm 5.1.

Various stopping criteria can be used in Step 3 of the k-means algorithm. They include:

- let $\varepsilon > 0$ be a given tolerance and m_t be a number of data points changing their clusters at the tth iteration of Algorithm 5.1. If

$$\frac{m_t}{m} \leq \varepsilon,$$

  then the algorithm terminates with $\boldsymbol{x}_t = (\boldsymbol{x}_{t,1}, \ldots, \boldsymbol{x}_{t,k})$ as a solution to the clustering problem;
- when no data point changes its clusters, then Algorithm 5.1 terminates. This corresponds to the previous stopping criterion when $\varepsilon = 0$; and
- let $\varepsilon > 0$ be a given tolerance and $\boldsymbol{x}_{t-1} = (\boldsymbol{x}_{t-1,1}, \ldots, \boldsymbol{x}_{t-1,k})$ and $\boldsymbol{x}_t = (\boldsymbol{x}_{t,1}, \ldots, \boldsymbol{x}_{t,k})$ be solutions found at iterations $t-1$ and t, $t > 1$. If

$$\frac{\zeta_k(\boldsymbol{x}_{t-1}) - \zeta_k(\boldsymbol{x}_t)}{\zeta_k(\boldsymbol{x}_{t-1})} \leq \varepsilon,$$

  where ζ_k is the objective function in the clustering problem (4.2), then Algorithm 5.1 terminates with $\boldsymbol{x}_t = (\boldsymbol{x}_{t,1}, \ldots, \boldsymbol{x}_{t,k})$ as a solution to the clustering problem.

It should be noted that the second stopping criterion works best in small data sets, although, it can be used also in larger data sets. The first criterion works best in medium sized and large data sets, and finally, the third stopping criterion works best in large and very large data sets.

In [263], conditions under which the k-means algorithm converges in a finite number iterations to the solution of the MSSC problem are established.

Proposition 5.1 *Algorithm 5.1 converges to an optimal solution of the clustering problem in a finite number of iterations.*

Proof It is obvious that the maximum number of subsets of the set A is 2^m, where m is the number of data points in A. Particularly, the maximum number of combinations in which a set of m data points can be partitioned into k non-empty groups is $S(m,k)$ given by (5.1). As mentioned above each iteration of the k-means algorithm consists of two main steps as follows:

(i) reassigning data points to the current cluster centers; and
(ii) updating of the cluster centers using the new distribution of points.

As the new centers provide minimum of the clustering function for the redistributed clusters and the objective function is strongly convex for each cluster, the value of the clustering objective function strictly decreases at each iteration of the k-means algorithm. That is, the k-means algorithm generates the sequence of combinations of points where the value of the clustering function decreases and therefore, all these combinations are different. Since the number of such combinations is finite the k-means algorithm terminates after finite number of iterations. □

It is easy to see that the mixed integer programming model (4.2) with the similarity measure d_2 can be reformulated as

$$\begin{cases} \text{minimize} & f_k(\boldsymbol{x}) \\ \text{subject to} & \boldsymbol{x} = (\boldsymbol{x}_1, \ldots, \boldsymbol{x}_k) \in \mathbb{R}^{nk}, \end{cases}$$

where

$$f_k(\boldsymbol{x}) = \frac{1}{m} \sum_{j=1}^{k} \sum_{i=1}^{m_j} d_2(\boldsymbol{x}_j, \boldsymbol{a}_i^j).$$

Here, k is the number of clusters, m_j is the number of objects in the cluster A^j, $j = 1, \ldots, k$, $\boldsymbol{a}_i^j \in A$ is the ith element of the cluster A^j, $i = 1, \ldots, m_j$, and $\boldsymbol{x}_j$ is the centroid of the jth cluster obtained by

$$\boldsymbol{x}_j = \frac{1}{m_j} \sum_{i=1}^{m_j} \boldsymbol{a}_i^j.$$

5.2.2 Variants of k-Means Algorithm

The k-means algorithm suffers from being sensitive to the selection of the initial clustering partition or cluster centers [12]. It converges to a local solution which can significantly differ from the global solution, especially in large data sets. Various versions of the k-means algorithm have been proposed in the literature, many of them focussing on the selection of a good initial partition (see, for example, [4, 64, 153, 154, 245]). Below we list some of these algorithms.

- *Forgy algorithm [108]:* the algorithm randomly chooses k points from the data set and uses them as initial centers. The idea behind this selection is that when choosing points randomly we are more likely to select a point from a region with the highest density of points. However, there is no guarantee that we will not select some poorly located outliers [12]. This algorithm is also called the h-means clustering algorithm. Application of this algorithm may lead to obtaining empty clusters [272].
- *MacQueen algorithm [204]:* in this algorithm, points in a data set are ordered. To solve the k-partition problem, first, one takes the first k points in the data set

A as the initial cluster centers. Then a point is assigned to a cluster with the least squared Euclidian distance between the point and the cluster center. After the assignment of each point its previous and new cluster centers are updated. This can be done easily. For example, assume that the data point $\bar{\boldsymbol{a}} \in A$ is moved from the cluster A^t to the cluster A^j, $t, j \in \{1, \ldots, k\}$. Let $\boldsymbol{x}_t$ and $\boldsymbol{x}_j$ be the centers of clusters A^t and A^j, respectively. Then these centers will be updated as

$$\boldsymbol{x}_t' = \frac{1}{m_t - 1}\Big(m_t \boldsymbol{x}_t - \bar{\boldsymbol{a}}\Big), \quad \text{and}$$

$$\boldsymbol{x}_j' = \frac{1}{m_j + 1}\Big(m_j \boldsymbol{x}_j + \bar{\boldsymbol{a}}\Big),$$

where m_t and m_j are the number of data points in clusters with centers $\boldsymbol{x}_t$ and $\boldsymbol{x}_j$, respectively, and $\boldsymbol{x}_t'$ and $\boldsymbol{x}_j'$ are the updated cluster centers. The outcome of this algorithm depends on the order of points in a data set. The number of clusters found by the MacQueen algorithm cannot change because each cluster should contain at least one data point. If a cluster contains only one data point, then this point cannot be assigned to a different cluster.

- *Ball and Hall's algorithm [41]:* for a given number k of clusters, the Ball and Hall's algorithm determines the starting cluster centers in k steps. First, some distance threshold T is defined. The first center is computed as the center of the whole data set A as

$$\boldsymbol{x}_1 = \frac{1}{m}\sum_{i=1}^{m} \boldsymbol{a}_i.$$

Assume that l, $(1 < l < k)$ centers are computed. In order to find the $(l + 1)$th initial cluster center, the algorithm chooses the first data point whose distance from all the previously found centers is no less than a given threshold T. This process continues until all k starting cluster centers are obtained. The usage of the distance threshold T allows one to ensure that the starting points are well separated. Nevertheless, it may be difficult to define an appropriate value for T. In addition, the algorithm is sensitive to ordering of points in a data set.

- *Maximin algorithm [126, 172]:* the original maximin algorithm chooses the first starting cluster center $\boldsymbol{x}_1$ arbitrarily. In some variants of this algorithm, a data point with the greatest Euclidean norm is selected as the first cluster center instead of an arbitrary selection. Then, the lth starting cluster center $\boldsymbol{x}_l$, $(1 < l \leq k)$ is chosen to be the data point that has the greatest minimum distance to the previously selected centers $\boldsymbol{x}_1, \ldots, \boldsymbol{x}_{l-1}$. More precisely, first for each data point $\boldsymbol{a} \in A$ we calculate

$$d_{\min}(\boldsymbol{a}) = \min_{t=1,\ldots,l-1} d_2(\boldsymbol{a}, \boldsymbol{x}_t)$$

and then define $\boldsymbol{x}_l$ as

$$\boldsymbol{x}_l = \operatorname*{argmax}_{\boldsymbol{a} \in A} d_{\min}(\boldsymbol{a}).$$

This process continues until all k starting cluster centers are obtained.

- *Lloyd algorithm:* it is believed that this algorithm is one of the oldest versions of the k-means clustering algorithm introduced in 1957. However, the algorithm was published only in 1982 [200]. For solving the k-partition problem, the Lloyd algorithm starts with an arbitrary (or random) set of starting cluster centers $X = \{\boldsymbol{x}_1, \ldots, \boldsymbol{x}_k\}$. Then for each $\boldsymbol{a} \in A$ it computes the closest center $\boldsymbol{y}_{\boldsymbol{a}} \in X$ to $\boldsymbol{a}$. At the final step it updates the cluster centers as

$$\boldsymbol{x}_j = \frac{1}{|\mathcal{I}_j|} \sum_{i \in \mathcal{I}_j} \boldsymbol{a}_i, \quad \mathcal{I}_j = \Big\{ i \in \{1, \ldots, m\} : \boldsymbol{y}_{\boldsymbol{a}_i} = \boldsymbol{x}_j \Big\}.$$

  This process continues until the set X is not changed in two successive iterations.
- *Hartigan and Wong algorithm [144]:* this algorithm is considered as an alternative heuristic to the Lloyd algorithm. Given a partition $A^1, \ldots, A^k$, the algorithm randomly selects a single point $\boldsymbol{a}$ from its cluster A^j, $j \in \{1, \ldots, k\}$. This point is considered as a singleton cluster with the center $\boldsymbol{a}$. Then the algorithm updates the center of the cluster $A^j \setminus \{\boldsymbol{a}\}$ and finds the closest cluster to which $\boldsymbol{a}$ should be reassigned by minimizing the clustering objective function.
- *X-means algorithm [234]:* this algorithm is different to other variants of the k-means algorithm since it produces not only the set of clusters, but also the optimal (true) number k of clusters. In the X-means algorithm, instead of predefining k, the user specifies a range $[k_{\min}, k_{\max}]$ for the number of clusters where $k \in [k_{\min}, k_{\max}]$. The Bayesian information criterion (BIC) score is used to identify k in this algorithm. The algorithm starts with $k = k_{\min}$ and adds new cluster centers when necessary until the upper bound $k_{\max}$ is reached. Then the BIC scores are computed for all number of clusters in the range and the optimal number k of clusters is selected with the best score. Finally, the cluster distribution corresponding to the number k is chosen as the output of the algorithm. The X-means algorithm consists of two main operations: the *Improve-Params* and the *Improve-Structure*. The first operation is used to run the k-means algorithm until it converges. The second operation finds out if and where a new center should appear. This is achieved by splitting some clusters.
- *j-means algorithm [141]:* this algorithm is able to tackle degeneracy which may happen with the k-means (more specifically with the h-means) algorithm. If among obtained clusters only $k - k_1$ are non-degenerate (i.e., non-empty) for some $1 \leq k_1 < k$, then k_1 data points that are most distant from their cluster centers are selected. Considering these points as new ones, additional cluster centers are obtained and all points are reassigned.

- *k-means++ algorithm [14]:* this algorithm chooses the first starting cluster center $\boldsymbol{x}_1 \in A$ randomly. Assuming that $l-1$, $l \geq 2$ starting cluster centers $\boldsymbol{x}_1, \ldots, \boldsymbol{x}_{l-1}$ have been selected, the lth starting cluster center is chosen to be a data point $\boldsymbol{a} \in A$ with the probability

$$P_l(\boldsymbol{a}) = \frac{d_{\min}(\boldsymbol{a})}{\sum_{t=1}^{l-1} d_p(\boldsymbol{a}, \boldsymbol{x}_t)}.$$

Here, d_p is any distance function—usually the squared Euclidean distance function—and

$$d_{\min}(\boldsymbol{a}) = \min_{t=1,\ldots,l-1} d_p(\boldsymbol{a}, \boldsymbol{x}_t)$$

is the minimum distance between the data point $\boldsymbol{a}$ and the set of starting cluster centers chosen so far. The k-means++ algorithm probabilistically selects $\log(k)$ centers in each round, and then greedily selects the center that reduces the value of the cluster function the most. Such a modification allows one to avoid choosing two centers that are close to each other.

There are several other initialization algorithms for the k-means clustering algorithm (see, e.g. [4, 64, 237]). These methods are based on the approach on dividing the search space into disjoint subsets of simple structure (for example, hypercubes), using them to identify dense regions of data and choosing starting cluster centers from the densest regions.

5.2.3 Global k-Means Algorithm

The objective functions in all optimization models of the partitional clustering problem are nonconvex and they may have many local minimizers. Moreover, as the number of clusters increases, the number of local minimizers increases considerably. Nevertheless, global or nearly global minimizers of the clustering problem are of interest as they provide the best cluster structure of a data set with the least number of clusters.

Global minimizers (or global solutions of the mixed integer programming problem (4.2)) of the function ζ_k are points where the function attains its least value over the feasible set. Since the clustering problem is NP-hard global optimization algorithms are not always applicable to solve this problem and, even if they are, finding global minimizers may become very time-consuming in large data.

In the most variants of the k-means algorithm some procedures are introduced to improve the quality of the solution. These procedures, mainly, try to select a good initial partition with a given number k of clusters.

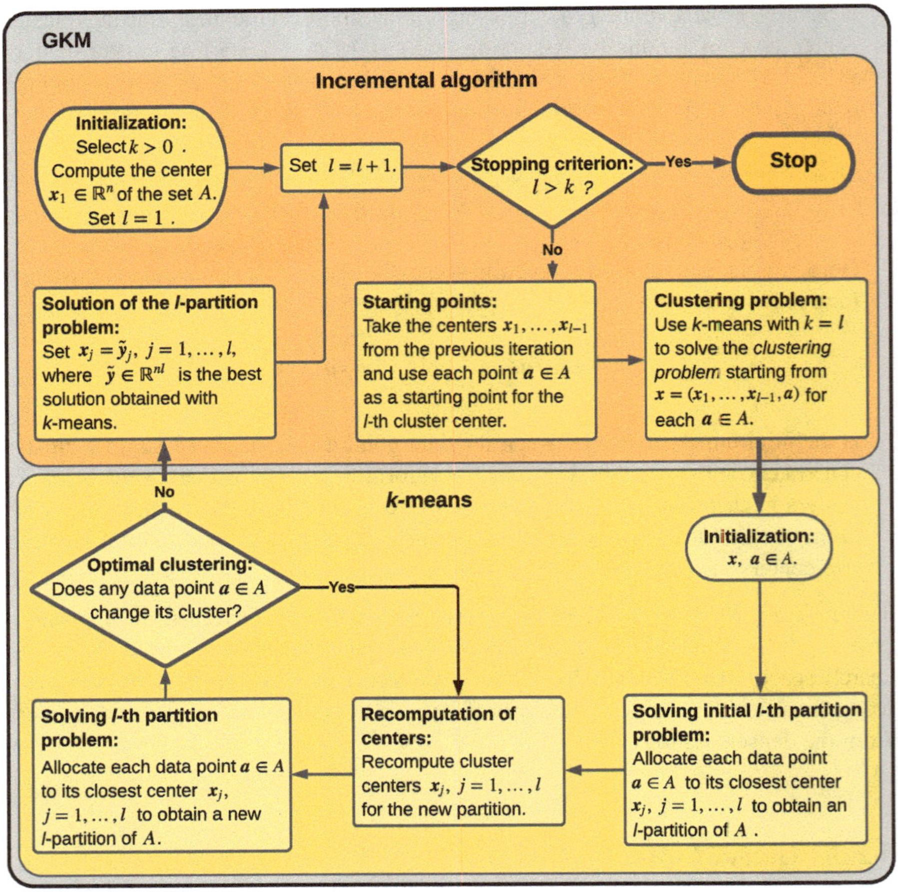

Fig. 5.2 Global k-means algorithm

Another approach is to compute clusters incrementally. Algorithms based on an *incremental approach* start with the calculation of one cluster center and gradually add a new cluster center at each iteration of the algorithm. More precisely, in order to compute k-partition, $k > 1$, of the data set A, incremental algorithms start from an initial state with the $k - 1$ centers for the $(k - 1)$-partition problem and the remaining kth center is placed in an appropriate position. The *global k-means algorithm* (GKM), introduced in [197], is one representative of these algorithms. The flowchart of this algorithm is given in Fig. 5.2.

The GKM is a significant improvement of the k-means algorithm [197]. It is an incremental algorithm where each data point is used as a starting point for the kth cluster center. Next, we give the GKM in step by step form.

The GKM applies the k-means algorithm m times at each iteration of the incremental algorithm, and therefore, it is not very efficient in large data sets. Two

Algorithm 5.2 Global k-means algorithm

Input: Data set A and the number k of clusters to be computed.
Output: Solution to the l-partition problem, $l = 1, \dots, k$.

1: *(Initialization)* Compute the centroid $\boldsymbol{x}_1$ of the data set A as

$$\boldsymbol{x}_1 = \frac{1}{m} \sum_{i=1}^{m} \boldsymbol{a}_i, \quad \boldsymbol{a}_i \in A, \ i = 1, \dots, m, \tag{5.2}$$

and set $l = 1$.

2: *(Stopping criterion)* Set $l = l + 1$. If $l > k$, then **stop**.

3: Take the centers $\boldsymbol{x}_1, \dots, \boldsymbol{x}_{l-1}$ from the $(l-1)$th iteration and consider each point $\boldsymbol{a} \in A$ as a starting point for the lth cluster center, thus obtaining m initial solutions with l points $(\boldsymbol{x}_1, \dots, \boldsymbol{x}_{l-1}, \boldsymbol{a})$. Apply the k-means algorithm with $k = l$ starting from each of them, and denote the obtained solution by $(\hat{\boldsymbol{y}}_1(\boldsymbol{a}), \dots, \hat{\boldsymbol{y}}_l(\boldsymbol{a}))$.

4: Compute the value of the function ζ_l, defined in (4.2), at the point $(\hat{\boldsymbol{y}}_1(\boldsymbol{a}), \dots, \hat{\boldsymbol{y}}_l(\boldsymbol{a}))$, find

$$\zeta_l^{\min} = \min_{\boldsymbol{a} \in A} \zeta_l\big(\hat{\boldsymbol{y}}_1(\boldsymbol{a}), \dots, \hat{\boldsymbol{y}}_l(\boldsymbol{a})\big),$$

and define the point $(\tilde{\boldsymbol{y}}_1, \dots, \tilde{\boldsymbol{y}}_l)$ such that

$$\zeta_l(\tilde{\boldsymbol{y}}_1, \dots, \tilde{\boldsymbol{y}}_l) = \zeta_l^{\min}.$$

5: Set $\boldsymbol{x}_j = \tilde{\boldsymbol{y}}_j, \ j = 1, \dots, l$ and go to Step 2.

different approaches were proposed to reduce the computational burden [197]. One approach is to compute the distance matrix $D = [d_{ij}], \ i, j = 1, \dots, m$ of the data set A, where $d_{ij} = d_2(\boldsymbol{a}_i, \boldsymbol{a}_j)$, before the application of the GKM. This reduces the number of distance function evaluations significantly. However, this approach has a limitation as the matrix D for large data sets (with tens of thousands of data points and more) cannot be stored in the memory of a computer.

The second approach is to use only one data point as a candidate for the next cluster center. More precisely, it selects a data point that provides the largest decrease of the cluster function, and this point is considered as the kth cluster center. This approach leads to the design of the fast global k-means algorithm (FGKM). Next, we give a very brief overview of this algorithm.

Let $\boldsymbol{x}_1, \dots, \boldsymbol{x}_{k-1}$ be a given solution to the $(k-1)$th clustering problem and $\zeta_{k-1}^* = \zeta_{k-1}(\boldsymbol{x}_1, \dots, \boldsymbol{x}_{k-1})$ be the corresponding value of the objective function given in (4.2). The FGKM computes an upper bound $\zeta_k^* \le \zeta_{k-1}^* - b_j$ on the ζ_k^* as

$$b_j = \sum_{i=1}^{m} \max\left\{0, r_{k-1}^i - d_2(\boldsymbol{a}_i, \boldsymbol{a}_j)\right\}, \quad j = 1, \dots, m, \tag{5.3}$$

where r_{k-1}^i is the squared distance between $\boldsymbol{a}_i$ and the closest center among $k-1$ cluster centers $\boldsymbol{x}_1, \ldots, \boldsymbol{x}_{k-1}$, defined in (4.27). Then a data point $\boldsymbol{a}_j \in A$ with the maximum value of b_j is chosen as a starting point for the kth cluster center. The FGKM can be applied to large data sets, however, it is usually not as accurate as the original GKM.

5.3 k-Medians Algorithm and Its Variants

There are some applications where clustering algorithms defined using the d_1 and d_∞ distance functions generate more meaningful results than those defined using the function d_2. Particularly, clustering algorithms with d_1 and d_∞ are more robust to outliers [312], and in high dimensional data mining applications the function d_1 is consistently more preferable than d_2 [2].

The distance function d_1 was used to define the similarity measure in clustering problems first time by Carmichael and Sneath in 1969 [59] (see, also [174]). In its current form the k-medians algorithm was introduced by Späth in 1976 [270]. Since then many variants of this algorithm have been proposed (see, e.g., [50, 82, 139, 234, 254, 288]. A comparison of clustering algorithms using the d_1 and d_∞ distance functions is given in [85].

5.3.1 *k-Medians Algorithm*

The *k-medians algorithm* aims to solve clustering problems where the similarity (dissimilarity) measure is defined using the L_1-norm, that is, the similarity measure is the distance function d_1 defined in (1.3). Otherwise, this algorithm is similar to the k-means algorithm. The flowchart of the k-medians algorithm is given in Fig. 5.3.

At each iteration of the k-medians algorithm we need to solve the following problem for each cluster $C = A^j,\ j = 1, \ldots, k$:

$$\begin{cases} \text{minimize} & \varphi(\boldsymbol{x}) \\ \text{subject to} & \boldsymbol{x} \in \mathbb{R}^n, \end{cases} \tag{5.4}$$

where

$$\varphi(\boldsymbol{x}) = \frac{1}{|C|} \sum_{\boldsymbol{c} \in C} d_1(\boldsymbol{x}, \boldsymbol{c}),$$

and $|C|$ is the cardinality of the cluster C. The coordinates of the solution $\boldsymbol{x}$ to this problem are medians of corresponding attributes.

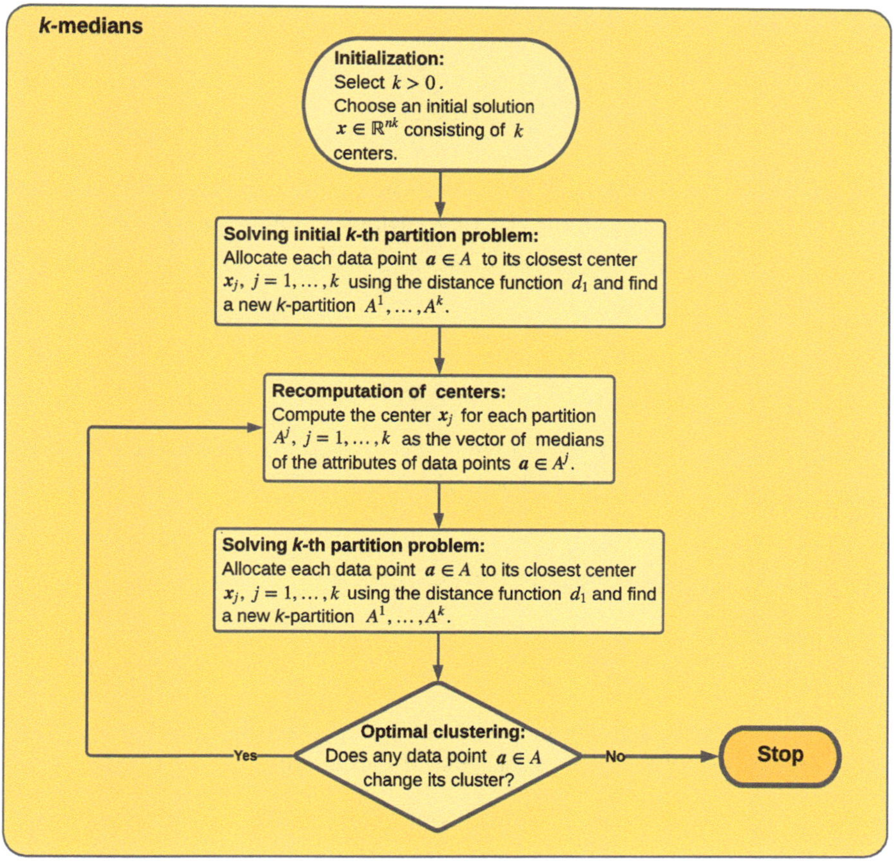

Fig. 5.3 k-medians algorithm

Definition 5.1 A point $\boldsymbol{x} \in \mathbb{R}^n$ whose coordinates are medians of attributes of the set C is called the median of this set.

Proposition 5.2 *Assume that for any $i \in \{1, \dots, n\}$ coordinates c_i are different for all $\boldsymbol{c} \in C$. Then the median of the set C is the solution to the problem (5.4).*

Proof The function φ can be written as

$$\varphi(\boldsymbol{x}) = \frac{1}{|C|} \sum_{c \in C} \sum_{i=1}^{n} |x_i - c_i| = \frac{1}{|C|} \sum_{i=1}^{n} \sum_{c \in C} |x_i - c_i|.$$

Consider functions

$$\psi_i(x_i) = \sum_{c \in C} |x_i - c_i| = \sum_{c \in C} \max\{x_i - c_i, c_i - x_i\}, \quad i = 1, \dots, n.$$

Then the function φ can be represented as

$$\varphi(\boldsymbol{x}) = \frac{1}{|C|} \sum_{i=1}^{n} \psi_i(x_i).$$

This means that the minimization of φ is equivalent to the minimization of functions $\psi_i,\ i = 1, \ldots, n$. For a given $i \in \{1, \ldots, n\}$ define the following sets:

$$C_i^- = \{\boldsymbol{c} \in C : c_i < x_i\},$$
$$C_i^+ = \{\boldsymbol{c} \in C : c_i > x_i\}, \quad \text{and}$$
$$C_i^0 = \{\boldsymbol{c} \in C : c_i = x_i\}.$$

Since all numbers c_i ($\boldsymbol{c} \in C$) are different it is obvious that for a given $\boldsymbol{x} \in \mathbb{R}^n$ the cardinality of the set C_i^0 is either 0 or 1. Then the subdifferential of the function ψ_i at x_i is

$$\partial\psi_i(x_i) = |C_i^-| - |C_i^+| + \Big[- |C_i^0|,\ |C_i^0| \Big]$$
$$= \Big[|C_i^-| - |C_i^+| - |C_i^0|,\ |C_i^-| - |C_i^+| + |C_i^0| \Big].$$

For a point x_i to be a global minimizer of the function ψ_i, it is necessary and sufficient that $0 \in \partial\psi_i(x_i)$. This means that at x_i we have

$$|C_i^-| - |C_i^+| - |C_i^0| \le 0, \quad \text{and}$$
$$|C_i^-| - |C_i^+| + |C_i^0| \ge 0.$$

Depending on the cardinality of the set C_i^0, we have two cases:

(i) for $|C_i^0| = 0$, we get $|C_i^-| - |C_i^+| = 0$, that is, $|C_i^-| = |C_i^+|$ and therefore, the total number of points $\boldsymbol{c} \in C$ with different ith coordinate is $2|C_i^-|$ (or $2|C_i^+|$). It means that this number is even and it is obvious that x_i is the median; and
(ii) for $|C_i^0| = 1$, we have $-1 \le |C_i^-| - |C_i^+| \le 1$. This leads to the following three options:

- if $|C_i^-| = |C_i^+| - 1$, then the number of points $\boldsymbol{c} \in C$ with different ith coordinates is even. Therefore, x_i is the median coinciding with one of c_i ($\boldsymbol{c} \in C$);
- if $|C_i^-| = |C_i^+|$, then the number of points $\boldsymbol{c} \in C$ with different ith coordinates is odd and x_i is the median coinciding with the coordinate which is exactly in the middle;

- if $|C_i^-| = |C_i^+| + 1$, then the number of points $\boldsymbol{c} \in C$ with different ith coordinates is even, and again x_i is the median coinciding with one of c_i ($\boldsymbol{c} \in C$).

This completes the proof. □

Remark 5.1 The assumption used in Proposition 5.2 is reasonable. If there is any two data points with the same ith coordinate, $i \in \{1, \ldots, n\}$, then one of them can be changed by adding a very small number to it.

In practice, the calculation of the median for each cluster A^j, $j = 1, \ldots, k$ can be time-consuming. One way to deal with this difficulty is to apply Weiszfeld's algorithm [293, 294] to find the medians. This algorithm proceeds as follows.

Algorithm 5.3 Weiszfeld's algorithm

Input: Finite point set $C \subset \mathbb{R}^n$ and a tolerance $\varepsilon > 0$.
Output: Median $\bar{\boldsymbol{c}}$ of the set C.

1: *(Initialization)* Compute the centroid $\boldsymbol{c}$ of the set C and set $\bar{\boldsymbol{c}} = \boldsymbol{c}$.

2: Compute

$$\boldsymbol{u} = \sum_{\boldsymbol{c} \in C} \frac{\boldsymbol{c}}{\|\boldsymbol{c} - \bar{\boldsymbol{c}}\|} \quad \text{and} \quad u_1 = \sum_{\boldsymbol{c} \in C} \frac{1}{\|\boldsymbol{c} - \bar{\boldsymbol{c}}\|}.$$

3: Compute $\bar{\boldsymbol{c}}_1 = \boldsymbol{u}/u_1$.

4: *(Stopping criterion)* If $\|\bar{\boldsymbol{c}}_1 - \bar{\boldsymbol{c}}\| < \varepsilon$, then **stop**. Otherwise, set $\bar{\boldsymbol{c}} = \bar{\boldsymbol{c}}_1$ and go to Step 2.

The Weiszfeld's algorithm may fail to converge when one of its estimates $\bar{\boldsymbol{c}}$ falls on one of the points $\boldsymbol{c} \in C$.

In addition, since (5.4) is a convex NSO problem one can apply any NSO algorithms, given in Chap. 3, to solve it and most of these methods will find the median in a finite number of steps.

The step by step form of the k-medians algorithm is given next.

In Step 4 of Algorithm 5.4, one can apply stopping criteria used in the k-means algorithm (see Sect. 5.2.1).

5.3.2 *Variants of k-Medians Algorithm*

As mentioned before, various versions of the k-medians algorithm have been proposed. Next, we describe the most important—or at least the most well-known—ones:

Algorithm 5.4 k-medians algorithm

Input: Data set A and the number of clusters k to be computed.
Output: Solution to the k-partition problem.

1: *(Initialization)* Select initial cluster centers $(\boldsymbol{x}_1, \ldots, \boldsymbol{x}_k) \in \mathbb{R}^{nk}$.
2: Allocate data points to the closest cluster center using the distance function d_1 and find the cluster partition $A^1, \ldots, A^k$.
3: Compute the center $\boldsymbol{x}_j$ of the cluster A^j as a vector of medians of attributes using data points from the cluster A^j, $j = 1, \ldots, k$.
4: *(Stopping criterion)* Repeat Steps 2 and 3 until a predefined stopping criterion is met.

- *ISODATA clustering algorithm [92, 157, 288]:* this algorithm does not require the number of clusters to be known a priori but only a user-defined threshold for the cluster separation. It uses splitting and merging to find clusters. First, the ISODATA algorithm places some initial cluster centers randomly with an initial number of clusters. Then it assigns data points to these centers using the d_1 distance function and obtains the initial cluster distribution of the data set. For each cluster, a new cluster center is computed as its median. Then the standard deviations within each cluster and also the distances between the new centers are calculated. Next, the following two operations are applied to obtain a new cluster distribution:

  - a cluster is split if its standard deviation is greater than the user-defined threshold; and
  - two clusters are merged if the distance between their centers is less than the user-defined threshold.

  These iterations continue until one of the following stopping criteria met:

  - the average inter-center distance falls below the user-defined threshold;
  - the average change in the inter-center distance between iterations is less than a threshold; or
  - the maximum number of iterations is reached.

  The outcome of the ISODATA algorithm strongly depends on the choice of starting cluster centers. In addition, the algorithm may become time-consuming for clustering in highly unstructured data sets. The strength of the ISODATA algorithm is that it requires limited information from the user.

- *X-medians algorithm:* this algorithm is an improvement of the original k-medians algorithm [234], and can be considered as a version of the X-means algorithm with the similarity measure defined using the L_1-norm. The X-medians algorithm does not require the number of clusters to be provided, instead lower and upper bounds for this number are required. The details of the X-means algorithm are given in Sect. 5.2.2.

In addition, versions of the k-medians algorithm for solving fuzzy clustering problems were proposed in [50, 116, 306]. These algorithms are similar to the fuzzy c-means algorithm to be described in Sect. 5.5.

5.4 k-Medoids Algorithm

In the MSSC problems the calculation of centroids or in the clustering problems with the d_1 and d_∞ distance functions, the calculation of cluster centers may yield points that are not in a data set A. The *medoid* is defined as the point of a cluster, whose average dissimilarity to all points in the cluster is the lowest in comparison with any other point from that cluster, that is, it is the most centrally located data point in the cluster. The *k-medoids algorithm* aims to find such points in clusters. A flowchart of this algorithm is given in Fig. 5.4.

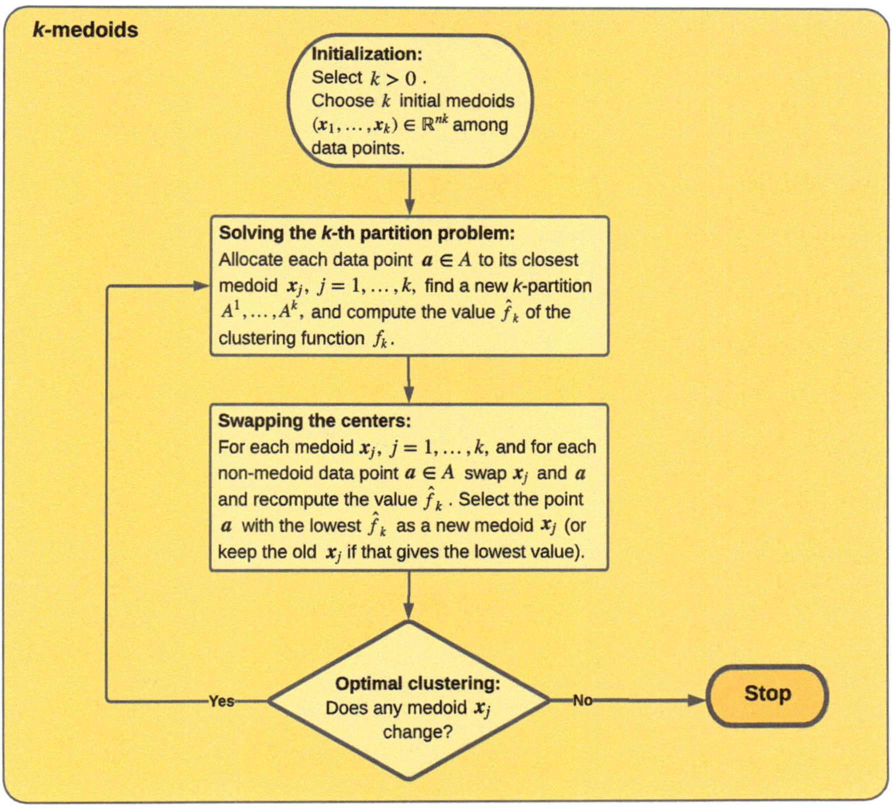

Fig. 5.4 k-medoids algorithm

The k-medoids algorithm is a partitional clustering algorithm. This algorithm is similar to the k-means algorithm but it calculates medoids instead of means. Therefore, it is considered to be more resilient to outliers compared to k-means. Different similarity measures using various distance functions can be used within the k-medoids algorithm.

The problem of finding k medoids $\boldsymbol{x} = (\boldsymbol{x}_1, \ldots, \boldsymbol{x}_k) \in \mathbb{R}^{nk}$ with $k > 1$ can be formulated as the constrained minimization problem

$$\begin{cases} \text{minimize} & f_k(\boldsymbol{x}) \\ \text{subject to} & \varphi(\boldsymbol{x}_j) = \min_{i=1,\ldots,m} \|\boldsymbol{x}_j - \boldsymbol{a}_i\| = 0, \quad j = 1, \ldots, k, \\ & \boldsymbol{x} = (\boldsymbol{x}_1, \ldots, \boldsymbol{x}_k) \in \mathbb{R}^{nk}, \end{cases} \tag{5.5}$$

where

$$f_k(\boldsymbol{x}) = \frac{1}{m} \sum_{i=1}^{m} \min_{j=1,\ldots,k} d_p(\boldsymbol{x}_j, \boldsymbol{a}_i).$$

Here, the constraints $\varphi(\boldsymbol{x}_j) = 0, \ j = 1, \ldots, k$ guarantee that the solution points $\boldsymbol{x}_j, \ j = 1, \ldots, k$ are medoids, that is, they belong to A. Applying the penalty function method, the problem (5.5) is reduced to the unconstrained minimization problem

$$\begin{cases} \text{minimize} & F_k(\boldsymbol{x}) \\ \text{subject to} & \boldsymbol{x} = (\boldsymbol{x}_1, \ldots, \boldsymbol{x}_k) \in \mathbb{R}^{nk}, \end{cases}$$

where

$$F_k(\boldsymbol{x}) = f_k(\boldsymbol{x}) + \rho \sum_{j=1}^{k} |\varphi(\boldsymbol{x}_j)|,$$

and $\rho > 0$ is the penalty parameter.

The k-medoids algorithm was first introduced by Späth in 1985 [273]. This algorithm minimizes the objective function value by swapping data points from one cluster to another one. First, the k-medoids algorithm randomly generates starting medoids using data points. With these medoids as initial centers, each data point is assigned to its closest medoid and the cluster distribution is obtained. Then those data points whose movements from one cluster to another one result in the reduction of the objective function value are chosen as new cluster centers (medoids). This process is continued until no point moving results in the decrease of the value of the objective function F_k.

The widely used version of the k-medoids algorithm is the *partitioning around medoids* (PAM) algorithm. It was first introduced in [173] (see, also [174, 231, 286]) where the d_1 distance function was used to define the similarity measure. One version of the PAM algorithm is given below where we use the similarity measure d_p with $p = 1, 2, \infty$. Note that the outcome of this algorithm does not depend on the order of points in a data set.

Algorithm 5.5 Partitioning around medoids

Input: Data set A and the number of clusters k to be computed.
Output: Solution to the k-partition problem.

1: *(Initialization)* For each data point $\boldsymbol{a} \in A$ calculate

$$f_{1,\boldsymbol{a}} = \sum_{\boldsymbol{b} \in A, \boldsymbol{b} \neq \boldsymbol{a}} d_p(\boldsymbol{a}, \boldsymbol{b}),$$

find $f_1^{\min} = \min_{\boldsymbol{a} \in A} f_{1,\boldsymbol{a}}$ and identify a data point $\bar{\boldsymbol{a}} \in A$ such that $f_{1,\bar{\boldsymbol{a}}} = f_1^{\min}$. Set $\boldsymbol{x}_1 = \bar{\boldsymbol{a}},\ s = 1$ and define the *set of selected points* $S = \{\bar{\boldsymbol{a}}\}$ and the *set of unselected points* $U = A \setminus S$.

(i) Set $s = s + 1$. If $s > k$, then go to Step 2 since the initial medoids $(\boldsymbol{x}_1, \ldots, \boldsymbol{x}_k)$ have been found.

(ii) For each data point $\boldsymbol{a} \in U$ calculate the value

$$f_{s,\boldsymbol{a}} = \sum_{\boldsymbol{b} \in U, \boldsymbol{b} \neq \boldsymbol{a}} \min \left\{ d_p(\boldsymbol{x}_1, \boldsymbol{b}), \ldots, d_p(\boldsymbol{x}_{s-1}, \boldsymbol{b}), d_p(\boldsymbol{a}, \boldsymbol{b}) \right\}.$$

Compute $f_s^{\min} = \min_{\boldsymbol{a} \in U} f_{s,\boldsymbol{a}}$ and find a point $\bar{\boldsymbol{a}} \in U$ such that $f_{s,\bar{\boldsymbol{a}}} = f_s^{\min}$. Set $\boldsymbol{x}_s = \bar{\boldsymbol{a}}$, the set of selected points $S = S \bigcup \{\bar{\boldsymbol{a}}\}$, the set of unselected points $U = A \setminus S$ and go to Step 1(i).

2: Assign each data point to its closest medoid, find the cluster partition $A^1, \ldots, A^k$ and compute the value $\hat{f}_k$ of the objective function f_k, given in the problem (5.5). Set $l = 1$.

3: Take the medoid $\boldsymbol{x}_l$. For each $\boldsymbol{a} \in U$, calculate

$$f_{l,\boldsymbol{a}} = \frac{1}{m} \sum_{\boldsymbol{b} \in U, \boldsymbol{b} \neq \boldsymbol{a}} \min \Big\{ d_p(\boldsymbol{x}_1, \boldsymbol{b}), \ldots, d_p(\boldsymbol{x}_{l-1}, \boldsymbol{b}), d_p(\boldsymbol{a}, \boldsymbol{b}), d_p(\boldsymbol{x}_{l+1}, \boldsymbol{b}), \ldots$$
$$, d_p(\boldsymbol{x}_k, \boldsymbol{b}) \Big\}.$$

Compute

$$f_l^{\min} = \min_{\boldsymbol{a} \in U} f_{l,\boldsymbol{a}},$$

and find a data point $\bar{\boldsymbol{a}}$ such that $f_{l,\bar{\boldsymbol{a}}} = f_l^{\min}$.

4: If $f_l^{\min} < \hat{f}_k$, then set $\boldsymbol{x}_l = \bar{\boldsymbol{a}}$ and $\hat{f}_k = f_l^{\min}$. Update the sets $S = S \setminus \{\boldsymbol{x}_l\} \bigcup \{\bar{\boldsymbol{a}}\}$ and $U = U \setminus \{\bar{\boldsymbol{a}}\} \bigcup \{\boldsymbol{x}_l\}$.

5: If $l < k$, set $l = l + 1$ and go to Step 3.

6: *(Stopping criterion)* If $\hat{f}_k < f_k$, then go to Step 2. Otherwise **stop**.

5.5 Fuzzy c-Means Algorithm

Hard clustering approaches generate partitions or groups where each data point belongs to one and only one cluster. *Fuzzy clustering* extends the idea into the *multi-label* domain where data points may belong simultaneously to many clusters. In practice, fuzzy clustering associates each data point with every cluster using a membership function. The output of the fuzzy clustering algorithms is, therefore, a clustering rather than a partition.

Given the data set A, the problem of finding c fuzzy clusters is formulated as the optimization problem

$$\begin{cases} \text{minimize} & U_c(W) \\ \text{subject to} & w_{ij} \in [0, 1], \quad i = 1, \ldots, m, \; j = 1, \ldots, c, \\ & \sum_{j=1}^{c} w_{ij} = 1, \quad i = 1, \ldots, m, \end{cases} \tag{5.6}$$

where

$$U_c(W) = \sum_{i=1}^{m} \sum_{j=1}^{c} w_{ij}^q d_p(\boldsymbol{x}_j, \boldsymbol{a}_i).$$

Here, $q > 1$ is a predefined real number—the so-called fuzzifier—and $W = [w_{ij}]$, $i = 1, \ldots, m$, $j = 1, \ldots, c$ is the $m \times c$ membership matrix. The *fuzzy cluster centers* $\boldsymbol{x}_1, \ldots, \boldsymbol{x}_c$ are defined as

$$\boldsymbol{x}_j = \frac{\sum_{i=1}^{m} w_{ij}^q \boldsymbol{a}_i}{\sum_{i=1}^{m} w_{ij}^q}, \quad j = 1, \ldots, c. \tag{5.7}$$

The design of the *membership values* w_{ij}—and thus, the *membership matrix* W—is an important problem in fuzzy clustering. One widely used formula for computing w_{ij} is

$$w_{ij} = \frac{1}{\sum_{t=1}^{c} \left(\frac{\|\boldsymbol{a}_i - \boldsymbol{x}_j\|}{\|\boldsymbol{a}_i - \boldsymbol{x}_t\|} \right)^{\frac{2}{q-1}}}, \quad i = 1, \ldots, m, \; j = 1, \ldots, c. \tag{5.8}$$

The *fuzzifier* q determines the level of cluster fuzziness. Large values of q result in smaller membership values w_{ij}. If there is no any prior information, one can take $q = 2$.

A fuzzy clustering algorithm usually selects an initial fuzzy partition of m data points into c clusters by initializing the membership matrix W, computes the value of the fuzzy objective function $U_c(W)$, and reassigns data points to clusters to

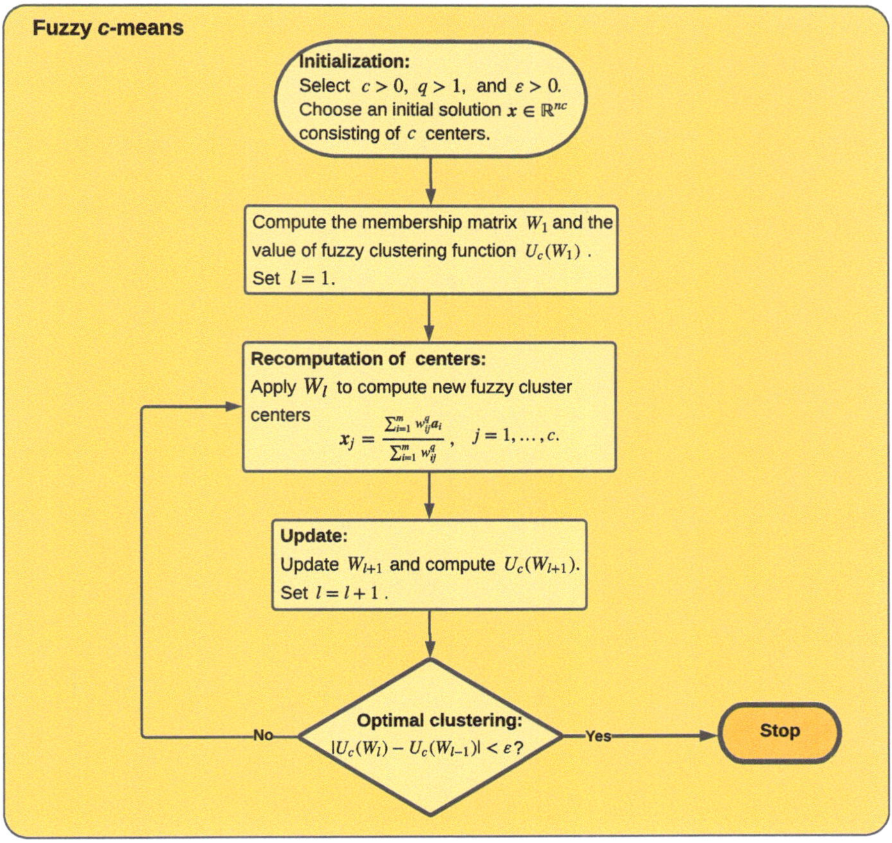

Fig. 5.5 Fuzzy c-means algorithm

reduce this objective function. A common fuzzy clustering algorithm is the *fuzzy c-means algorithm*. The flowchart of this algorithm is given in Fig. 5.5.

The fuzzy c-means algorithm is an extension of the k-means algorithm. It is also referred as the *soft clustering* or *soft k-means algorithm*. This algorithm was first introduced by Dunn in 1973 [93] and was modified by Bezdek in 1981 [47]. Similar to k-means, the d_2 similarity measure is usually used with the fuzzy c-means algorithm. In addition, the variants of the fuzzy c-means algorithm applying similarity measures with the L_1- and L_∞-norms are given in [50, 116, 306]. The fuzzy c-means algorithm is widely used, for instance, in pattern recognition. The step by step description of this algorithm is given next.

Algorithm 5.6 Fuzzy c-means algorithm

Input: Data set A, the number of clusters c to be computed and a tolerance $\varepsilon > 0$.
Output: Solution to the c-clustering problem and the membership matrix W.

1: *(Initialization)* Select c initial cluster centers $(\boldsymbol{x}_1, \dots, \boldsymbol{x}_c) \in \mathbb{R}^{nc}$. Apply (5.8) to compute the membership matrix W_1. Then compute the value of the objective function $U_c(W_1)$. Set $l = 1$.
2: Apply (5.7) and the membership matrix W_l to compute new cluster centers $\boldsymbol{x}_1, \dots, \boldsymbol{x}_c$.
3: Update the membership matrix W_{l+1} and compute the value of the objective function $U_c(W_{l+1})$.
4: *(Stopping criterion)* If

$$|U_c(W_{l+1}) - U_c(W_l)| < \varepsilon,$$

then **stop**. Otherwise set $l = l + 1$ and go to Step 2.

Note that, if we use d_2 as the similarity measure and $q \to 1$, then in the equation (5.8) for each data point $\boldsymbol{a}_i \in A$ the coefficients w_{ij} become either 1 or 0. This means that the fuzzy clustering problem becomes the hard clustering problem and the fuzzy c-means algorithm becomes the k-means algorithm. In addition, usually for any given data point the membership value for one cluster is significantly greater than its values for all other clusters. This shows a higher confidence in the assignment of that point to this cluster. Therefore, using the largest values of the membership function we can replace the fuzzy cluster distribution by the hard cluster distribution.

5.6 Clustering Algorithms Based on Mixture Models

Finite mixture models are a class of probability distribution formed by a convex combination of two or more probability density functions. They are initially developed by Newcomb in 1886 [226] and Pearson in 1894 [233], and later extended for solving regression [240] and clustering problems [43, 45, 49, 51, 104, 196, 211, 212, 296].

To some extent, partitional clustering algorithms based on the mixture models can be considered as fuzzy clustering algorithms. However, the probabilities of each data point being a member of a particular cluster are used to define the membership matrix in algorithms based on the mixture models.

5.6.1 *Mixture Models*

In the mixture model approach, it is assumed that data points arise from $k \geq 2$ distinct random processes. Each of these processes is modelled by a specific density function. Let z be a *random variable.* A *density function* φ is a mixture of k components $\psi_1, \ldots, \psi_k$ if

$$\varphi(z) = \sum_{j=1}^{k} \lambda_j \psi_j(z), \tag{5.9}$$

where λ_j are the *mixing weights* satisfying the conditions

$$\sum_{j=1}^{k} \lambda_j = 1, \quad 0 \le \lambda_j \le 1, \quad j = 1, \ldots, k.$$

In practice, it is usually assumed that the density functions ψ_j are of parametric form, that is they depend on some parameter $\theta_j,\ j = 1, \ldots, k$. In general, these parameters are unknown. Then (5.9) can be written as

$$\varphi(z, \boldsymbol{\theta}) = \sum_{j=1}^{k} \lambda_j \psi_j(z, \theta_j).$$

Here, ψ_j are called *probability density functions,* $j = 1, \ldots, k$ and the overall *parameter vector* is $\boldsymbol{\theta} = (\lambda_1, \ldots, \lambda_k, \theta_1, \ldots, \theta_k)$.

Clustering algorithms based on mixture models are partitional model-based algorithms. Assume that the number of clusters k is predefined. Then the data points to be clustered are drawn from a mixture of k clusters in some unknown proportions $\lambda_1, \ldots, \lambda_k$, that is, each data point $\boldsymbol{a} \in A$ is taken from a population whose probability density function is the *mixture probability density function* of the form

$$f(\boldsymbol{a}, \boldsymbol{\theta}) = \sum_{j=1}^{k} \lambda_j f_j(\boldsymbol{a}, \theta_j), \tag{5.10}$$

where $f_j(\boldsymbol{a}, \theta_j)$ is the probability density function of the jth component, $\boldsymbol{a}$ is a vector of input variables (data points), θ_j is the component specific parameter vector for the density function f_j, λ_j is the (unknown) mixing proportion—also known as a prior probability of the component j—and $\boldsymbol{\theta}$ is the vector of all parameters: $\boldsymbol{\theta} = (\lambda_1, \ldots, \lambda_k, \theta_1, \ldots, \theta_k)$.

The model (5.10) is considered as a finite mixture model density with the parameter vector $\boldsymbol{\theta}$. This parameter can be estimated, for instance, by the maximum likelihood method. Nevertheless, estimation of parameters $\theta_1, \ldots, \theta_k$ and coefficients $\lambda_1, \ldots, \lambda_k$, when $f_j,\ j = 1, \ldots, k$ are not the same parametric probability density functions, is a challenging problem. Therefore, from now on we assume that functions $f_j,\ j = 1, \ldots, k$ are represented using the same parametric distribution, that is, they are the same function $f_j \equiv \bar{f},\ j = 1, \ldots, k$ for some probability density function $\bar{f}$.

Various probability density functions can be used to design clustering algorithms based on the finite mixture model. The multivariate Gaussian mixtures are the most popular choices. In this case, parameters to be estimated are the mean value vector and the dispersion matrix. In addition, the beta and Bernoulli distributions have been used to design mixture models based clustering algorithms. Once the mixture model has been fitted, a probabilistic clustering of data into k clusters can be obtained in terms of the fitted posterior probabilities of component membership for data. An outright assignment of data into k clusters is achieved by assigning each data point to the component to which it has the highest estimated posterior probability of belonging.

5.6.2 Maximum Likelihood Estimation

The parameters $\theta_1, \ldots, \theta_k$ and coefficients $\lambda_1, \ldots, \lambda_k$ can be estimated using the *maximum likelihood* (ML) estimation by applying the expectation maximization algorithm. Given m independent points $\boldsymbol{a}_i \in A,\ i = 1, \ldots, m$, we can formulate a *likelihood function* as

$$L(\boldsymbol{\theta}) = \prod_{i=1}^{m} \left(\sum_{j=1}^{k} \lambda_j \bar{f}(\boldsymbol{a}_i, \theta_j) \right), \quad \text{or}$$

$$L_0(\boldsymbol{\theta}) \equiv \ln L(\boldsymbol{\theta}) = \sum_{i=1}^{m} \ln \left(\sum_{j=1}^{k} \lambda_j \bar{f}(\boldsymbol{a}_i, \theta_j) \right). \tag{5.11}$$

Now, the clustering problem becomes a ML estimation problem of the giving number of k clusters and the set A. The coefficients $\lambda_1, \ldots, \lambda_k$ and parameters $\theta_1, \ldots, \theta_k$ are estimated by maximizing the function L or, equivalently, the function L_0.

Functions L and L_0 are multi modal and may have many local maximizers. The standard procedure for finding the ML estimate—that is, to maximize the function L or L_0—is the EM algorithm. This algorithm is particularly applicable in the multi parameter situations.

5.6.3 Expectation Maximization Clustering Algorithm

The *expectation maximization* (EM) algorithm is the primary tool in finite mixture models and clustering algorithms based on these models [212]. The algorithm seeks to find the ML estimates iteratively applying two steps: expectation step (E-step) and maximization step (M-step). Then these estimates are used for computing weights

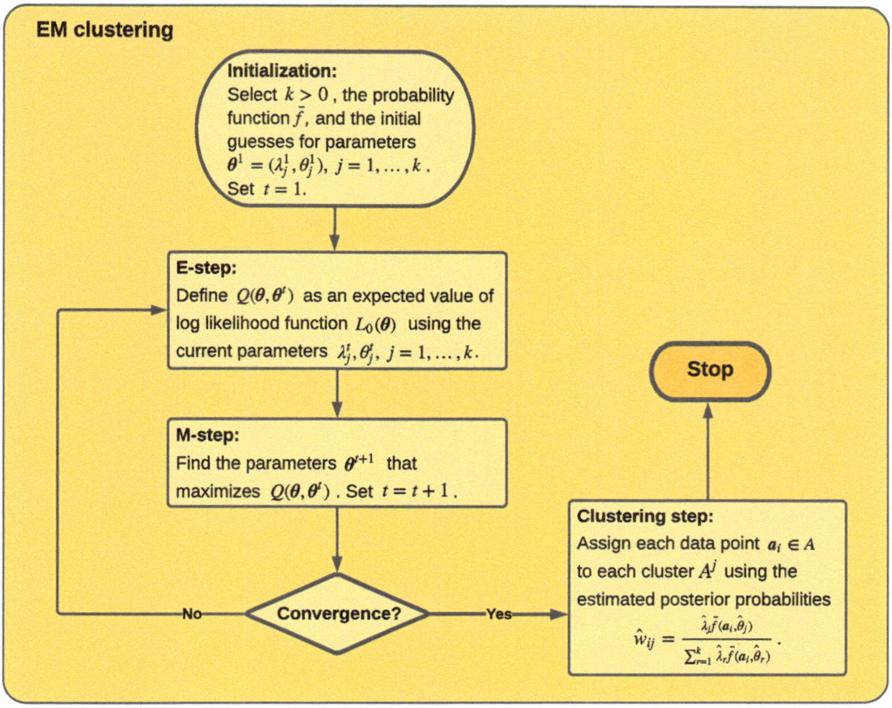

Fig. 5.6 Expectation maximization clustering algorithm

for cluster distribution. A flowchart of the EM clustering algorithm is given in Fig. 5.6.

The E-step estimates the expected value of the complete data log likelihood function (5.11) using the observed data $\boldsymbol{a}$ and the current parameter estimates $\lambda_j,\ \theta_j,\ j = 1, \ldots, k$. Let $\boldsymbol{\theta}^t = (\lambda_j^t,\ \theta_j^t),\ j = 1, \ldots, k$ be the parameters estimate at the tth iteration. At the next iteration, the EM algorithm calculates the function

$$Q(\boldsymbol{\theta}, \boldsymbol{\theta}^t) = \sum_{i=1}^{m} \ln \sum_{j=1}^{k} w_{ij}^t \lambda_j \bar{f}(\boldsymbol{a}_i, \theta_j), \tag{5.12}$$

where

$$w_{ij}^t = \frac{\lambda_j^t \bar{f}(\boldsymbol{a}_i, \theta_j^t)}{\sum_{r=1}^{k} \lambda_r^t \bar{f}(\boldsymbol{a}_i, \theta_r^t)}$$

is the *posterior probability* that the ith data point belongs to the jth component of the mixture after the tth iteration.

The M-step maximizes the expectation of log likelihood for each component separately using the posterior probabilities as weights. In the M-step the $Q(\boldsymbol{\theta}, \boldsymbol{\theta}^t)$ is maximized with respect to $\boldsymbol{\theta}$ and the $(t+1)$th iteration of the EM algorithm is defined as

$$\boldsymbol{\theta}^{t+1} = \underset{\boldsymbol{\theta} \in \Theta}{\operatorname{argmax}}\ Q(\boldsymbol{\theta}, \boldsymbol{\theta}^t).$$

Here, Θ denotes the set of parameters $\boldsymbol{\theta}$. The E-steps and M-steps are repeated until some prespecified stopping criterion is met. One criterion can be defined based on the convergence of the parameters $\boldsymbol{\theta}^t$; however, this might be too demanding if there is a large number of parameters. The other criterion—probably, the most usual stopping criterion—is to stop when the relative increase in the likelihood function is sufficiently small. In addition, the predefined maximum number of iterations can be used as a stopping criterion.

Once the estimates of λ_j and $\theta_j,\ j = 1, \ldots, k$ are obtained—we denote them as $\hat{\lambda}_j$ and $\hat{\theta}_j$, respectively—each data point $\boldsymbol{a}_i \in A$ can be assigned to the cluster A^j (using Bayes rule) via the estimated posterior probability

$$\hat{w}_{ij} = \frac{\hat{\lambda}_j \bar{f}(\boldsymbol{a}_i, \hat{\theta}_j)}{\sum_{r=1}^k \hat{\lambda}_r \bar{f}(\boldsymbol{a}_i, \hat{\theta}_r)}.$$

This process is considered as a fuzzy clustering of a point $\boldsymbol{a}_i$. In addition, we can form a deterministic clustering by applying the rule

- assign $\boldsymbol{a}_i$ to A^j, if $\hat{w}_{ij} > \hat{w}_{ir}$ for all $r = 1, \ldots, k, \quad r \neq j$.

Note that the EM algorithm is a local search algorithm and can converge only to local maximizers of the functions L and L_0 [76, 210]. Thus, there is no guarantee of finding the best cluster structure.

5.7 Self-Organizing Map Algorithm

Self-organizing map (SOM) is an unsupervised neural network [185] (see also [184]) that usually contains a 2-dimensional array of neurons. The SOM algorithm is widely used since it generates an intuitive two-dimensional map of a multidimensional data set. The flowchart of the method is given in Fig. 5.7.

Assume that we are given a set of input data vectors $A = \{\boldsymbol{a}_1, \ldots, \boldsymbol{a}_m\}$ $(\boldsymbol{a}_i \in \mathbb{R}^n)$ and a set of k *neurons* that are represented as k *weights* $W = \{\boldsymbol{w}_1, \ldots, \boldsymbol{w}_k\}$ $(\boldsymbol{w}_j \in \mathbb{R}^n)$. The data points $\boldsymbol{a}_i,\ i \in \{1, \ldots, m\}$ are presented to the network one at a time. The point $\boldsymbol{a}_i$ is compared with all weight vectors $\boldsymbol{w}_j,\ j = 1, \ldots, k$, and the nearest $\boldsymbol{w}_j$ is selected as the *best matching unit* (BMU) for this point. We say that the data

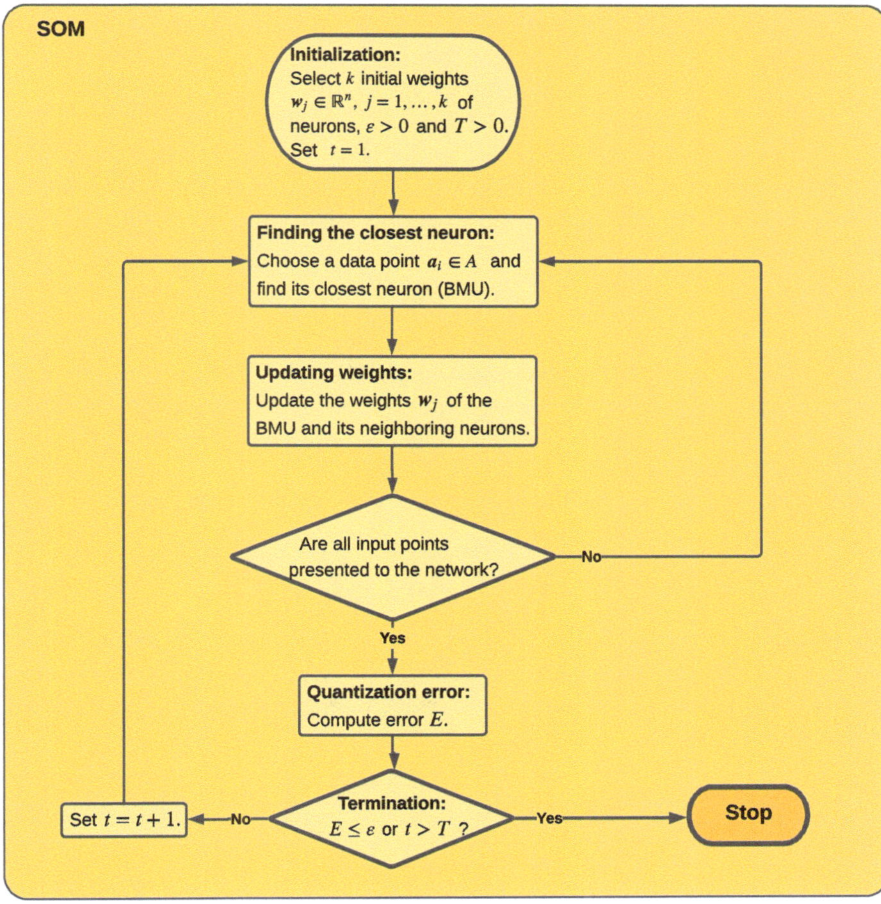

Fig. 5.7 Self-organizing map algorithm

point $\boldsymbol{a}_i$ is mapped to the best matching neuron $\boldsymbol{c}$ and denote the corresponding weight by $\boldsymbol{w_c}$.

The weights of the BMU are adjusted by

$$\boldsymbol{w}_j = \boldsymbol{w}_j + \alpha(t)\beta(t)(\boldsymbol{a}_i - \boldsymbol{w}_j), \quad j = 1, \ldots, k, \tag{5.13}$$

where β is a neighborhood function and α is a learning rate at the iteration t. Usually β is a decreasing exponential function of t. For instance, it can be defined as

$$\beta(t) = \exp\Big(-\frac{r^2}{2\sigma(t)^2}\Big),$$

where

$$\sigma(t) = \eta \frac{T - t}{T}, \quad \eta \geq 1.$$

The value of the function β depends on the iteration $t > 0$, the maximum number of iterations T given for the algorithm, and the distance r in the output space of each neuron in the set of neighborhood weights N_c. The set N_c around the BMU are selected such that

$$N_c = \{\boldsymbol{w}_l : d_{nt}(c, l) \leq r, \ l \neq c\},$$

where $d_{nt}(c, l) \in \mathbb{N}$ is the distance between the BMU and a neuron l in 2-dimensional coordinates of the network topology and $r > 0$ is the predefined radius. The learning rate α is a decreasing linear function of t that reduces the effect of the neighborhood function β as $t \to T$.

The quality of the map is usually measured by the quantization error

$$E = \frac{1}{m} \sum_{i=1}^{m} \|\boldsymbol{a}_i - \boldsymbol{w}_c\|, \tag{5.14}$$

where $\boldsymbol{w}_c$ is the weight of the BMU of $\boldsymbol{a}_i, \ i = 1, \dots, m$. The overall goal of the SOM algorithm is to minimize this error. A general description of the SOM algorithm is as follows.

Algorithm 5.7 Self-organizing map algorithm

Input: Data set A and the number of clusters k to be computed.
Output: Set of k weights $\boldsymbol{w}_j, \ j = 1, \dots, k$ of neurons.

1: *(Initialization)* Initialize the maximum number of iterations T, a radius r of the network and weight vectors $\boldsymbol{w}_j, j = 1, \dots, k$. Set stopping tolerance $\varepsilon > 0$ and the iteration counter $t = 1$.
2: Select a data point $\boldsymbol{a}_i, \ i = 1, \dots, m$ and find its closest neuron $\boldsymbol{c}$ (BMU), where $\boldsymbol{c}$ is

$$\boldsymbol{c} = \operatorname*{argmin}_{j=1,\dots,k} \|\boldsymbol{a}_i - \boldsymbol{w}_j\|.$$

Denote the corresponding weight by $\boldsymbol{w}_c$.
3: Set $\boldsymbol{w}_j = \boldsymbol{w}_c$. Update the weights of the BMU and its neighboring neurons using

$$\boldsymbol{w}_j = \boldsymbol{w}_j + \alpha(t)\beta(t)(\boldsymbol{a}_i - \boldsymbol{w}_j),$$

where β is a neighborhood function and α is a learning rate at the iteration t.
4: If all input data are presented to the network go to Step 5, otherwise go to Step 2.
5: *(Stopping criterion)* Calculate E using (5.14). If $E \leq \varepsilon$ or $t > T$, then **stop**. Otherwise set $t = t + 1$ and go to Step 2.

Algorithm 5.7, in general, generates a suboptimal partition if the initial weights are not properly selected. Therefore, the choice of initial weights is very important. Several algorithms have been introduced for initialization of weights in the SOM algorithm [217–219]. In addition, a global optimization approach for the determination of weights of layered feed-forward networks is introduced in [265]. The description of other neural network architectures for the learning of recognition categories can be found in [60, 62].

Chapter 6
Metaheuristic Clustering Algorithms

6.1 Introduction

Partitional clustering is a global optimization problem. However, traditional clustering algorithms such as k-means and other heuristics described in the previous chapter can only guarantee convergence to a local solution which may not reflect an existing cluster structure of a data set. There have been some attempts to reformulate the clustering problem in order to solve it globally. For instance, in [261] the clustering function is reformulated as an equivalent implicit concave function. Then the concept of convexity cuts is implemented to obtain the optimal solution. This approach leads to an exact algorithm for clustering; however, the algorithm is very time-consuming, requires a large memory, and is not applicable in large data sets. In addition, clustering algorithms based on the dynamic programming [159, 244] and branch and bound techniques [84, 186] are applicable only in small data sets [140].

Metaheuristic algorithms are well-known optimization tools for global optimization. They can handle both discrete and continuous variables, and they have been widely applied for solving clustering problems. In this chapter, we consider both single point-based and population-based—also known as evolutionary algorithms—metaheuristics.

The single point-based metaheuristics applied to solve clustering problems include tabu search and simulated annealing. These algorithms start from an initial solution, use some probabilities to escape from local solutions, and keep record on the best solution obtained. The tabu search and simulated annealing algorithms for clustering are described in Sects. 6.2 and 6.3, respectively.

In the rest of this chapter, we consider evolutionary algorithms. These algorithms are population-based metaheuristics. That is instead of just one initial solution, we have a set of initial solutions. Among the evolutionary algorithms, the genetic algorithm, the artificial bee colony optimization, the particle swarm optimization,

A. M. Bagirov et al., *Partitional Clustering via Nonsmooth Optimization*,
Unsupervised and Semi-Supervised Learning,
https://doi.org/10.1007/978-3-030-37826-4_6

and the ant colony optimization algorithms are frequently used in clustering. The descriptions of these algorithms for clustering are given in Sects. 6.4–6.7.

Evolutionary algorithms are easy to implement and well suited for clustering. These algorithms have both local and global search abilities. The evolutionary algorithms are based on optimization of an objective function (fitness function) that guides the evolutionary search. In the case of clustering, different fitness functions can be formulated using various models of the clustering problems.

Evolutionary algorithms can be applied either directly to the clustering problem, considering it as a global optimization problem or in combination with other clustering algorithms to improve the quality of solutions obtained by the latter algorithms. However, the evolutionary clustering algorithms become inaccurate and may require large computational effort in large data sets. In these cases, they can still be applied to generate good starting cluster centers to be used by other clustering algorithms which may lead to a design of efficient and accurate clustering algorithms also for large data sets.

The comparison of clustering algorithms based on different metaheuristics is presented in [11]. Survey of these algorithms can be found, for instance, in [148, 252]. The description of metaheuristics for clustering not included in this book can be found, for example, in [74, 125, 137, 138, 276]. The theoretical issues of convergence of evolutionary algorithms are discussed in [107].

6.2 Tabu Search Clustering Algorithm

Tabu search (TS) is a metaheuristic algorithm that explores the solution space beyond the local optimality and is not trapped in local solutions. The TS method was introduced in 1986 [118] and then modified in 1989 by Glover [119–122]. It can handle both discrete and continuous variables and, therefore, it is applicable to solve combinatorial optimization problems. The TS based clustering algorithm was first developed in 1995 [9], and then it was extended to solve the fuzzy clustering problem in 1997 [10]. The basic idea of the TS clustering algorithm is given in Fig. 6.1.

The TS algorithm includes the following elements. It starts from a single starting point. *A move* is a procedure which generates a new trial point using the current point. The set of all possible moves out of the current point constitutes the *set of candidate moves*. It is possible to make a move out of the current iteration point to a point where the value of the objective function is greater (or equal) than its value at the current point. This property of the TS method makes it suitable for solving nonconvex clustering problems. The *tabu restrictions* are certain conditions imposed on the set of candidate moves to make some of them forbidden. All such moves constitute the *tabu list*. On the other hand, *aspiration criteria* are rules that override tabu restrictions, that is, if a certain move is on the tabu list, then the aspiration criterion, when satisfied, can make this move allowable.

Thus, the TS (clustering) algorithm starts with an initial point and computes the value of the objective function at this point. Then, it generates a set of possible

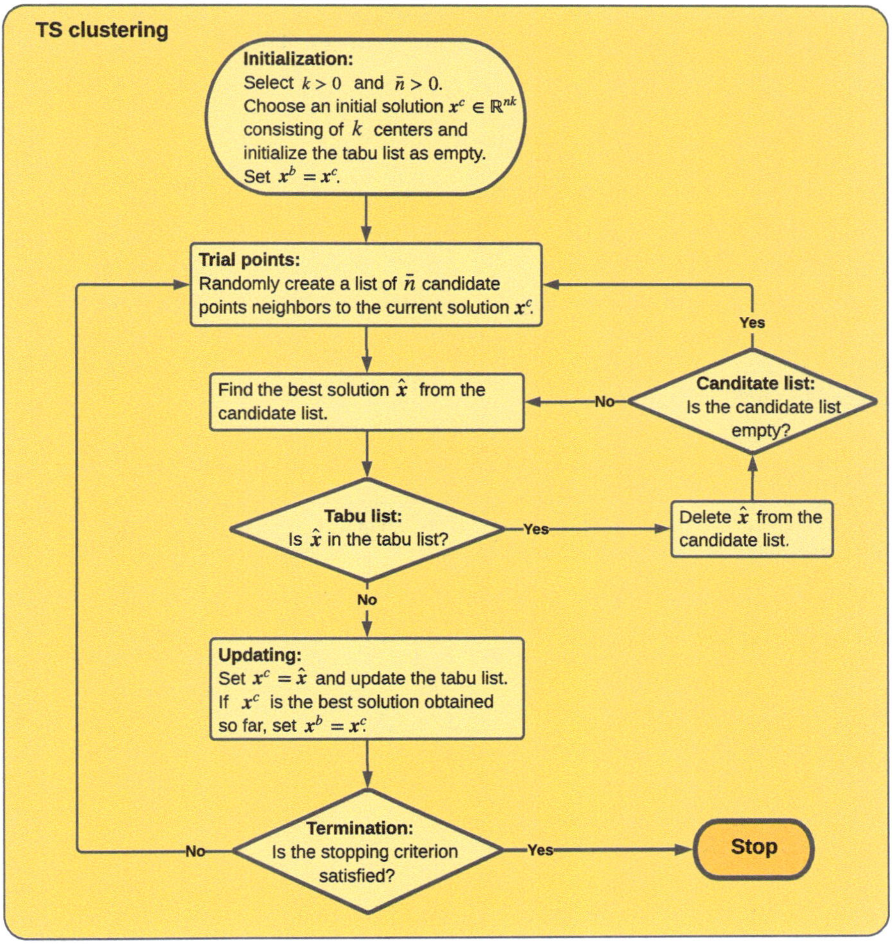

Fig. 6.1 Tabu search clustering algorithm

moves and their tabu list. If the best of these moves is not tabu—or if the best is tabu, but satisfies the aspiration criterion—then the algorithm uses that move to determine a new point. Otherwise, the algorithm selects the best move which is not on the tabu list and finds a new point using it. The algorithm repeats this procedure until the number of iterations reaches its maximum value. Then the best solution obtained so far is accepted as the solution for the clustering problem.

The move picked at a given iteration is added to the tabu list to avoid possible reverse moves. The maximum number of moves in the tabu list is restricted above, and when this number is reached the first move is removed from the list. Usually, the objective function value is used to define the aspiration criterion. For instance, if a tabu move leads to a better solution than the best solution so far, the aspiration criterion is satisfied and the tabu restriction is overridden.

The TS clustering algorithm uses the mixed integer formulation (4.2) with the squared Euclidean distance d_2 as an objective function. Therefore, the centroids $\boldsymbol{x}_j$ can be computed as

$$\boldsymbol{x}_j = \frac{\sum_{i=1}^{m} w_{ij}\boldsymbol{a}_i}{\sum_{i=1}^{m} w_{ij}}, \quad j = 1, \ldots, k,$$

where $w_{ij} = 1$ if the data point $\boldsymbol{a}_i$ is allocated to the cluster A^j and $w_{ij} = 0$, otherwise.

Next, we give the step by step form of the TS clustering algorithm. In what follows the current iteration point, trial point, and the best solution are denoted by $\boldsymbol{x}^c$, $\boldsymbol{x}^t$, and $\boldsymbol{x}^b$, respectively. Consequently, ζ_k^c, ζ_k^t, and ζ_k^b are the values of the objective function ζ_k at these points.

Algorithm 6.1 Tabu search clustering algorithm

Input: Data set A and the number of clusters k to be computed.
Output: Solution to the k-partition problem.

1: *(Initialization)* Select the length of the tabu list n_{tl}, the number of trial points $\bar{n}$, the probability threshold P and the maximum number of iterations $j_{\max}$.

2: Select an arbitrary starting cluster center $\boldsymbol{x}_1^c = (\boldsymbol{x}_{1,1}^c, \ldots, \boldsymbol{x}_{1,k}^c)$ and compute the value $\zeta_{k,1}^c$ of the objective function ζ_k at $\boldsymbol{x}_1^c$. Set $\boldsymbol{x}^b = \boldsymbol{x}_1^c$ and $\zeta_k^b = \zeta_{k,1}^c$. Set the initial value for the length of the tabu list $t_{\text{tl}} = 0$ and $j = 1$.

3: *(Set of candidate moves)* Using $\boldsymbol{x}_j^c$ and P randomly generate $\bar{n}$ trial points $\boldsymbol{x}_1^t, \ldots, \boldsymbol{x}_{\bar{n}}^t$ (see Remark 6.1). Compute the values $\zeta_{k,1}^t, \ldots, \zeta_{k,\bar{n}}^t$ of the objective function ζ_k at these points.

4: Arrange these values in an ascending order. Denote this order as $\zeta_{[1]}^t, \ldots, \zeta_{[\bar{n}]}^t$ and their corresponding points as $\boldsymbol{x}_{[1]}^t, \ldots, \boldsymbol{x}_{[\bar{n}]}^t$.

5: *(Tabu restrictions and aspiration rule)* If the trial point corresponding to the value $\zeta_{[1]}^t$ is not on the tabu list or it is on the tabu list but $\zeta_{[1]}^t < \zeta_k^b$, then set $\boldsymbol{x}_{j+1}^c = \boldsymbol{x}_{[1]}^t$, $\zeta_{k,j+1}^c = \zeta_{[1]}^t$ and go to Step 7.

6: Select $l,\ l \in \{1, \ldots, \bar{n}\}$ such that $\zeta_{[l]}^t$ is the best objective function value which is not on the tabu list. If all $\zeta_{[1]}^t, \ldots, \zeta_{[\bar{n}]}^t$ are on the tabu list, then go to Step 3. Otherwise, set $\boldsymbol{x}_{j+1}^c = \boldsymbol{x}_{[l]}^t$, $\zeta_{k,j+1}^c = \zeta_{[l]}^t$.

7: Add $\boldsymbol{x}_{j+1}^c$ to the bottom of the tabu list and set $t_{\text{tl}} = t_{\text{tl}} + 1$. If $t_{\text{tl}} > n_{\text{tl}}$, then delete the first element from the tabu list and set $t_{\text{tl}} = t_{\text{tl}} - 1$. If $\zeta_k^b > \zeta_{k,j+1}^c$, then set $\boldsymbol{x}^b = \boldsymbol{x}_{j+1}^c$ and $\zeta_k^b = \zeta_{k,j+1}^c$.

8: *(Stopping criterion)* If $j = j_{\max}$, then **stop**. $\boldsymbol{x}^b$ is the best solution found and ζ_k^b is the corresponding best value. Otherwise, set $j = j + 1$ and go to Step 3.

Remark 6.1 Different strategies can be used to generate points $\boldsymbol{x}_1^t, \ldots, \boldsymbol{x}_{\bar{n}}^t$ in Step 3 of Algorithm 6.1. One strategy is as follows. Let $\mathcal{A}$ be an array of dimension m whose ith element $\mathcal{A}_i$ is an index j representing the cluster A^j, $j = 1, \ldots, k$ to which $\boldsymbol{a}_i$ is allocated. Clearly, we have $w_{ij} = 1$, if $\mathcal{A}_i = j$ and 0, otherwise. Denote by $\mathcal{A}^c$ the current solution and by $\mathcal{A}^t$ the trial point. For all $i = 1, \ldots, m$ generate a random number u from the uniform distribution $U[0, 1]$. If $u < P$, then set $\mathcal{A}_i^t = \mathcal{A}_i^c$. Otherwise, draw randomly an integer $\bar{l}$ from the set $\{l : l \in \{1, \ldots, k\},\ l \neq \mathcal{A}_i^c\}$ and set $\mathcal{A}_i^t = \bar{l}$. Generate a new trial point $\boldsymbol{x}^t$ and repeat the process $\bar{n}$ times.

6.3 Simulated Annealing Clustering Algorithm

Simulated annealing (SA) is a well-known stochastic technique for approximating the global optimum of a given function in a large search space. Its name and inspiration come from the annealing process in metallurgy. Annealing is the process of heating up a solid to a high temperature followed by slow cooling achieved by decreasing of the temperature. This process consists of several steps, and at each step the temperature is kept a constant for a period of time sufficient to reach the thermal equilibrium. At equilibrium the solid may have many configurations, each corresponding to a specific energy level.

The SA method for optimization was introduced in 1983 by Kirkpatrick et.al. [178] and then modified in 1985 by Cerny [65] for solving discrete optimization problems. Later, it was extended to solve global optimization problems with continuous variables [1]. Under some mild conditions it is proved that the SA method converges to the global solution with the probability one [1, 201].

The SA method is well suited for solving clustering problems. In this context it was first introduced in 1991 by Selim and Al-Sultan [262] (see also [182]) for solving the MSSC problems using their integer programming model. Later, different versions of the SA clustering method are developed, for instance, in [54, 275]. The basic idea of the SA clustering algorithm is given in Fig. 6.2.

Next, we describe the SA clustering algorithm in more detail. We use the nonsmooth nonconvex optimization model of the clustering problem (4.3) with the squared Euclidean distance d_2. The SA clustering algorithm randomly selects an initial solution $\boldsymbol{x}^c \in \mathbb{R}^{nk}$ consisting of k cluster centers. This algorithm is designed to escape from solutions which are only local minima of the objective function. This is achieved by accepting—with some probability—a new solution with a lower quality. The probability of acceptance is governed by a parameter called the *temperature*.

The SA clustering algorithm consists of *inner* and *outer* iterations. The outer iterations can be considered as a global search phase of the method whereas the inner iterations are its local search phase. In the outer iteration the temperature T, which is analogous to the temperature in the physical process of annealing, is updated. With this aim, an arbitrary initial value T_1 for the temperature and a temperature multiplier $r \in (0, 1)$ are taken, and the temperature is updated by the formula $T_{j+1} = rT_j,\ j = 1, 2, \ldots$. Thus, the temperature starts with its highest value at the first iteration and reduces to its final value at the last iteration.

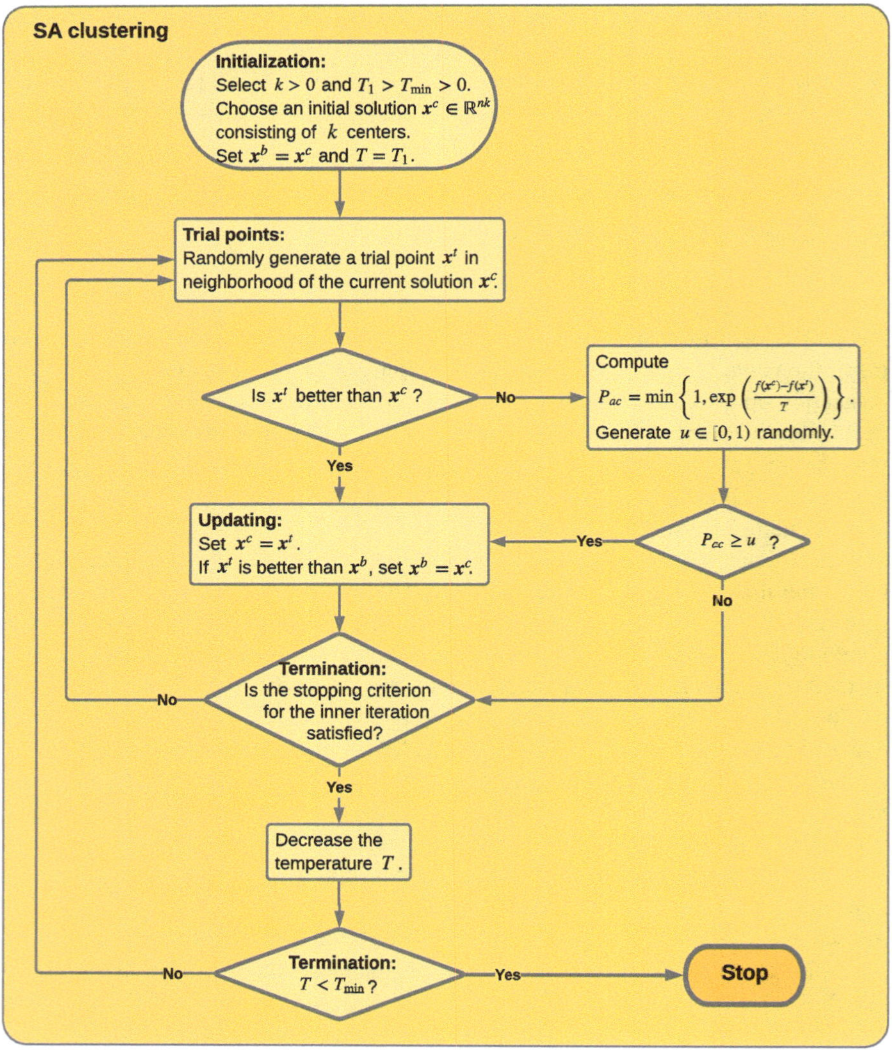

Fig. 6.2 Simulated annealing clustering algorithm

In the inner iteration, the current solution x^c is modified to generate a new trial solution x^t. If the trial solution reduces the value of the objective, then it is accepted as the new solution. If it increases the value of the objective, then the trial solution can still be accepted with an *acceptance probability*

$$P_{ac} = \min\left\{1, \exp\left(\frac{f_k(x^c) - f_k(x^t)}{T}\right)\right\}.$$

More precisely, a random number u from the uniform distribution $U[0, 1]$ is generated. If $P_{ac} \geq u$, then the trial solution is accepted as a new solution; otherwise the inner iterations are repeated. The SA clustering algorithm keeps record on the best solution $\boldsymbol{x}^b$ obtained so far and the corresponding value of the objective function f_k^b.

Remark 6.2 There are various ways for generating points $\boldsymbol{x}^t$ from some neighborhood of the point $\boldsymbol{x}^c$ in Step 3 of Algorithm 6.2. Nevertheless, it is worth of noting that after few iterations of the algorithm, it can be expected that the current solution $\boldsymbol{x}^c$ will provide much lower objective function value than a random point from the search space. Therefore, as a general rule, the trial points should be skewed toward the current solution as the solution process proceeds.

Algorithm 6.2 Simulated annealing clustering algorithm

Input: Data set A and the number of clusters k to be computed.
Output: Solution to the k-partition problem.

1: *(Initialization)* Select the initial value T_1 and the final value $T_{\min} < T_1$ of the temperature, a multiplier $r \in (0, 1)$ for scheduling the temperature, and the maximum number $j_{\max}$ of steps for each inner iteration.

2: Select an arbitrary starting cluster center $\boldsymbol{x}^c = (\boldsymbol{x}_1^c, \ldots, \boldsymbol{x}_k^c) \in \mathbb{R}^{nk}$ and compute the value f_k^c of the objective function f_k at $\boldsymbol{x}^c$. Set $\boldsymbol{x}^b = \boldsymbol{x}^c$, $f_k^b = f_k^c$, $T = T_1$ and $j = 1$.

3: Randomly generate a trial point $\boldsymbol{x}^t$ in some neighborhood of the current iteration point $\boldsymbol{x}^c$, and compute the value f_k^t of the function f_k at this point.

4: If $f_k^t \geq f_k^c$, then go to Step 5. Otherwise, set $\boldsymbol{x}^c = \boldsymbol{x}^t$ and $f_k^c = f_k^t$. If $f_k^t \geq f_k^b$, then set $j = j + 1$ and go to Step 7. Otherwise, set $\boldsymbol{x}^b = \boldsymbol{x}^t$, $f_k^b = f_k^t$, $j = 1$ and go to Step 3.

5: Generate a random number u from $U[0, 1]$ and compute

$$P_{ac} = \min\left\{1, \exp\left(\frac{f_k^c - f_k^t}{T}\right)\right\}.$$

If $P_{ac} \geq u$, then set $\boldsymbol{x}^c = \boldsymbol{x}^t$ and $f_k^c = f_k^t$.

6: Set $j = j + 1$.

7: *(Inner iteration termination)* If $j \leq j_{\max}$, then go to Step 3.

8: *(Outer iteration termination)* Set $T = rT$. If $T < T_{\min}$, then **stop** with $\boldsymbol{x}^b$. Otherwise go to Step 3.

The choice of parameters in the SA clustering algorithm is very important to obtain high quality solutions. These parameters include the initial value T_1 of the temperature, the temperature multiplier $r \in (0, 1)$, and the maximum number $j_{\max}$ of steps in inner iterations. The parameter T_1 is usually chosen large in order to accept many solutions at the early stage of the solution process and to explore the search space. The number r should not be very close to 1 to avoid the large number of outer iterations. For instance, $r = 0.5$ can be a reasonably good choice. Finally, the number $j_{\max}$ should be large enough to allow the method to intensively explore the neighborhood of the current solution. This number depends on the number n of decision variables and $j_{\max} = 2n$ is a reasonable choice.

6.4 Genetic Algorithm for Clustering

Among all evolutionary algorithms *the genetic algorithm* (GA) has been most widely applied to design clustering algorithms [72, 102, 148, 199, 209, 224, 242, 252]. The GA was introduced by Goldberg in 1980s [123], although, some of its ideas appeared already in 1970s [146]. The GA is based on the Darwin theory rule: "the strongest species that survives" and "the survival of an organism can be maintained through the process of reproduction, crossover and mutation".

The GA uses a number of approaches, known as *evolutionary approaches*, which are motivated by the ideas of natural selection and evolution. An important component of the GA is a *fitness function.* This function is used to measure the suitability of a solution generated by the GA. In most cases the fitness function may coincide with the objective function of an optimization problem under consideration. Evolutionary approaches include operators such as *selection, recombination,* and *mutation* on a population—usually encoded as *chromosomes*—to obtain a global solution to an optimization problem [115, 123, 124].

The GA starts by generating an initial population. The solutions in this population—chromosomes (individuals) made up from *genes*—are represented as strings using binary, integer, or real encodings. The *selection* operator propagates solutions from a generation to the next one based on their fitness. The selection usually uses a probabilistic scheme where chromosomes with higher fitness have a higher probability of being reproduced in the next generation. The *crossover* operator as a form of recombination exchanges subsequences of the bit strings between parents and is effective at exploring the search space. Finally, the *mutation* acts as a fine tuning for the exploration process. This means that the population is evolving toward better solutions.

When applying the GA to clustering, the population consists of different partitions to the clustering problem (chromosomes), and each chromosome consists of the cluster representatives (genes). The general scheme of the GA for clustering is given in Fig. 6.3.

The GA for clustering starts with the initialization of the starting population—a set of candidate solutions (partitions) to the clustering problem. This population can be generated both deterministically and randomly, or as a combination of these two. The size of the population (i.e., the number of chromosomes) depends on the number of clusters, the number of data points, and other characteristics of a data set. However, this number cannot be chosen very large as this may make solving the clustering problem very time-consuming. Then by using the current population the algorithm applies the selection, crossover, and mutation operations to generate the next population. At the same time, the algorithm stores the best solution found so far and injects it into the population after each operation. This procedure is called

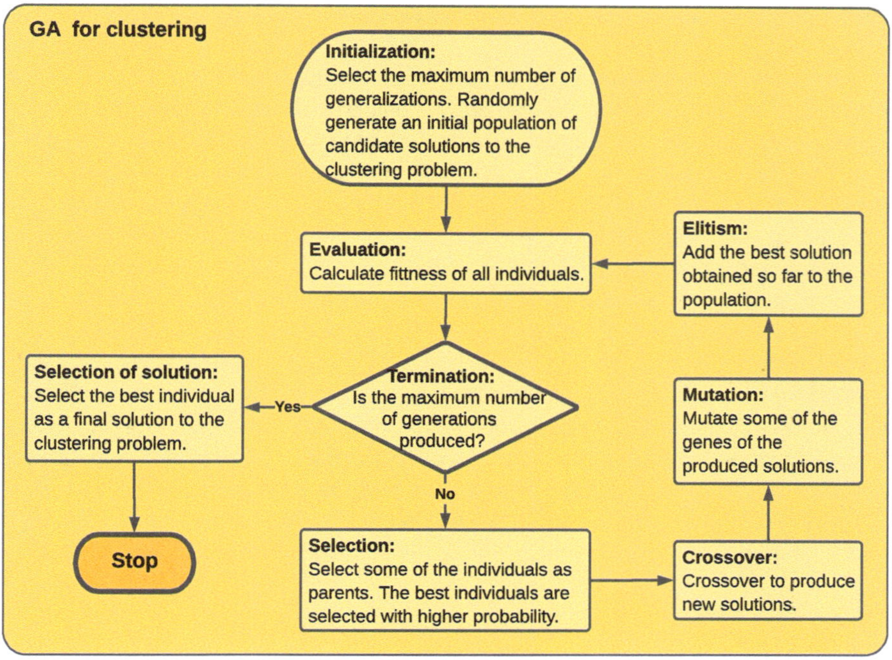

Fig. 6.3 Genetic algorithm for clustering

the *elitism* and it guarantees that the solution quality obtained by the GA will not decrease from a generation to the next one. The algorithm terminates when the number of iterations reaches its maximum value.

The GA can be applied to solve different types of clustering problems such as centroid based and medoid based clustering. Depending on the type of the clustering problem, different fitness functions can be used in the GA. For example, for the medoid based clustering problem we can use the nonsmooth clustering function (4.4) with the squared Euclidean distance d_2 and restrict the genes to be data points (i.e., possible medoids). On the other hand, for the MSSC problem the same fitness function can be used with no such a restriction.

The GA is able to solve clustering problems in data sets with both numeric and categorial attributes [69]. In some versions of the GA for clustering the number of clusters needs to be given and fixed beforehand, while some other versions define this number during the solution process.

In many cases the GA performs better than the k-means algorithm and its variations. Unlike most traditional clustering algorithms, the GA for clustering performs a global search and theoretically is able to find a global solution. Thus, we could expect that the GA for clustering is more accurate than the most traditional algorithms. However, this is not always the case as we need to restrict the number of iterations as well as the size of the initial population. This means that the GA for clustering may find suboptimal solutions to clustering problems, especially in large data sets.

Another challenge when using the GA for clustering is the representation of the problem using the bit strings. There are various approaches [48, 162, 241] to this representation and also to address crossover problems. Such difficulties restrict the use of the GA for clustering to a small number of clusters. The other drawback of the GA for clustering is its sensitivity to the selection of the population size, and the mutation and crossover probabilities. Parameters of the GA need to be fine-tuned for each specific clustering problem type and, to some extent, to a data set. This task has been studied, for instance, in [127]; however, there are still difficulties and open questions to be answered in order to achieve good results on specific problems.

In addition to using the GA for solving entire clustering problem, it can be applied together with some local search clustering algorithms. In [162], it is shown that the hybrid GA incorporating problem specific heuristics is good for clustering. For instance, one approach is to use the GA to generate good initial cluster centers and then to apply the k-means algorithm to find the final partition [16]. Another approach is to apply the GA to identify both the number of clusters and the high quality initial cluster centers which can be used by the k-means algorithm [242].

6.5 Artificial Bee Colony Clustering Algorithm

Artificial bee colony (ABC) algorithm is a metaheuristic for solving global optimization problems. It was first introduced in 2005 by Karaboga [44, 164] and then modified and improved in [165]. The ABC algorithm can be used to solve both unconstrained and constrained optimization problems.

The ABC algorithm simulates the intelligent foraging behavior of honey bee swarms. In the artificial bee swarm there are three types of bees: employed bees, onlookers, and scouts. Onlookers and scouts are also called the *unemployed bees*. An *employed bee* takes a particular food source to exploit and shares its information with onlookers in the nest. An *onlooker* waits in the nest and evaluates a food source using the information shared by the employed bee. Finally, a *scout* looks for a new food source in the search space. A *food source* means a possible solution for the optimization problem. Therefore, in the ABC algorithm it is important to estimate the quality (fitness)—the *nectar amount*—of a food source. In the algorithm, the employed bees and onlookers implement the exploitation process while the scouts execute the exploration process.

It is assumed in the ABC algorithm that for each food source there is only one employed bee and, thus, the number of employed bees is the same as the number of food sources around the hive. In addition, the number of onlookers is the same as the number of employed bees. An employed bee generates an alteration on the position of the food source in her memory and checks the nectar amount of the new source. If this amount is higher than that of the previous one, the bee memorizes the new position and disregards the old one. Otherwise, she holds the previous position in her memory.

An onlooker bee evaluates the nectar information obtained from all employed bees and selects a food source with a probability depending on its nectar amount. When the nectar of a food source is exhausted it is *abandoned*. Then the corresponding employed bee becomes a scout and starts for the search of a new food source.

Applying these principles one can construct the ABC clustering algorithm. The ABC based clustering algorithms are developed in [160, 166, 311]. First, we present the general scheme of the ABC clustering algorithm in Fig. 6.4. Then we describe it in detail and give its step by step algorithm.

Let s be the size of the population (number of food sources), M be the maximum number of cycles, and L be the limit for abandonment. First, an initial population of food sources $\boldsymbol{u}_i$, $i = 1, \dots, s$, is randomly generated. Each food source $\boldsymbol{u}_i$ is an nk-dimensional vector, where n is the number of attributes in the data set A. These food sources can be considered as candidate solutions for the cluster partition. The population is subjected to repeat the cycles for $i_C = 1, \dots, M$ of the search processes of the employed bees, onlookers, and scouts, respectively. Note that the cluster partition is considered as the food source.

In order to formulate the *fitness function* in the ABC clustering algorithm, we use the clustering objective function f_k of the nonsmooth optimization model given in (4.4). Then the fitness function can be defined as

$$F_k(\boldsymbol{u}_i) = \frac{1}{1 + f_k(\boldsymbol{u}_i)}. \tag{6.1}$$

The value of this function corresponds to the nectar amount in the food source $\boldsymbol{u}_i$. An onlooker bee selects the ith food source $\boldsymbol{u}_i$ with the *probability* P_i defined as

$$P_i = \frac{F_k(\boldsymbol{u}_i)}{\sum_{l=1}^{s} F_k(\boldsymbol{u}_l)}, \quad i = 1, \dots, s. \tag{6.2}$$

Therefore, the more nectar amount at the $\boldsymbol{u}_i$th food source means the higher probability of selecting this source. A *candidate food source*, based on the old one in the memory, is computed as

$$v_{ij} = u_{ij} + c_{ij}(u_{ij} - u_{lj}), \tag{6.3}$$

where $l \in \{1, \dots, s\}$ and $j \in \{1, \dots, nk\}$ are randomly chosen indices. Nevertheless, the number l should be different from i. The numbers c_{ij} are randomly generated from $[-1, 1]$.

If a food source $\boldsymbol{u}_i$ cannot be improved in L cycles, then it is abandoned and the scout discovers a new food source to replace it. Assume that the abandoned source is $\boldsymbol{u}_i$ and $j \in \{1, , \dots, nk\}$ are the indices of the elements of the food source vector. In order to generate the new food source $\boldsymbol{v}_i$, the maximum and minimum values $u_{\max,j}$

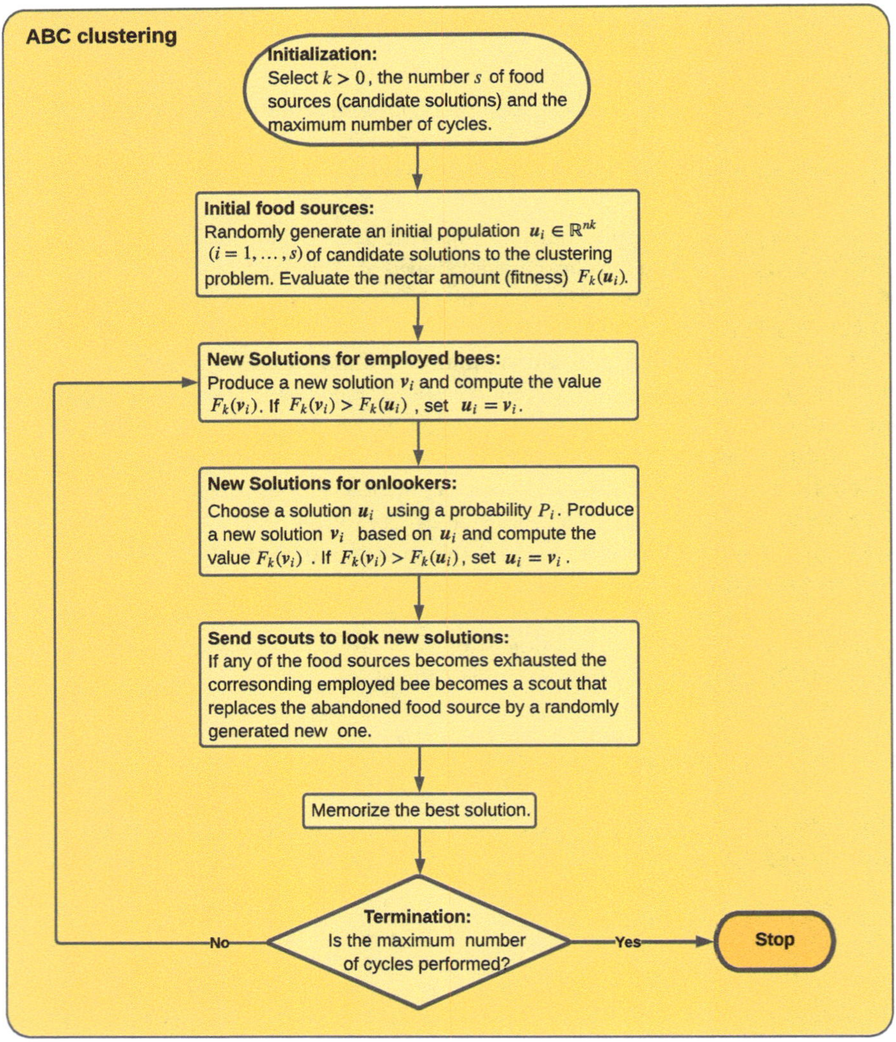

Fig. 6.4 Artificial bee colony clustering algorithm

and $u_{\min,j}$ for each $j \in \{1, \ldots, nk\}$ over the current population are computed. Then the new food source is defined as

$$v_{ij} = u_{\min,j} + \lambda_{ij}(u_{\max,j} - u_{\min,j}). \tag{6.4}$$

Here, λ_{ij} is a random number from (0, 1). After its calculation the fitness of $\boldsymbol{v}_i$ is compared with that of $\boldsymbol{u}_i$. If the new food source has an equal or better nectar than the old source, it replaces the old one. Otherwise, the old one is retained. One can see that the local search in the ABC algorithm is performed by employed and onlooker bees whereas the global search is performed by scouts.

Algorithm 6.3 Artificial bee colony clustering algorithm

Input: Data set A and the number of clusters k to be computed.
Output: Solution to the k-partition problem.

1: *(Initialization)* Select the number of food sources (solution candidates) s. Generate (randomly) the initial population $\boldsymbol{u}_i,\ i = 1, \ldots, s$ of food sources and set $e_i = 0$ for all $i = 1, \ldots, s$. Select the maximum number of cycles M and the limit for abandonment L, and set $i_C = 0$.

2: Evaluate the nectar amount (fitness) $F_k(\boldsymbol{u}_i)$ of the food sources using (6.1) ($i = 1, \ldots, s$).

3: Set $i_C = i_C + 1$.

4: For each employed bee

- (i) produce a new solution $\boldsymbol{v}_i$ using (6.3) and set $e_i = e_i + 1$;
- (ii) calculate the value of $F_k(\boldsymbol{v}_i)$;
- (iii) if $F_k(\boldsymbol{v}_i) > F_k(\boldsymbol{u}_i)$, then the current solution $\boldsymbol{u}_i$ is replaced with $\boldsymbol{v}_i$ and we set $e_i = 0$. Otherwise, the current solution $\boldsymbol{u}_i$ is retained;
- (iv) calculate probabilities P_i for each $\boldsymbol{u}_i$ by applying (6.2).

5: For each onlooker bee

- (i) choose a solution $\boldsymbol{u}_i$ using probabilities P_i;
- (ii) generate a new solution $\boldsymbol{v}_i$ and set $e_i = e_i + 1$;
- (iii) calculate the value of $F_k(\boldsymbol{v}_i)$;
- (iv) if $F_k(\boldsymbol{v}_i) > F_k(\boldsymbol{u}_i)$, then the current solution $\boldsymbol{u}_i$ is replaced with $\boldsymbol{v}_i$ and we set $e_i = 0$. Otherwise, the current solution $\boldsymbol{u}_i$ is retained;
- (v) update the probability for each solution using (6.2).

6: For each food source $\boldsymbol{u}_i,\ i = 1, \ldots, s$, if the exploitation number $e_i > L$, this food source is abandoned, and the corresponding employed bee becomes a scout.

7: For each scout

- (i) replace the abandoned food source $\boldsymbol{u}_i, i = 1, \ldots, s$ by $\boldsymbol{v}_i$ using (6.4);
- (ii) set $e_i = 0$;
- (iii) calculate the value of $F_k(\boldsymbol{v}_i)$;
- (iv) if $F_k(\boldsymbol{v}_i) > F_k(\boldsymbol{u}_i)$, then the current solution $\boldsymbol{u}_i$ is replaced with $\boldsymbol{v}_i$. Otherwise, the current solution $\boldsymbol{u}_i$ is retained.

8: Memorize the best solution.

9: *(Stopping criterion)* If $i_C < M$, then go to Step 3. Otherwise, accept the best solution as the solution to the clustering problem and **stop**.

6.6 Particle Swarm Optimization Clustering Algorithm

Particle swarm optimization (PSO) is a population-based evolutionary algorithm designed to solve general optimization problems. It was developed by Eberhart and Kennedy in 1995 [175]. This algorithm simulates social behavior of bird flocking or fish schooling. Details of the PSO algorithm can be found in [95] (see also, [39]). The PSO based clustering algorithms are introduced in [73, 163, 287]. The basic idea of these algorithms is given in Fig. 6.5.

In comparison with other evolutionary algorithms, the PSO algorithm is easy to implement and has only few parameters to adjust. In this algorithm, the population is called a *swarm*. Therefore, the swarm consists of a set of potential solutions to the optimization (or clustering) problem. Each candidate solution in the swarm is called a *particle*. In the clustering framework this means that each particle is a different cluster distribution. The particles move (or fly) in the search space following the current optimal particle. They have memory and are able to retain part of their previous state.

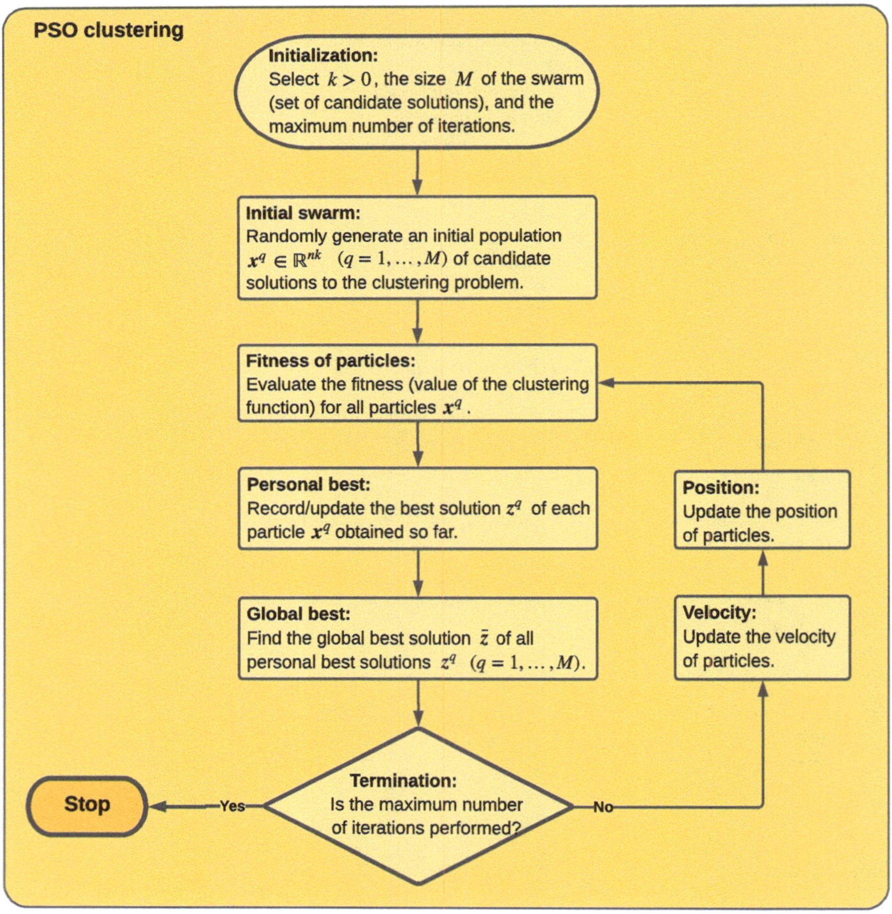

Fig. 6.5 Particle swarm optimization clustering algorithm

A *fitness function*, which is problem specific, is used to evaluate fitness values of particles. Here, the nonsmooth clustering function (4.4) with the squared Euclidean distance d_2 is used. This formulation of the clustering problem has an advantage that it leads to the unconstrained optimization problem. Other clustering models involve some constraints and in order to utilize them the PSO clustering algorithms should contain some mechanisms to take into account these constraints.

In the PSO based clustering algorithm particles are cluster centers, that is, each particle has the representation $\boldsymbol{x}^q = (\boldsymbol{x}_1^q, \ldots, \boldsymbol{x}_k^q)$, where $\boldsymbol{x}_j^q \in \mathbb{R}^n,\ j = 1, \ldots, k$, $q = 1, \ldots, M$, and M is the size of the swarm. The qth particle at the tth iteration is associated by its *current position* $\boldsymbol{x}_t^q$, the *personal best position* $\boldsymbol{z}_t^q$, and the *velocity* $\boldsymbol{v}_t^q$. The new position $\boldsymbol{x}_{t+1}^q$ of this particle is computed as

$$\boldsymbol{x}_{t+1}^q = \boldsymbol{x}_t^q + \boldsymbol{v}_{t+1}^q, \tag{6.5}$$

where the ith component of the velocity vector $\boldsymbol{v}_{t+1}^q$ is given by

$$v_{t+1,i}^q = w v_{t,i}^q + c_1 r_{1i}\left(z_{t,i}^q - x_{t,i}^q\right) + c_2 r_{2i}\left(\bar{z}_{t,i} - x_{t,i}^q\right). \tag{6.6}$$

Here, w is the inertia weight, c_1 and c_2 are the acceleration constants, r_{1i} and r_{2i} are random numbers from $U[0, 1]$, and $\bar{\boldsymbol{z}}_t$ is the *best position*. It follows from (6.6) that the velocity of the qth particle is calculated based on the following three contributions:

- a fraction of the previous velocity;
- a difference between the best and the current personal positions of the qth particle (the cognitive component); and
- a difference between the best particle found so far (the best of the personal bests) and the current personal position of the qth particle.

If $f(\boldsymbol{x}_{t+1}^q) < f(\boldsymbol{z}_t^q)$, then the new best position is updated as $\boldsymbol{z}_{t+1}^q = \boldsymbol{x}_{t+1}^q$, otherwise the best position is not updated. In (6.6), the best position $\bar{\boldsymbol{z}}_t$ can be used in two different ways: in the first case the position is determined using the whole swarm—the g_{best} (global best) version—and in the second case the swarm is divided into overlapping neighborhoods, and the best particle is determined for each neighborhood—the l_{best} (local best) version. In the latter case, the third term in (6.6) is rewritten in the form

$$c_2 r_{2i}\left(\bar{z}_{t,i}^s - x_{t,i}^q\right),$$

where $\bar{\boldsymbol{z}}_t^s$ is the best particle in the sth neighborhood to which the qth particle belongs.

The PSO algorithm starts with the randomly generated initial swarm and apply (6.5) and (6.6) to update particles. At each iteration, the personal best for each particle and the global (or local) best for the whole swarm are stored. The algorithm proceeds until a predefined stopping criterion is met. In the literature two different stopping criteria have been used: one criterion is to define the maximum number

$t_{\max}$ of iterations and to terminate the algorithm when the number of iterations exceeds $t_{\max}$. Alternatively, the algorithm can be terminated when the difference between the velocity values in two successive iterations or over several successive iterations is close to zero.

Here, we present the version of the PSO clustering algorithm with the g_{best}. The algorithm with the l_{best} version can be described in a similar way.

Algorithm 6.4 Particle swarm optimization clustering algorithm

---

Input: Data set A and the number k of clusters to be computed.
Output: Solution to the k-partition problem.

1: *(Initialization)* Select the size of the swarm M and the maximum number of iterations $t_{\max}$. Randomly generate the initial population (swarm) X_1 containing particles $\boldsymbol{x}_1^q = (\boldsymbol{x}_{1,1}^q, \dots, \boldsymbol{x}_{1,k}^q) \in \mathbb{R}^{nk}$, $q = 1, \dots, M$. Set $t = 1$.
2: For each particle from the population X_t compute the corresponding cluster distribution.
3: Evaluate the fitness function f_k for all particles from X_t using their corresponding cluster distributions.
4: Update the best position z_t^q for all particles $\boldsymbol{x}_t^q$ from X_t.
5: Update the best solution $\bar{z}_t$ over the whole swarm X_t.
6: Update the velocity by applying (6.6) and positions of all particles by applying (6.5).
7: *(Stopping criterion)* Set $t = t + 1$. If $t \le t_{\max}$, then go to Step 2. Otherwise, accept the best solution $\bar{z}_{t-1}$ as the solution to the clustering problem and **stop**.

6.7 Ant Colony Optimization Clustering Algorithm

Ant colony optimization (ACO) was introduced by Dorigo and Maniezzo in 1996 [89]. The ACO algorithm was originally designed to solve some challenging problems in optimization such as the traveling salesman and the asymmetric traveling salesman as well as the quadratic assignment and the job-shop scheduling problems [88, 90]. The ACO algorithm is a population-based metaheuristics. The idea of this algorithm comes from mimicking the behavior of real ants in the search of food.

Individually, each ant is blind and its perceptive capabilities are very limited. However, cooperating the colony of ants is capable of finding the shortest path from the nest to the food source and back. This is achieved by laying down a special substance called *pheromones* during the search for food. The concentration of the pheromone on the paths helps to direct the colony to the food sources. Furthermore, an increase in the concentration of pheromones in a trail attracts more ants to it.

Ants use the following three mechanisms to build pheromone trails along the shortest path:

- ants move randomly, but prefer locations with higher pheromone concentrations;
- depending on the length of the trail ants deposit a certain amount of pheromones on their trails, the shorter the trail, the more pheromones are deposited; and

- the pheromones evaporate over time, hence those paths that were preferred earlier can be abandoned if the amount of pheromones decreases significantly.

Therefore, pheromone trails are used for communication between ants. Using this communication they exchange information about the path which should be followed. Ants moving in the shorter path return to the nest earlier than those moving in the longer path. Thus, the amount of pheromone deposited in the shorter path is always more than that of in the longer path. This, in turn, means that the probability of choosing the shorter path by other ants is greater than that of the longer path. As the number of ants following a given path increases, the path becomes more attractive and, consequently, more ants follow this path and deposit their pheromones. This cooperative behavior results in finding the shortest path and forms the intelligent swarm behavior.

Similar to the most other evolutionary algorithms, the fitness function in the ACO algorithm is problem specific. As far as the clustering problem is concerned the fitness function can be defined as the objective function in the mixed integer formulation of the clustering problem (4.2) or as the nonsmooth nonconvex clustering function (4.4). The use of the latter function allows us to avoid the application of special techniques for handling constraints.

Next, we give the flowchart of the ACO clustering algorithm in Fig. 6.6 and describe the algorithm in detail.

Let M be the number of ants in the population. In the ACO clustering algorithm, each ant represents one clustering solution, that is, each ant is represented as $\boldsymbol{x} = (\boldsymbol{x}_1, \ldots, \boldsymbol{x}_k) \in \mathbb{R}^{nk}$. In order to update cluster centers, one needs to define the *pheromone trail matrix* as

$$\tau = [\tau_{ij}], \quad i = 1, \ldots, m, \ j = 1, \ldots, k.$$

At the iteration $t + 1$, the following formula is used to update τ_{ij}:

$$\tau_{ij}^{t+1} = (1 - u)\, \tau_{ij}^t + \sum_{l=1}^{M} \Delta\tau_{ij}^l. \tag{6.7}$$

Here, $u \in (0, 1)$ is the *persistence of the trail* while $1 - u$ is the *evaporation rate*. If u is close to 1, then the past information is forgotten faster. The amount $\Delta\tau_{ij}^l$ is defined as

$$\Delta\tau_{ij}^l = \frac{1}{f_k^l} \quad \text{for all} \quad l = 1, \ldots, M, \tag{6.8}$$

where f_k^l is the value of the clustering function f_k at the solution provided by the lth ant. This means that ants update the pheromone according to the objective function value at the solution they generate.

The assignment of a data point $\boldsymbol{a}_i$ to the cluster A^j is determined according to the amount of pheromone using the probability P_{ij}:

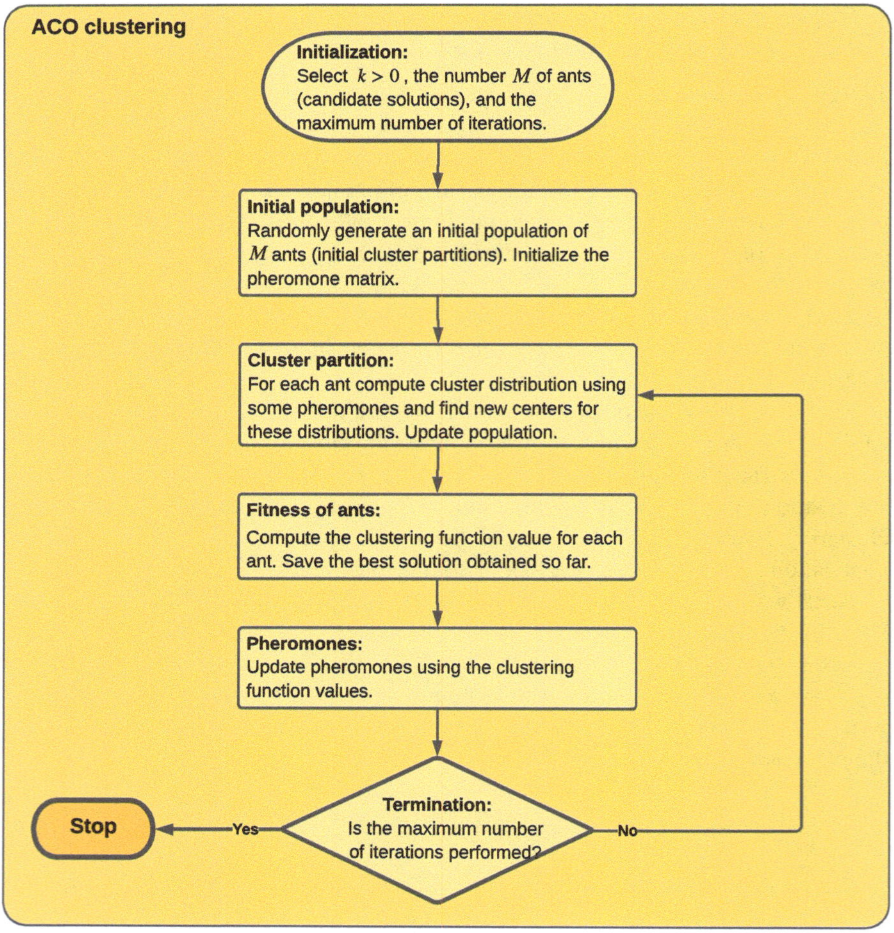

Fig. 6.6 Ant colony optimization clustering algorithm

$$P_{ij} = \frac{\tau_{ij}}{\sum_{q=1}^{k} \tau_{iq}} \quad \text{for all} \quad i = 1, \dots, m, \; j = 1, \dots, k. \tag{6.9}$$

The ith point is assigned to the cluster $A^{\bar{j}}, \; \bar{j} \in \{1, \dots, k\}$ if we have

$$\bar{j} \in \text{Argmax}\left\{P_{ij} : j \in \{1, \dots, k\}\right\}.$$

If there are several such indices $\bar{j}$, we can choose one of them.

Now we are ready to give the ACO clustering algorithm in its step by step form.

Algorithm 6.5 Ant colony optimization clustering algorithm

Input: Data set A and the number of clusters k to be computed.
Output: Solution to the k-partition problem.

1: *(Initialization)* Select the maximum number of iterations $t_{\max}$, the persistence of the trail $u \in (0, 1)$ and the number M of ants in the population.
2: Randomly generate the initial population X_1 containing M ants. Initialize the pheromone matrix τ and set $t = 1$.
3: Calculate the cluster distribution using probabilities P_{ij}, defined in (6.9), and calculate cluster centers for each solution from X_t.
4: *(Global search)* Compute the best solution providing the least value of the clustering function.
5: *(Local search)* Perform a local search to improve current solutions and to generate a new population X_{t+1}.
6: Update the pheromone matrix by applying (6.7) and (6.8).
7: *(Stopping criterion)* Set $t = t + 1$. If $t \leq t_{\max}$, then go to Step 3. Otherwise, accept the best solution found so far as the solution to the clustering problem and **stop**.

Most ACO algorithms contain local search component which is used to improve the solution obtained by the global (ACO) search algorithm. One such local search algorithm, introduced in [266], proceeds as follows.

Algorithm 6.6 Local search algorithm for ant colony optimization clustering

Input: Set of partitions $\{\hat{A}_l,\ l = 1, \ldots, M\}$ and the probability threshold $P_t \in [0, 1]$.
Output: Set of improved partitions $\{\bar{A}_l,\ l = 1, \ldots, M\}$.

1: *(Initialization)* Set $t = 1$.
2: Compute the clustering objective function value $\hat{f}_k^t$ associated with the partition $\hat{A}_t$.
3: For each $\boldsymbol{a}_i \in \hat{A}_t,\ i = 1, \ldots, m$, draw a random number r from $U[0, 1]$. If $r \leq P_t$, randomly select an integer j between 1 and k such that $\boldsymbol{a}_i \notin \hat{A}_t^j$. Reassign $\boldsymbol{a}_i$ to $\hat{A}_t^j$ and denote the new partition with this assignment by $\tilde{A}_t$.
4: Calculate new cluster centers and the objective function value $\tilde{f}_k^t$ associated with $\tilde{A}_t$. If $\tilde{f}_k^t < \hat{f}_k^t$, then set $\bar{A}_t = \tilde{A}_t$. Otherwise, set $\bar{A}_t = \hat{A}_t$.
5: *(Stopping criterion)* Set $t = t + 1$. If $t \leq M$, then go to Step 2. Otherwise **stop**.

Note that here, $\hat{A}_l$ is the set $\hat{A}_l = \{\hat{A}_l^1, \ldots, \hat{A}_l^k\}$.

Different versions of the ant colony optimization clustering algorithm can be found in [111, 152, 198, 253].

Chapter 7
Incremental Clustering Algorithms

7.1 Introduction

As we mentioned in Chap. 4, the clustering problem (4.3) is a nonsmooth global optimization problem and may have many local minimizers. Applying the conventional global optimization techniques is not always a good choice since they are time-consuming for solving such problems, particularly in large data sets. The local methods are fast, however, depending on the choice of starting cluster centers they may end up at the closest local minimizer. Therefore, the success of these methods in solving clustering problems heavily depends on the choice of initial centers.

Since the second half of 1980s, several algorithms have been introduced to choose *favorable* starting cluster centers for local search clustering algorithms, especially for the k-means algorithm [4, 14, 16, 19, 64, 190, 197]. In some of these algorithms, starting points are generated randomly using certain procedures. The use of the incremental approach allows us to choose good starting points in a deterministic way from different parts of the search space. The paper [106] is among the first introducing the incremental algorithm.

The existing incremental algorithms in cluster analysis can be divided, without any loss of generality, into the following classes:

- algorithms where new data points are added at each iteration and cluster centers are refined accordingly. Such algorithms are called *single pass incremental clustering algorithms;* and
- algorithms where clusters are built incrementally adding one cluster center at a time. This type of algorithms are called *sequential clustering algorithms*.

In the single pass incremental algorithms, new data points are presented as a sequence of items and can be examined only in a few passes (usually just one). At each iteration of these algorithms clusters are updated according to newly arrived

A. M. Bagirov et al., *Partitional Clustering via Nonsmooth Optimization*,
Unsupervised and Semi-Supervised Learning,
https://doi.org/10.1007/978-3-030-37826-4_7

data. These algorithms require limited memory and also limited processing time per item (see [130] and references therein).

In the second type of incremental algorithms, the data set is considered as static and clusters are computed incrementally. Such algorithms compute clusters step by step starting with one cluster for the whole data set and gradually adding one cluster center at each iteration [19, 26, 29, 142, 197]. In this book, we consider this type of incremental clustering algorithms.

There are following three optimization problems to be solved at each iteration of incremental clustering algorithms [229]:

- problem of finding a center of one cluster;
- auxiliary clustering problem, defined in (4.29), to obtain starting points for cluster centers; and
- clustering problem, given in (4.3), to determine all cluster centers.

In this chapter, we discuss different approaches for solving each of these problems. In Sect. 7.2, we describe how a center of one cluster can be found. The general incremental clustering algorithm is given in Sect. 7.3. This algorithm involves solving of the auxiliary clustering problem (4.29).

Since both the cluster and the auxiliary cluster functions are nonconvex they may have a large number of local minimizers. Therefore, having favorable starting points will help us to obtain either global or nearly global solutions to clustering problems. We describe the algorithm for finding such starting points for cluster centers in Sect. 7.4. This algorithm generates a set of starting points for the cluster centers, where the points guarantee the decrease of the cluster function at each iteration of the incremental algorithm. Section 7.5 presents the multi-start incremental clustering algorithm. This algorithm is an improvement of the general incremental algorithm that applies the algorithm for finding a set of starting cluster centers.

Finally, the incremental k-medians algorithm and the discussion on the decrease of its computational complexity are given in Sect. 7.6. This algorithm is a modification of the k-medians algorithm, where the latter algorithm is used at each iteration of the multi-start incremental algorithm to solve the clustering problem (4.3).

7.2 Finding a Center of One Cluster

In Chap. 5, the problem of finding a center of a cluster is formulated as an optimization problem. Considering a cluster C, the problem of finding its center $\boldsymbol{x} \in \mathbb{R}^n$ can be reformulated as follows:

$$\begin{cases} \text{minimize} & \varphi(\boldsymbol{x}) \\ \text{subject to} & \boldsymbol{x} \in \mathbb{R}^n, \end{cases} \tag{7.1}$$

where

$$\varphi(\boldsymbol{x}) = \frac{1}{|C|} \sum_{\boldsymbol{c} \in C} d_p(\boldsymbol{x}, \boldsymbol{c}).$$

If the similarity measure d_2 is used, then the centroid of the cluster C is the solution to the problem (7.1) which can be easily computed. If the distance function d_1 is applied, then according to Proposition 5.2 the median of the set C is a solution to this problem. This means that there is no need to solve the problem (7.1) when the similarity functions d_1 and d_2 are applied in the clustering problem.

Next, we consider the problem (7.1) when the function d_∞ is used. Unlike the functions d_1 and d_2, there is no explicit formula for finding a solution to this problem with the function d_∞, and one needs to apply some optimization methods to solve it. In this case, we have

$$\varphi(\boldsymbol{x}) = \frac{1}{|C|} \sum_{\boldsymbol{c} \in C} d_\infty(\boldsymbol{x}, \boldsymbol{c}),$$

and the subdifferential of the function φ at $\boldsymbol{x} \in \mathbb{R}^n$ is

$$\partial \varphi(\boldsymbol{x}) = \frac{1}{|C|} \sum_{\boldsymbol{c} \in C} \partial d_\infty(\boldsymbol{x}, \boldsymbol{c}),$$

where the subdifferential $\partial d_\infty(\boldsymbol{x}, \boldsymbol{c})$ is given in (4.10) and (4.11). Recall that the necessary and sufficient condition for a point $\boldsymbol{x}$ to be a minimum is $\mathbf{0} \in \partial \varphi(\boldsymbol{x})$.

For a moderately large number of points in the set C, the subdifferential $\partial \varphi(\boldsymbol{x})$ may have a huge number of extreme points and therefore, the computation of the whole subdifferential is not an easy task. To solve the problem (7.1) in this case, we can apply versions of the bundle method which are finite convergent for minimizing convex piecewise linear functions [32].

Another option is to use smoothing techniques to approximate the function d_∞ by the smooth functions to replace the problem (7.1) by the sequence of smooth optimization problems. Then we can apply any smooth optimization method to solve these problems.

7.3 General Incremental Clustering Algorithm

As we mentioned, the incremental approach provides an efficient way to generate starting cluster centers. In this section, we describe a general scheme of the *incremental clustering algorithm* (INC-CLUST) using the nonconvex nonsmooth optimization model of the clustering problem. Recall the clustering problem (4.3)

$$\begin{cases} \text{minimize} & f_k(\boldsymbol{x}) \\ \text{subject to} & \boldsymbol{x} = (\boldsymbol{x}_1, \ldots, \boldsymbol{x}_k) \in \mathbb{R}^{nk}, \end{cases} \tag{7.2}$$

where the function f_k, given in (4.4), is

$$f_k(\boldsymbol{x}) = \frac{1}{m}\sum_{i=1}^{m} \min_{j=1,\ldots,k} d_p(\boldsymbol{x}_j, \boldsymbol{a}_i). \tag{7.3}$$

We also recall the auxiliary clustering problem (4.29)

$$\begin{cases} \text{minimize} & \bar{f}_k(\boldsymbol{y}) \\ \text{subject to} & \boldsymbol{y} \in \mathbb{R}^n, \end{cases} \tag{7.4}$$

where the function $\bar{f}_k$, defined in (4.28), is

$$\bar{f}_k(\boldsymbol{y}) = \frac{1}{m}\sum_{i=1}^{m} \min\left\{r_{k-1}^i, d_p(\boldsymbol{y}, \boldsymbol{a}_i)\right\}, \tag{7.5}$$

and r_{k-1}^i, given in (4.27), is the distance between the data point $\boldsymbol{a}_i$, $i = 1, \ldots, m$ and its cluster center:

$$r_{k-1}^i = \min_{j=1,\ldots,k-1} d_p(\boldsymbol{x}_j, \boldsymbol{a}_i). \tag{7.6}$$

The general scheme of the INC-CLUST for solving the k-partition problem (7.2) is given in Fig. 7.1 and Algorithm 7.1.

Algorithm 7.1 Incremental clustering algorithm (INC-CLUST)

Input: Data set A and the number of clusters k to be computed.
Output: The l-partition of the set A with $l = 1, \ldots, k$.

1: *(Initialization)* Compute the center $\boldsymbol{x}_1 \in \mathbb{R}^n$ of the set A. Set $l = 1$.
2: *(Stopping criterion)* Set $l = l + 1$. If $l > k$, then **stop**—the k-partition problem has been solved.
3: *(Computation of the next cluster center)* Find a starting point $\bar{\boldsymbol{y}} \in \mathbb{R}^n$ for the lth cluster center by solving the auxiliary clustering problem (7.4).
4: *(Refinement of all cluster centers)* Select $(\boldsymbol{x}_1, \ldots, \boldsymbol{x}_{l-1}, \bar{\boldsymbol{y}})$ as a starting point to solve the clustering problem (7.2) and find a solution $(\tilde{\boldsymbol{y}}_1, \ldots, \tilde{\boldsymbol{y}}_l)$.
5: *(Solution to the lth partition problem)* Set $\boldsymbol{x}_j = \tilde{\boldsymbol{y}}_j$, $j = 1, \ldots, l$ as a solution to the lth partition problem and go to Step 2.

Remark 7.1 Algorithm 7.1 in addition to the k-partition problem solves also all intermediate l-partition problems, where $l = 1, \ldots, k - 1$.

Steps 3 and 4 are the most important steps of Algorithm 7.1, where both the auxiliary clustering problem (7.4) and the clustering problem (7.2) are solved. Since these problems are nonconvex they may have a large number of local minimizers.

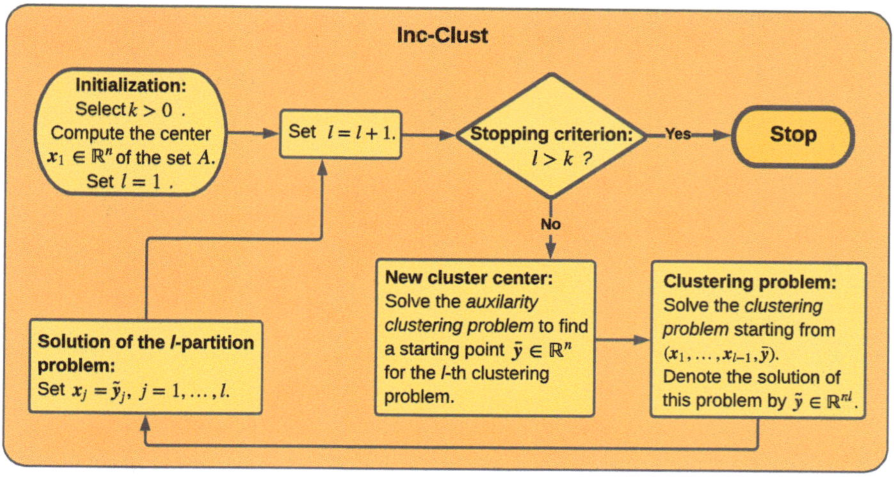

Fig. 7.1 Incremental clustering algorithm (INC-CLUST)

In the next section, we describe a special procedure to generate favorable starting points for solving these problems. Such an approach allows us to find high quality solutions to the clustering problem using local search methods.

7.4 Computation of Set of Starting Cluster Centers

In this section, first we describe an algorithm for finding starting points for solving the auxiliary clustering problem (7.4). We assume that for some $l > 1$, the solution $(\boldsymbol{x}_1, \ldots, \boldsymbol{x}_{l-1})$ to the $(l-1)$-clustering problem is known. Consider the sets

$$\bar{S}_1 = \{\boldsymbol{y} \in \mathbb{R}^n : r^{\boldsymbol{a}}_{l-1} \le d_p(\boldsymbol{y}, \boldsymbol{a}) \quad \text{for all} \quad \boldsymbol{a} \in A\}, \quad \text{and} \tag{7.7}$$

$$\bar{S}_2 = \{\boldsymbol{y} \in \mathbb{R}^n : r^{\boldsymbol{a}}_{l-1} > d_p(\boldsymbol{y}, \boldsymbol{a}) \quad \text{for some} \quad \boldsymbol{a} \in A\}. \tag{7.8}$$

Here, $r^{\boldsymbol{a}}_{l-1}$, $\boldsymbol{a} \in A$ is defined by (4.27). It is obvious that cluster centers $\boldsymbol{x}_1, \ldots, \boldsymbol{x}_{l-1} \in \bar{S}_1$. The set $\bar{S}_2$ contains all points $\boldsymbol{y} \in \mathbb{R}^n$ which are not cluster centers and attract at least one point from the data set A.

Since the number $l-1$ of clusters is less than the number m of data points in the set A all points which are not cluster centers belong to the set $\bar{S}_2$ (because such points attract at least themselves) and therefore, the set $\bar{S}_2$ is not empty. Obviously

$$\bar{S}_1 \cap \bar{S}_2 = \emptyset \quad \text{and} \quad \bar{S}_1 \cup \bar{S}_2 = \mathbb{R}^n.$$

Figure 7.2 illustrates the sets $\bar{S}_1$ and $\bar{S}_2$ where the similarity measure d_2 is applied to find cluster centers. There are three clusters in this figure. Their centers are shown by "red" circles. The set $\bar{S}_2$ consists of all points inside three balls except cluster

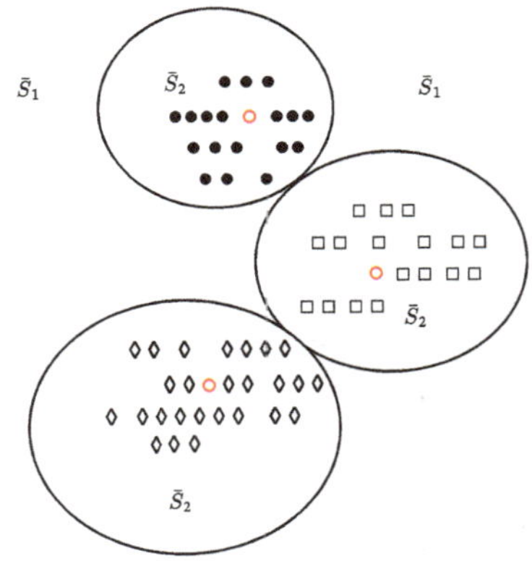

Fig. 7.2 Illustration of sets $\bar{S}_1$ and $\bar{S}_2$

centers and the set $\bar{S}_1$ contains three cluster centers and the part of the space outside balls.

Note that

$$\bar{f}_l(\mathbf{y}) \leq \frac{1}{m} \sum_{\mathbf{a} \in A} r_{l-1}^{\mathbf{a}} \quad \text{for all} \quad \mathbf{y} \in \mathbb{R}^n, \quad \text{and}$$

$$\bar{f}_l(\mathbf{y}) = f_{l-1}(\mathbf{x}_1, \ldots, \mathbf{x}_{l-1}) = \frac{1}{m} \sum_{\mathbf{a} \in A} r_{l-1}^{\mathbf{a}} \quad \text{for all} \quad \mathbf{y} \in \bar{S}_1.$$

This means that the lth auxiliary cluster function $\bar{f}_l$ is constant on the set $\bar{S}_1$, and any point from this set is a global maximizer of this function. In general, a local search method terminates at any of these points. Therefore, starting points for solving the auxiliary clustering problem (7.4) should not be chosen from the set $\bar{S}_1$.

We introduce a special procedure which allows one to select starting points from the set $\bar{S}_2$. Take any $\mathbf{y} \in \bar{S}_2$ and consider the sets $B_i(\mathbf{y}),\ i = 1, 2, 3$ defined in (4.30). Then the set A can be divided into two subsets $\bar{B}_{12}(\mathbf{y})$ and $\bar{B}_3(\mathbf{y})$, where

$$\bar{B}_{12}(\mathbf{y}) = B_1(\mathbf{y}) \cup B_2(\mathbf{y}) \quad \text{and} \quad \bar{B}_3(\mathbf{y}) = B_3(\mathbf{y}). \tag{7.9}$$

The set $\bar{B}_3(\mathbf{y})$ contains all data points $\mathbf{a} \in A$ which are closer to the point $\mathbf{y}$ than to their cluster centers, and the set $\bar{B}_{12}(\mathbf{y})$ contains all other data points. Since $\mathbf{y} \in \bar{S}_2$ the set $\bar{B}_3(\mathbf{y}) \neq \emptyset$. Furthermore,

$$\bar{B}_{12}(\mathbf{y}) \cap \bar{B}_3(\mathbf{y}) = \emptyset \quad \text{and} \quad A = \bar{B}_{12}(\mathbf{y}) \cup \bar{B}_3(\mathbf{y}).$$

Figure 7.3 depicts the set $\bar{B}_3(\mathbf{y})$ for a given $\mathbf{y}$ (black ball). There are two clusters in this data set and their centers are shown by "red" circles. The set $\bar{B}_3(\mathbf{y})$ contains all "yellow" data points and the set $\bar{B}_{12}(\mathbf{y})$ contains the rest of the data set.

At a point $\mathbf{y} \in \mathbb{R}^n$ using the sets $\bar{B}_{12}(\mathbf{y})$ and $\bar{B}_3(\mathbf{y})$, the lth auxiliary cluster function $\bar{f}_l$ can be written as

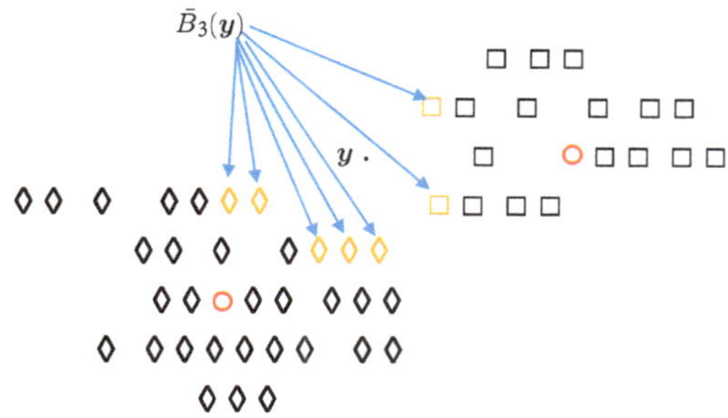

Fig. 7.3 Illustration of sets $\bar{B}_{12}(y)$ and $\bar{B}_3(y)$

$$\bar{f}_l(\mathbf{y}) = \frac{1}{m}\Big(\sum_{\mathbf{a}\in \bar{B}_{12}(\mathbf{y})} r_{l-1}^{\mathbf{a}} + \sum_{\mathbf{a}\in \bar{B}_3(\mathbf{y})} d_p(\mathbf{y}, \mathbf{a})\Big).$$

The difference between the values of $\bar{f}_l(\mathbf{y})$ and $f_{l-1}(\mathbf{x}_1, \ldots, \mathbf{x}_{l-1})$ is

$$z_l(\mathbf{y}) = \frac{1}{m} \sum_{\mathbf{a}\in \bar{B}_3(\mathbf{y})} \Big(r_{l-1}^{\mathbf{a}} - d_p(\mathbf{y}, \mathbf{a})\Big),$$

which can be rewritten as

$$z_l(\mathbf{y}) = \frac{1}{m} \sum_{\mathbf{a}\in A} \max \Big\{0, r_{l-1}^{\mathbf{a}} - d_p(\mathbf{y}, \mathbf{a})\Big\}. \tag{7.10}$$

The difference $z_l(\mathbf{y})$ shows the decrease of the value of the lth cluster function f_l comparing with the value $f_{l-1}(\mathbf{x}_1, \ldots, \mathbf{x}_{l-1})$ if the points $\mathbf{x}_1, \ldots, \mathbf{x}_{l-1}, \mathbf{y}$ are chosen as the cluster centers for the lth clustering problem.

It is reasonable to choose a point $\mathbf{y} \in \mathbb{R}^n$ that provides the largest decrease $z_l(\mathbf{y})$ of the clustering function as the starting point for minimizing the auxiliary clustering function. Since it is not easy to choose such a point from the whole space $\mathbb{R}^n$ we restrict ourselves to the data set A.

If a data point $\mathbf{a} \in A$ is a cluster center, then this point belongs to the set $\bar{S}_1$, otherwise it belongs to the set $\bar{S}_2$. Therefore, we choose points $\mathbf{y}$ from the set $\bar{A}_0 = A \setminus \bar{S}_1$. Obviously, $\bar{A}_0 \neq \emptyset$. Take any $\mathbf{y} = \mathbf{a} \in \bar{A}_0$, compute $z_l(\mathbf{a})$ and define the number

$$z_{\max}^1 = \max_{\mathbf{a}\in \bar{A}_0} z_l(\mathbf{a}). \tag{7.11}$$

The number $z_{\max}^1$ represents the largest decrease of the cluster function which can be provided by any data point. Let $\gamma_1 \in [0, 1]$ be a given number. Compute the following subset of $\bar{A}_0$:

$$\bar{A}_1 = \big\{\mathbf{a} \in \bar{A}_0 : \ z_l(\mathbf{a}) \ge \gamma_1 z_{\max}^1\big\}. \tag{7.12}$$

The set $\bar{A}_1$ contains all data points that provide the decrease of the cluster function no less than the threshold $\gamma_1 z^1_{\max}$. This set is obtained from the set $\bar{A}_0$ by removing data points that do not provide sufficient decrease of the cluster function. Apparently, $\bar{A}_1 \neq \emptyset$ for any $\gamma_1 \in [0, 1]$. If $\gamma_1 = 0$, then $\bar{A}_1 = \bar{A}_0$ and if $\gamma_1 = 1$, then the set $\bar{A}_1$ contains data points providing the largest decrease $z^1_{\max}$.

For each point $\boldsymbol{a} \in \bar{A}_1$ compute the set $\bar{B}_3(\boldsymbol{a})$ and its center $\boldsymbol{c}(\boldsymbol{a})$. Replace the point $\boldsymbol{a}$ by $\boldsymbol{c}(\boldsymbol{a})$ since the center $\boldsymbol{c}(\boldsymbol{a})$ is a better representative of the set $\bar{B}_3(\boldsymbol{a})$ than the point $\boldsymbol{a}$. If the similarity measure d_p is defined using the L_2-norm, then $\boldsymbol{c}(\boldsymbol{a})$ is the centroid of the set $\bar{B}_3(\boldsymbol{a})$. In other cases, $\boldsymbol{c}(\boldsymbol{a})$ is found as a solution to the problem (7.1) where

$$\varphi(\boldsymbol{x}) = \frac{1}{|\bar{B}_3(\boldsymbol{a})|} \sum_{\boldsymbol{b} \in \bar{B}_3(\boldsymbol{a})} d_p(\boldsymbol{x}, \boldsymbol{b}).$$

Let

$$\bar{A}_2 = \left\{\boldsymbol{c} \in \mathbb{R}^n : \text{ there exists } \boldsymbol{a} \in \bar{A}_1 \text{ such that } \boldsymbol{c} = \boldsymbol{c}(\boldsymbol{a})\right\}$$

be the set of such solutions. It is obvious that $\bar{A}_2 \neq \emptyset$. For each $\boldsymbol{c} \in \bar{A}_2$, compute the number $z_l(\boldsymbol{c})$ using (7.10) and find the number

$$z^2_{\max} = \max_{\boldsymbol{c} \in \bar{A}_2} z_l(\boldsymbol{c}). \tag{7.13}$$

The number $z^2_{\max}$ represents the largest value of the decrease

$$f_{l-1}(\boldsymbol{x}_1, \dots, \boldsymbol{x}_{l-1}) - f_l(\boldsymbol{x}_1, \dots, \boldsymbol{x}_{l-1}, \boldsymbol{c})$$

among all centers $\boldsymbol{c} \in \bar{A}_2$.

For a given number $\gamma_2 \in [0, 1]$, define the following subset of $\bar{A}_2$:

$$\bar{A}_3 = \left\{\boldsymbol{c} \in \bar{A}_2 : z_l(\boldsymbol{c}) \geq \gamma_2 z^2_{\max}\right\}. \tag{7.14}$$

The set $\bar{A}_3$ contains all points $\boldsymbol{c} \in \bar{A}_2$ that provide the decrease of the cluster function no less than the threshold $\gamma_2 z^2_{\max}$. This set is obtained from the set $\bar{A}_2$ by removing centers which do not provide the sufficient decrease of the cluster function. It is clear that the set $\bar{A}_3 \neq \emptyset$ for any $\gamma_2 \in [0, 1]$. If $\gamma_2 = 0$, then $\bar{A}_3 = \bar{A}_2$ and if $\gamma_2 = 1$, then the set $\bar{A}_3$ contains only centers $\boldsymbol{c}$ providing the largest decrease of the cluster function f_l.

All points from the set $\bar{A}_3$ are considered as starting points for solving the auxiliary clustering problem (7.4). Since all data points are used for the computation of the set $\bar{A}_3$, it contains starting points from different parts of the data set. Such a strategy allows us to find either global or nearly global solutions to the problem (7.2) (as well as to the problem (7.4)) using local search methods.

Applying a local search algorithm, the auxiliary clustering problem (7.4) is solved using starting points from $\bar{A}_3$. A local search algorithm generates the same number of solutions as the number of starting points. The set of these solutions is denoted by $\bar{A}_4$. This set is a non-empty subset of the set of stationary points of the auxiliary cluster function $\bar{f}_l$.

A local search algorithm starting from different points may arrive to the same stationary point or stationary points which are close to each other. To identify such stationary points we define a tolerance $\varepsilon > 0$. If the distance between any two points from the set $\bar{A}_4$ is less than this tolerance, then we keep a point with the lower value of the function $\bar{f}_l$ and remove another point from the set $\bar{A}_4$.

Next, we define

$$\bar{f}_l^{\min} = \min_{\mathbf{y} \in \bar{A}_4} \bar{f}_l(\mathbf{y}). \tag{7.15}$$

The number $\bar{f}_l^{\min}$ is the lowest value of the auxiliary cluster function $\bar{f}_l$ over the set $\bar{A}_4$. Let $\gamma_3 \in [1, \infty)$ be a given number. Introduce the following set:

$$\bar{A}_5 = \big\{\mathbf{y} \in \bar{A}_4 : \bar{f}_l(\mathbf{y}) \le \gamma_3 \bar{f}_l^{\min}\big\}. \tag{7.16}$$

The set $\bar{A}_5$ contains all stationary points where the value of the function $\bar{f}_l$ is no more than the threshold $\gamma_3 \bar{f}_l^{\min}$. Note that the set $\bar{A}_5 \neq \emptyset$. If $\gamma_3 = 1$, then $\bar{A}_5$ contains the best local minimizers of the function $\bar{f}_l$ obtained using starting points from the set $\bar{A}_3$. If γ_3 is sufficiently large, then $\bar{A}_5 = \bar{A}_4$. Points from the set $\bar{A}_5$ are used as a set of starting points for the lth cluster center to solve the lth clustering problem (7.2).

Summarizing all described above, the algorithm for finding starting points to solve the problem (7.2) proceeds as follows [228].

Algorithm 7.2 allows us to use more than one starting point to solve the clustering problem (7.2) in Step 4 of Algorithm 7.1. Moreover, these points always guarantee the decrease of the cluster function value at each iteration of the incremental algorithm and are distinct from each other in the search space. Such an approach

---

Algorithm 7.2 Finding set of starting points for the lth cluster center

Input: Data set A and the solution $(\mathbf{x}_1, \ldots, \mathbf{x}_{l-1})$ to the $(l-1)$-clustering problem, $l \ge 2$.
Output: Set of starting points for the lth cluster center.

1: *(Initialization)* Select numbers $\gamma_1, \gamma_2 \in [0, 1]$ and $\gamma_3 \in [1, \infty)$.
2: Compute $z_{\max}^1$ using (7.11) and the set $\bar{A}_1$ using (7.12).
3: Compute $z_{\max}^2$ using (7.13) and the set $\bar{A}_3$ using (7.14).
4: Compute the set $\bar{A}_4$ of stationary points of the auxiliary clustering problem (7.4) applying a local search algorithm and using starting points from the set $\bar{A}_3$.
5: Compute $\bar{f}_l^{\min}$ using (7.15) and the set $\bar{A}_5$ using (7.16). $\bar{A}_5$ is the set of starting points for the lth cluster center.

Algorithm 7.3 Multi-start incremental clustering algorithm (MSINC-CLUST)

Input: Data set A and the number of clusters k to be computed.
Output: The l-partition of the set A with $l = 1, \dots, k$.

1: *(Initialization)* Compute the center $\boldsymbol{x}_1 \in \mathbb{R}^n$ of the set A. Set $l = 1$.

2: *(Stopping criterion)* Set $l = l + 1$. If $l > k$, then **stop**—the k-partition problem has been solved.

3: *(Computation of a set of starting points for the lth cluster center)* Apply Algorithm 7.2 to compute the set $\bar{A}_5$ defined by (7.16).

4: *(Computation of a set of cluster centers)* For each $\bar{\boldsymbol{y}} \in \bar{A}_5$, select $(\boldsymbol{x}_1, \dots, \boldsymbol{x}_{l-1}, \bar{\boldsymbol{y}})$ as a starting point to solve the lth clustering problem (7.2) and find its solution $(\hat{\boldsymbol{y}}_1, \dots, \hat{\boldsymbol{y}}_l)$. Denote by $\bar{A}_6$ a set of all such solutions.

5: *(Computation of the best solution)* Compute

$$f_l^{\min} = \min\left\{ f_l(\hat{\boldsymbol{y}}_1, \dots, \hat{\boldsymbol{y}}_l) : (\hat{\boldsymbol{y}}_1, \dots, \hat{\boldsymbol{y}}_l) \in \bar{A}_6 \right\},$$

and the collection of cluster centers $(\tilde{\boldsymbol{y}}_1, \dots, \tilde{\boldsymbol{y}}_l)$ such that

$$f_l(\tilde{\boldsymbol{y}}_1, \dots, \tilde{\boldsymbol{y}}_l) = f_l^{\min}.$$

6: *(Solution to the lth partition problem)* Set $\boldsymbol{x}_j = \tilde{\boldsymbol{y}}_j, \; j = 1, \dots, l$ as a solution to the lth partition problem and go to Step 2.

allows us to apply local search methods to obtain a high quality solution to the global optimization problem (7.2).

7.5 Multi-Start Incremental Clustering Algorithm

In this section, we present the *multi-start incremental clustering algorithm* (MSINC-CLUST) for solving the problem (7.2). This algorithm is an improvement of Algorithm 7.1 where in Step 3, Algorithm 7.2 is applied. Similar to Algorithm 7.1, the MSINC-CLUST builds clusters dynamically adding one cluster center at a time by solving the auxiliary clustering problem (7.4).

The MSINC-CLUST applies Algorithm 7.2 to compute a set of starting cluster centers. Using these centers as initial points, the lth clustering problem (7.2) is solved ($l = 2, \dots, k$). Then a solution with the least cluster function value, defined in (7.3), is accepted as the solution to the clustering problem. The flowchart of the MSINC-CLUST is given in Fig. 7.4 and its step by step description is presented in Algorithm 7.3.

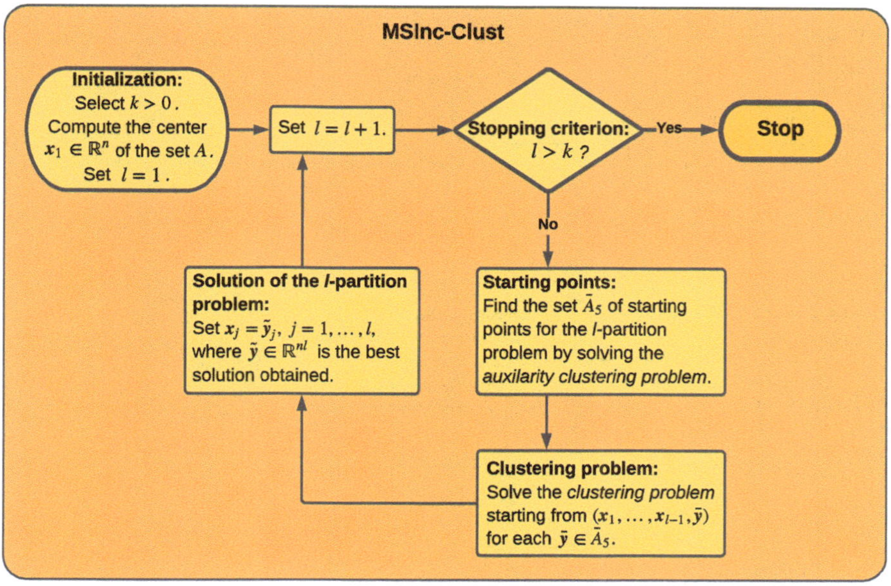

Fig. 7.4 Multi-start incremental clustering algorithm (MSINC-CLUST)

Remark 7.2 Similar to Algorithm 7.1, this algorithm solves all intermediate l-partition problems ($l = 1, \ldots, k - 1$) in addition to the k-partition problem. However, Algorithm 7.1 can find only stationary points of the clustering problem, while Algorithm 7.3 is able to find either global or nearly global solutions.

Note that the most important steps in Algorithm 7.3 are Step 3, where the auxiliary clustering problem (4.29) is solved to find starting points, and Step 4, where the clustering problem (7.2) is solved for each starting point. To solve these problems, we will introduce different algorithms in this and the following two chapters.

7.6 Incremental k-Medians Algorithm

In this section, we design the *incremental k-medians algorithm* (IKMED) as an application of Algorithm 7.3. The k-medians algorithm (Algorithm 5.4), presented in Chap. 5, is simple and easy to implement. However, this algorithm is sensitive to the choice of starting points and finds only local solutions that can be significantly different from the global solution in large data sets. The IKMED overcomes these

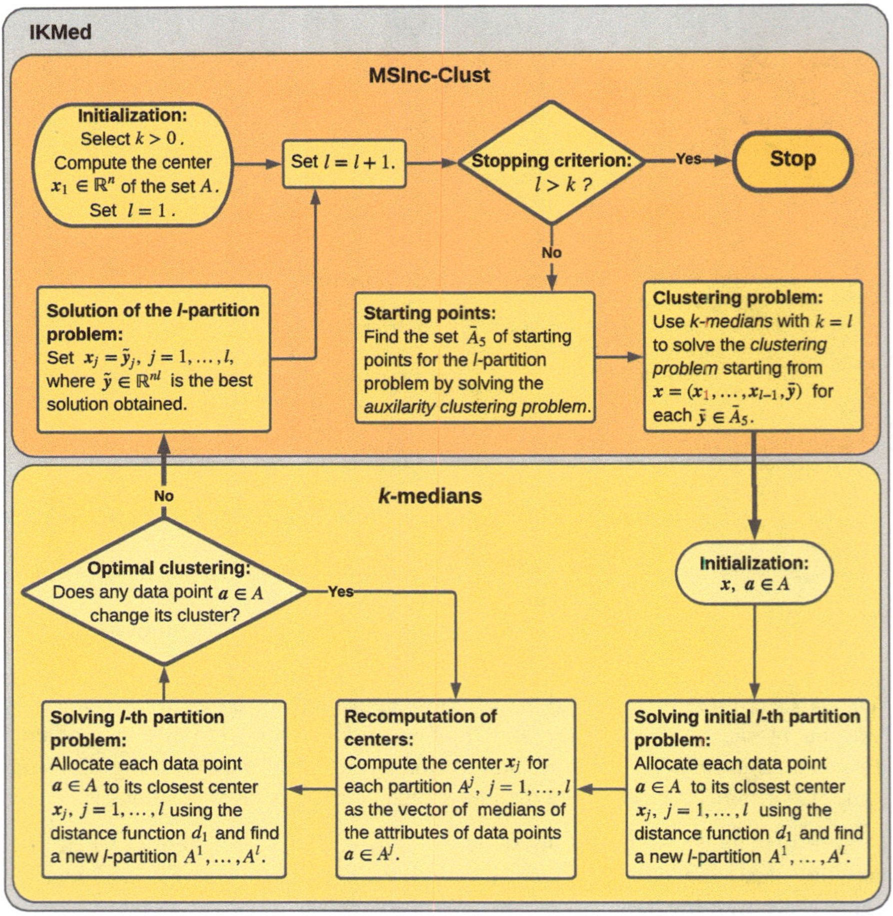

Fig. 7.5 Incremental k-medians algorithm (IKMED)

drawbacks by applying Algorithm 7.2. Characteristically for k-medians, the distance function d_1 is used to define the similarity measure in the IKMED. Fig. 7.5 illustrates the flowchart of this algorithm.

The IKMED first calculates the center of the whole data set as its median. Then it applies Algorithm 7.2 to compute the set of initial cluster centers by solving the auxiliary clustering problem (7.4). Using these centers, the clustering problem (7.2) is solved. Note that Algorithm 5.4 is utilized to solve both problems (7.2) and (7.3).

The following algorithm describes the IKMED in step by step.

Algorithm 7.4 Incremental k-medians algorithm (IKMED)

Input: Data set A and the number of clusters k to be computed.
Output: The l-partition of the set A with $l = 1, \dots, k$.

1: *(Initialization)* Compute the center $\boldsymbol{x}_1 \in \mathbb{R}^n$ of the set A. Set $l = 1$.

2: *(Stopping criterion)* Set $l = l + 1$. If $l > k$, then **stop**—the k-partition problem has been solved.

3: *(Computation of a set of starting points for the auxiliary clustering problem)* Apply Algorithm 7.2 to compute the set $\bar{A}_3$ of starting points for solving the auxiliary clustering problem (7.4).

4: *(Computation of a set of starting points for the lth cluster center)* Apply Algorithm 5.4 to solve the auxiliary clustering problem (7.4) starting from each point $\boldsymbol{y} \in \bar{A}_3$. This algorithm generates a set $\bar{A}_5$ of starting points for the lth cluster center.

5: *(Computation of a set of cluster centers)* For each $\bar{\boldsymbol{y}} \in \bar{A}_5$ apply Algorithm 5.4 for $k = l$ to solve the clustering problem (7.2) starting from the point $(\boldsymbol{x}_1, \dots, \boldsymbol{x}_{l-1}, \bar{\boldsymbol{y}})$ and find its solution $(\hat{\boldsymbol{y}}_1, \dots, \hat{\boldsymbol{y}}_l)$. Denote by $\bar{A}_6$ a set of all such solutions.

6: *(Computation of the best solution)* Compute

$$f_l^{\min} = \min \left\{ f_l(\hat{\boldsymbol{y}}_1, \dots, \hat{\boldsymbol{y}}_l) : (\hat{\boldsymbol{y}}_1, \dots, \hat{\boldsymbol{y}}_l) \in \bar{A}_6 \right\},$$

and the collection of cluster centers $(\tilde{\boldsymbol{y}}_1, \dots, \tilde{\boldsymbol{y}}_l)$ such that

$$f_l(\tilde{\boldsymbol{y}}_1, \dots, \tilde{\boldsymbol{y}}_l) = f_l^{\min}.$$

7: *(Solution to the lth partition problem)* Set $\boldsymbol{x}_j = \tilde{\boldsymbol{y}}_j,\ j = 1, \dots, l$ as a solution to the lth partition problem and go to Step 2.

In Step 4 of Algorithm 7.4 one can apply the modified version of the k-medians Algorithm 5.4 to solve the auxiliary clustering problem and to find starting points for the lth cluster center. In this version, cluster centers $\boldsymbol{x}_1, \dots, \boldsymbol{x}_{l-1}$ are fixed and the algorithm updates only the lth center. Therefore, we call it the *partial k-medians algorithm*. The description of this algorithm is given below.

We use the sets $\bar{S}_2$ and $\bar{B}_3(\boldsymbol{y}),\ \boldsymbol{y} \in \mathbb{R}^n$, defined in (7.8) and (7.9), respectively. Note that we employ the distance function d_1 in computing these sets.

Algorithm 7.5 Partial k-medians algorithm

Input: Data set A and the solution $(\boldsymbol{x}_1, \ldots, \boldsymbol{x}_{l-1}) \in \mathbb{R}^{n(l-1)}$ to the $(l-1)$-partition problem.
Output: Solution $\bar{\boldsymbol{y}} \in \mathbb{R}^n$ to the auxiliary clustering problem (7.4).

1: *(Initialization)* Select a starting point $\boldsymbol{y}_1 \in \bar{S}_2$. Set $h = 1$.
2: Compute the set $\bar{B}_3(\boldsymbol{y}_h)$.
3: *(Stopping criterion)* If $\bar{B}_3(\boldsymbol{y}_h) = \bar{B}_3(\boldsymbol{y}_{h-1})$ for $h > 1$, then **stop** with the solution $\bar{\boldsymbol{y}} = \boldsymbol{y}_h$ to the auxiliary clustering problem.
4: Find a center $\bar{\boldsymbol{c}}$ of the set $\bar{B}_3(\boldsymbol{y}_h)$ by computing its coordinates as the medians of the corresponding attributes.
5: Set $\boldsymbol{y}_{h+1} = \bar{\boldsymbol{c}},\ h = h + 1$ and go to Step 2.

Remark 7.3 The set $\bar{S}_2$ contains all data points $\boldsymbol{a} \in A$ which are not cluster centers and therefore, in Step 1 one can choose the point $\boldsymbol{y}_1$ among such data points. More specifically, we can choose $\boldsymbol{y}_1 \in A \setminus \bar{S}_1$ where the set $\bar{S}_1$ is given in (7.7). Furthermore, since for any $\boldsymbol{y} \in \bar{S}_2$ the set $\bar{B}_3(\boldsymbol{y})$ is not empty and the value of the auxiliary cluster function decreases at each iteration h the problem of finding the center of the sets $\bar{B}_3(\boldsymbol{y}_h),\ h \geq 1$ in Step 4 is well defined.

Note that the stopping criterion in Step 3 means that the algorithm terminates when no data point changes its cluster.

The most time-consuming steps in Algorithm 7.4 are Steps 3, 4, and 5. To reduce the computational effort required in these steps, we discuss three different approaches as follows:

1. *Reduction of the number of starting cluster centers.* As mentioned above, starting points for solving the auxiliary clustering problem (7.4) can be chosen from the set $A \setminus \bar{S}_1$. At the lth iteration ($l \geq 2$) of Algorithm 7.4, we can remove points that are close to cluster centers $\boldsymbol{x}_1, \ldots, \boldsymbol{x}_{l-1}$. For each cluster $A^q,\ 1 \leq q \leq l-1$, compute its average radius

$$r_{av}^q = \frac{1}{|A^q|} \sum_{\boldsymbol{a} \in A^q} d_1(\boldsymbol{x}_q, \boldsymbol{a}),$$

and define the subset $\hat{A}^q \subseteq A^q$ as

$$\hat{A}^q = \left\{\boldsymbol{a} \in A^q : r_{av}^q \leq d_1(\boldsymbol{x}_q, \boldsymbol{a})\right\}.$$

Note that if the cluster A^q is not empty, then the set $\hat{A}^q$ is also non-empty. Consider the following subset of the set A:

$$\hat{A} = \bigcup_{q=1}^{l-1} \hat{A}^q.$$

Replacing the set $A \setminus \bar{S}_1$ by the set $\hat{A} \setminus \bar{S}_1$ allows us to reduce—in some cases significantly—the number of starting cluster centers and to remove those points which do not provide the sufficient decrease of the cluster function.

2. *Exclusion of some stationary points of the auxiliary clustering problem (7.4).* If any two stationary points from the set $\bar{A}_4$ are close to each other with respect to some predefined tolerance, then one of them is removed while another one is kept. In order to do so we define a tolerance $\varepsilon = \hat{f}_1/ml$, where $\hat{f}_1$ is the optimal value of the cluster function f_1. If $d_1(\boldsymbol{y}_1, \boldsymbol{y}_2) \leq \varepsilon$ for two points $\boldsymbol{y}_1, \boldsymbol{y}_2 \in \bar{A}_4$, then the point with the lowest value of the auxiliary cluster function is kept in $\bar{A}_4$ and another point is removed.
3. *Use of the triangle inequality to reduce the number of distance calculations.* Since d_1 is the distance function it satisfies the triangle inequality. This can be used to reduce the number of distance function calculations of Algorithm 5.4 in solving both the clustering and the auxiliary clustering problems. First, we consider the auxiliary clustering problem (7.4). Assume that $(\boldsymbol{x}_1, \ldots, \boldsymbol{x}_{l-1})$ is the solution to the $(l-1)$-partition problem. Recall that the distance between the data point $\boldsymbol{a} \in A$ and its cluster center is denoted by

$$r_{l-1}^{\boldsymbol{a}} = \min_{j=1,\ldots,l-1} d_1(\boldsymbol{x}_j, \boldsymbol{a}).$$

Let $\bar{\boldsymbol{y}}$ be a current approximation to the solution of the problem (7.4). Compute distances $d_1(\bar{\boldsymbol{y}}, \boldsymbol{x}_j),\ j = 1, \ldots, l-1$. Assume that $\boldsymbol{a} \in A^j$ for some $j \in \{1, \ldots, l-1\}$. According to the triangle inequality we have

$$d_1(\bar{\boldsymbol{y}}, \boldsymbol{x}_j) \leq d_1(\boldsymbol{a}, \bar{\boldsymbol{y}}) + d_1(\boldsymbol{a}, \boldsymbol{x}_j) = d_1(\boldsymbol{a}, \bar{\boldsymbol{y}}) + r_{l-1}^{\boldsymbol{a}}, \quad \text{or}$$
$$d_1(\boldsymbol{a}, \bar{\boldsymbol{y}}) \geq d_1(\bar{\boldsymbol{y}}, \boldsymbol{x}_j) - r_{l-1}^{\boldsymbol{a}}.$$

This means that if $d_1(\bar{\boldsymbol{y}}, \boldsymbol{x}_j) > 2r_{l-1}^{\boldsymbol{a}}$, then $d_1(\boldsymbol{a}, \bar{\boldsymbol{y}}) > r_{l-1}^{\boldsymbol{a}}$ and therefore, there is no need to calculate the distance $d_1(\boldsymbol{a}, \bar{\boldsymbol{y}})$ as the point $\boldsymbol{a}$ does not belong to the cluster with the center $\bar{\boldsymbol{y}}$.

Similar approach can be considered for the clustering problem (7.2). Let $(\bar{\boldsymbol{x}}_1, \ldots, \bar{\boldsymbol{x}}_l)$ be a current approximation to the solution of the lth partition problem. Compute distances $d_1(\bar{\boldsymbol{x}}_i, \bar{\boldsymbol{x}}_j)$ for $i, j = 1, \ldots, l$. Assume that for a given point $\boldsymbol{a} \in A$, the distances $d_1(\boldsymbol{a}, \bar{\boldsymbol{x}}_i),\ i = 1, \ldots, j$ have been calculated or estimated for some $j \in \{1, \ldots, l-1\}$. Let $\tilde{\boldsymbol{x}} \in \{\bar{\boldsymbol{x}}_1, \ldots, \bar{\boldsymbol{x}}_j\}$ be such that

$$d_1(\boldsymbol{a}, \tilde{\boldsymbol{x}}) = \min_{i=1,\ldots,j} d_1(\boldsymbol{a}, \bar{\boldsymbol{x}}_i).$$

According to the triangle inequality we have

$$d_1(\tilde{\boldsymbol{x}}, \bar{\boldsymbol{x}}_{j+1}) \le d_1(\boldsymbol{a}, \tilde{\boldsymbol{x}}) + d_1(\boldsymbol{a}, \bar{\boldsymbol{x}}_{j+1}), \quad \text{or}$$
$$d_1(\boldsymbol{a}, \bar{\boldsymbol{x}}_{j+1}) \ge d_1(\tilde{\boldsymbol{x}}, \bar{\boldsymbol{x}}_{j+1}) - d_1(\boldsymbol{a}, \tilde{\boldsymbol{x}}).$$

If $d_1(\tilde{\boldsymbol{x}}, \bar{\boldsymbol{x}}_{j+1}) > 2d_1(\boldsymbol{a}, \tilde{\boldsymbol{x}})$, then there is no need to calculate the distance $d_1(\boldsymbol{a}, \bar{\boldsymbol{x}}_{j+1})$ as the point $\boldsymbol{a}$ does not belong to the cluster A^{j+1} with the center $\bar{\boldsymbol{x}}_{j+1}$. The last approach allows us to significantly reduce the number of distance function evaluations as the number of clusters increases.

Chapter 8
Nonsmooth Optimization Based Clustering Algorithms

8.1 Introduction

In Chap. 4, we formulated different optimization models of the clustering problem. Using these models, one can apply various heuristics or optimization methods to solve clustering problems. Algorithms considered in this chapter are based on the NSO model of these problems. More specifically, we consider incremental clustering algorithms where at each iteration the clustering and the auxiliary clustering problems are solved by applying either heuristics like the k-means or NSO algorithms [19, 22, 24, 26, 29, 33, 36, 170, 171].

We start with the description of the modified global k-means algorithm in Sect. 8.2. This algorithm is an improvement of the GKM. The main difference between these two algorithms is that the GKM uses only data points to find starting cluster centers whereas the modified global k-means algorithm solves the auxiliary clustering problem to compute them. In Sect. 8.3, we describe a further improvement of the modified global k-means algorithm called the fast modified global k-means algorithm. In addition, we discuss various procedures to reduce the computational complexity of the modified global k-means algorithm.

Then, we introduce three incremental clustering algorithms where the LMBM, the DGM, and the HSM are used to solve the clustering and the auxiliary clustering problems. More precisely, the limited memory bundle method for clustering is described in Sect. 8.4; the discrete gradient clustering algorithm is presented in Sect. 8.5; and the smooth incremental clustering algorithm is given in Sect. 8.6.

A. M. Bagirov et al., *Partitional Clustering via Nonsmooth Optimization*,
Unsupervised and Semi-Supervised Learning,
https://doi.org/10.1007/978-3-030-37826-4_8

8.2 Modified Global k-Means Algorithm

In this section, we present the *modified global k-means algorithm* (MGKM) to solve the clustering problem (7.2) where the similarity measure d_2 is used [19, 21]. The algorithm is the modified version of the GKM, and it is based on the incremental approach. The flowchart of the MGKM is illustrated in Fig. 8.1.

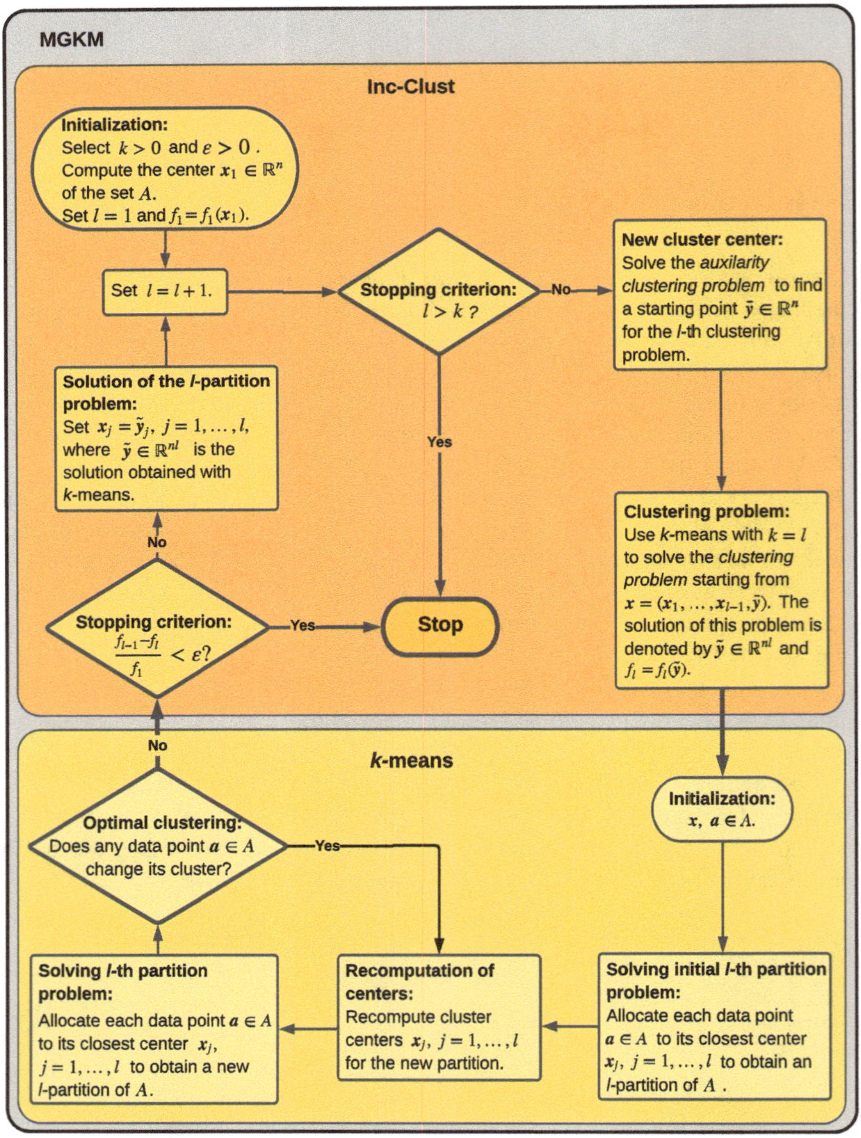

Fig. 8.1 Modified global k-means algorithm (MGKM)

The MGKM starts with the computation of the centroid of the whole data set. Then a new cluster center is added at each iteration. More precisely, the auxiliary clustering problem (7.4) is solved to compute a starting point for the lth center. The new center together with the previous $l-1$ cluster centers is taken as a starting point for solving the lth partition problem. The k-means Algorithm 5.1 is applied starting from this point to find the lth partition of the data set.

Assume that $(\boldsymbol{x}_1, \dots, \boldsymbol{x}_{l-1})$, $l \geq 2$, be a solution to the $(l-1)$th clustering problem. Let $p = 2$. Recall the sets $\bar{S}_1$ and $\bar{S}_2$ defined in (7.7) and (7.8), respectively. Take any $\boldsymbol{y} \in \bar{S}_2$ and consider the sets $\bar{B}_{12}(\boldsymbol{y})$ and $\bar{B}_3(\boldsymbol{y})$ given in (7.9). The algorithm for finding a starting point for the lth cluster center involves Algorithm 5.1 and proceeds as follows.

Algorithm 8.1 Finding a starting point

Input: Data set A and the solution $(\boldsymbol{x}_1, \dots, \boldsymbol{x}_{l-1})$ to the $(l-1)$th clustering problem, $l \geq 2$.
Output: Starting point for the lth cluster center.

1: For each $\boldsymbol{a} \in \bar{S}_2 \cap A$, compute the set $\bar{B}_3(\boldsymbol{a})$, its center $\boldsymbol{c}$ and the value $\bar{f}_{l,\boldsymbol{a}} = \bar{f}_l(\boldsymbol{c})$ of the auxiliary cluster function $\bar{f}_l$ at the point $\boldsymbol{c}$.

2: Compute

$$\bar{f}_{l,\min} = \min_{\boldsymbol{a} \in \bar{S}_2 \cap A} \bar{f}_{l,\boldsymbol{a}}, \quad \bar{\boldsymbol{a}} = \operatorname{argmin}_{\boldsymbol{a} \in \bar{S}_2 \cap A} \bar{f}_{l,\boldsymbol{a}},$$

and select the corresponding center $\bar{\boldsymbol{c}}$.

3: Compute the set $\bar{B}_3(\bar{\boldsymbol{c}})$ and its center.

4: Reassign data points and update the center of the set $\bar{B}_3(\bar{\boldsymbol{c}})$ until no more points escape or return to this set. Let $\bar{\boldsymbol{y}}$ be the final value of $\bar{\boldsymbol{c}}$. Then $\bar{\boldsymbol{y}}$ is a starting point for the lth cluster center.

In Steps 1 and 2 of Algorithm 8.1, a starting point is found to minimize the auxiliary cluster function $\bar{f}_l$, given in (7.5). This point is chosen among all data points that can attract at least one data point (see Step 1). For each such data point $\boldsymbol{a}$, the set $\bar{B}_3(\boldsymbol{a})$ and its center are computed. Then the function $\bar{f}_l$ is evaluated at these centers, and the center that provides the lowest value of the function $\bar{f}_l$ is selected as a starting point to minimize the function $\bar{f}_l$.

In Step 4 of Algorithm 8.1, we apply Algorithm 5.1 to minimize the auxiliary cluster function $\bar{f}_l$. In this case the first $l-1$ cluster centers are fixed and only the lth cluster center is updated at each iteration.

Remark 8.1 Algorithm 8.1 is a special case of Algorithm 7.2 when we select $\gamma_1 = 0$ and $\gamma_2 = 1$.

Proposition 8.1 *Let $\bar{\boldsymbol{y}}$ be a cluster center generated by Algorithm 8.1. Then this point is a local minimizer of the auxiliary cluster function $\bar{f}_l$.*

Proof Recall the sets $B_i(\mathbf{y}),\ i = 1, 2, 3$ defined in (4.30). Since $\bar{\mathbf{y}}$ is a cluster center we have $B_2(\bar{\mathbf{y}}) = \emptyset$. This is due to the fact that in the hard clustering problem, each data point belongs to only one cluster. Then the function (7.5) can be rewritten as

$$\bar{f}_l(\bar{\mathbf{y}}) = \frac{1}{m}\left(\sum_{\mathbf{a}\in B_1(\bar{\mathbf{y}})} r_{l-1}^{\mathbf{a}} + \sum_{\mathbf{a}\in B_3(\bar{\mathbf{y}})} d_2(\bar{\mathbf{y}}, \mathbf{a})\right).$$

It is clear that $\bar{\mathbf{y}}$ is a minimum point of the convex function

$$\varphi(\mathbf{x}) = \frac{1}{m}\sum_{\mathbf{a}\in B_3(\bar{\mathbf{y}})} d_2(\mathbf{x}, \mathbf{a}),$$

that is $\varphi(\bar{\mathbf{y}}) \le \varphi(\mathbf{x})$ for all $\mathbf{x} \in \mathbb{R}^n$. There exists $\varepsilon > 0$ such that

$$d_2(\mathbf{y}, \mathbf{a}) > r_{l-1}^{\mathbf{a}} \quad \text{for all} \quad \mathbf{a} \in B_1(\bar{\mathbf{y}}) \quad \text{and} \quad \text{for all} \quad \mathbf{y} \in B(\bar{\mathbf{y}}; \varepsilon), \quad \text{and}$$
$$d_2(\mathbf{y}, \mathbf{a}) < r_{l-1}^{\mathbf{a}} \quad \text{for all} \quad \mathbf{a} \in B_3(\bar{\mathbf{y}}) \quad \text{and} \quad \text{for all} \quad \mathbf{y} \in B(\bar{\mathbf{y}}; \varepsilon).$$

Then for any $\mathbf{y} \in B(\bar{\mathbf{y}}; \varepsilon)$ we have

$$\begin{aligned}
\bar{f}_l(\mathbf{y}) &= \frac{1}{m}\left(\sum_{\mathbf{a}\in B_1(\bar{\mathbf{y}})} r_{l-1}^{\mathbf{a}} + \sum_{\mathbf{a}\in B_3(\bar{\mathbf{y}})} d_2(\mathbf{y}, \mathbf{a})\right) \\
&= \frac{1}{m}\sum_{\mathbf{a}\in B_1(\bar{\mathbf{y}})} r_{l-1}^{\mathbf{a}} + \varphi(\mathbf{y}) \\
&\ge \frac{1}{m}\sum_{\mathbf{a}\in B_1(\bar{\mathbf{y}})} r_{l-1}^{\mathbf{a}} + \varphi(\bar{\mathbf{y}}) \\
&= \bar{f}_l(\bar{\mathbf{y}}).
\end{aligned}$$

Therefore, $\bar{f}_l(\mathbf{y}) \ge \bar{f}_l(\bar{\mathbf{y}})$ for all $\mathbf{y} \in B(\bar{\mathbf{y}}; \varepsilon)$. This completes the proof. □

Next, we give the step by step form of the MGKM.

Algorithm 8.2 Modified global k-means algorithm (MGKM)

Input: Data set A, the number of clusters k to be computed and a tolerance $\varepsilon \geq 0$.
Output: The l-partition of the set A with $l = 1, \ldots, k$.

1: *(Initialization)* Compute the center $\boldsymbol{x}_1 \in \mathbb{R}^n$ of the set A. Let f_1 be the corresponding value of the clustering function (7.3). Set $l = 1$.
2: *(Stopping criterion)* Set $l = l + 1$. If $l > k$, then **stop**—the k-partition problem has been solved.
3: *(Computation of the next cluster center)* Let $\boldsymbol{x}_1, \ldots, \boldsymbol{x}_{l-1}$ be the cluster centers for the $(l-1)$th partition problem. Apply Algorithm 8.1 to find a starting point $\bar{\boldsymbol{y}} \in \mathbb{R}^n$ for the lth cluster center.
4: *(Refinement of all cluster centers)* Select $(\boldsymbol{x}_1, \ldots, \boldsymbol{x}_{l-1}, \bar{\boldsymbol{y}})$ as a new starting point, apply the k-means Algorithm 5.1 to solve the clustering problem (7.2) for $k = l$. Let $(\tilde{\boldsymbol{y}}_1, \ldots, \tilde{\boldsymbol{y}}_l)$ be a solution to this problem and f_l be the corresponding value of the clustering function.
5: *(Stopping criterion)* If

$$\frac{f_{l-1} - f_l}{f_1} \leq \varepsilon,$$

then **stop**, otherwise set $\boldsymbol{x}_j = \tilde{\boldsymbol{y}}_j, \ j = 1, \ldots, l$ and go to Step 2.

Algorithm 8.2 has two stopping criteria. The algorithm stops when either it solves all l-partition problems, $l = 1, \ldots, k$ or the stopping criterion in Step 5 is satisfied. Note that $f_l^* = \inf\{f_l(\boldsymbol{x}), \ \boldsymbol{x} \in \mathbb{R}^{nk}\} \geq 0$ for all $l \geq 1$ and the sequence $\{f_l^*\}$ is decreasing, that is,

$$f_{l+1}^* \leq f_l^* \quad \text{for all} \quad l \geq 1.$$

This means that the stopping criterion in Step 5 will be satisfied after a finite number of iterations and therefore, Algorithm 8.2 computes as many clusters as the data set A contains with respect to the tolerance $\varepsilon > 0$. Note that the choice of this tolerance is crucial for Algorithm 8.2: large values of ε can result in the appearance of large clusters whereas small values can lead to small and artificial clusters.

In Step 3 of Algorithm 8.2, a starting point for the lth cluster center is computed. This is done by applying Algorithm 8.1 and minimizing the auxiliary cluster function. This algorithm requires the calculation of the distance or affinity matrix of the data set A. The matrix can be computed before the application of Algorithm 8.1 in small and medium sized data sets. However, it cannot be done for large data sets as such a matrix is too big to be stored in the memory of a computer. This means that in the latter case, the affinity matrix is computed at each iteration of the MGKM.

8.3 Fast Modified Global k-Means Algorithm

Algorithm 8.2 is time-consuming in large data sets as it requires the computation of the affinity matrix at each iteration. The *fast modified global k-means algorithm* (FMGKM) [29] is an improved version of Algorithm 8.2 and does not rely on the affinity matrix to compute starting points. Instead, the FMGKM uses some weights within the auxiliary cluster function for generating starting points from different parts of the data set. This leads to the elimination of computing or sorting the whole affinity matrix and therefore, to the reduction of computational effort and the memory usage. The flowchart of the FMGKM is similar to that of the MGKM given in Fig. 8.1.

Next, we describe the FMGKM. Let

$$U = \{u_1, \ldots, u_s\}$$

be a finite set of positive numbers. For $u \in U$, the auxiliary cluster function $\bar{f}_l$, given in (7.5), is modified as follows:

$$\bar{f}_l^u(\mathbf{y}) = \frac{1}{m} \sum_{\mathbf{a} \in A} \min \left\{ r_{l-1}^{\mathbf{a}}, u\, d_2(\mathbf{y}, \mathbf{a}) \right\}. \tag{8.1}$$

If $u = 1$, then $\bar{f}_l^u(\mathbf{y}) = \bar{f}_l(\mathbf{y})$ for all $\mathbf{y} \in \mathbb{R}^n$. Take $u \in U$ and define the set

$$\tilde{S}_2^u = \left\{ \mathbf{y} \in \mathbb{R}^n : r_{l-1}^{\mathbf{a}} > u\, d_2(\mathbf{y}, \mathbf{a}) \ \text{ for some } \mathbf{a} \in A \right\},$$

and for any $\mathbf{y} \in \tilde{S}_2^u$ consider the set

$$\tilde{B}_3^u(\mathbf{y}) = \left\{ \mathbf{a} \in A : r_{l-1}^{\mathbf{a}} > u\, d_2(\mathbf{y}, \mathbf{a}) \right\}.$$

The set $\tilde{S}_2^u$ is similar to the set $\bar{S}_2$ defined in (7.8) and the set $\tilde{B}_3^u(\mathbf{y})$ is similar to the set $\bar{B}_3(\mathbf{y})$ described in (7.9). The set $\tilde{B}_3^u(\mathbf{y})$ contains all data points attracted by the point $\mathbf{y} \in \tilde{S}_2^u$ with a given weight $u > 0$.

The following algorithm is a modified version of Algorithm 8.1 and computes a starting point for the lth cluster center.

Algorithm 8.3 Finding a starting point

Input: Data set A, the solution $(\boldsymbol{x}_1, \ldots, \boldsymbol{x}_{l-1})$ to the $(l-1)$th clustering problem, $l \geq 2$ and the set $U = \{u_1, \ldots, u_s\}$.
Output: Starting point for the lth cluster center.

1: Set $t = 1$.
2: Take $u_t \in U$. For each $\boldsymbol{a} \in \bar{S}_2^{u_t} \cap A$ compute the set $\bar{B}_3^{u_t}(\boldsymbol{a})$, its center $\boldsymbol{c}$ and the value $\bar{f}_{l,\boldsymbol{a}}^{u_t} = \bar{f}_l^{u_t}(\boldsymbol{c})$ of the function $\bar{f}_l^{u_t}$ at the point $\boldsymbol{c}$.
3: Compute

$$\bar{f}_{l,\min}^{u_t} = \min_{\boldsymbol{a} \in \bar{S}_2^{u_t} \cap A} \bar{f}_{l,\boldsymbol{a}}^{u_t} \quad \text{and} \quad \bar{\boldsymbol{a}} = \operatorname*{argmin}_{\boldsymbol{a} \in \bar{S}_2^{u_t} \cap A} \bar{f}_{l,\boldsymbol{a}}^{u_t},$$

and select the corresponding center $\bar{\boldsymbol{c}}_t$.
4: Compute the set $\bar{B}_3^{u_t}(\bar{\boldsymbol{c}}_t)$ and its center.
5: Reassign data points until no more points escape or return to this set. Let $\bar{\boldsymbol{y}}(u_t)$ be the final value for the center $\bar{\boldsymbol{c}}_t$. Compute the value $\bar{f}_{l,t}$ of the auxiliary function $\bar{f}_l$, given in (7.5), at the point $\bar{\boldsymbol{y}}(u_t)$.
6: Set $t = t + 1$. If $t \leq s$, then go to Step 2.
7: Compute

$$\bar{f}_{l,\min} = \min_{t=1,\ldots,s} \bar{f}_{l,t},$$

and let

$$\bar{f}_{l,t_0} = \bar{f}_{l,\min} \quad \text{for} \quad t_0 \in \{1, \ldots, s\}.$$

Set $\bar{\boldsymbol{y}} = \bar{\boldsymbol{y}}(u_{t_0})$ as a starting point for the lth cluster center.

In order to solve the auxiliary clustering problem (7.4) in Step 5 of Algorithm 8.3, we apply Algorithm 5.1. Here, only one cluster center is updated, other cluster centers are known from previous iterations and they are fixed. Since Algorithm 5.1 is only able to find local solutions to this problem more than one starting points are used to compute high quality solutions.

Starting points are computed using the function (8.1) with different values of u. If u is sufficiently small, then the starting point will be close to other cluster centers, most likely near the center of the largest cluster. If $u = 1$, then we get the same starting point as in the case of Algorithm 8.1. As the value of u is increased the starting points become more isolated data points. This leads to the finding of starting points from different parts of the data set.

The FMGKM is presented in step by step form as follows.

Algorithm 8.4 Fast modified global k-means algorithm (FMGKM)

Input: Data set A, the number of clusters k to be computed and a tolerance $\varepsilon > 0$.
Output: The l-partition of the set A with $l = 1, \ldots, k$.

1: *(Initialization)* Compute the center $\boldsymbol{x}_1 \in \mathbb{R}^n$ of the set A. Let f_1 be the corresponding value of the clustering function (7.3). Set $l = 1$.

2: *(Stopping criterion)* Set $l = l + 1$. If $l > k$, then **stop**—the k-partition problem has been solved.

3: *(Computation of the next cluster center)* Let $\boldsymbol{x}_1, \ldots, \boldsymbol{x}_{l-1}$ be the cluster centers for the $(l-1)$th partition problem. Apply Algorithm 8.3 to find a starting point $\bar{\boldsymbol{y}} \in \mathbb{R}^n$ for the lth cluster center.

4: *(Refinement of all cluster centers)* Select $(\boldsymbol{x}_1, \ldots, \boldsymbol{x}_{l-1}, \bar{\boldsymbol{y}})$ as a new starting point, apply Algorithm 5.1 to solve the clustering problem (7.2) for $k = l$. Let $(\tilde{\boldsymbol{y}}_1, \ldots, \tilde{\boldsymbol{y}}_l)$ be a solution to this problem and f_l be the corresponding value of the clustering function.

5: *(Stopping criterion)* If

$$\frac{f_{l-1} - f_l}{f_1} < \varepsilon,$$

then **stop**, otherwise set $\boldsymbol{x}_j = \tilde{\boldsymbol{y}}_j, \; j = 1, \ldots, l$ and go to Step 2.

The most time-consuming step in Algorithm 8.4 is Step 3, where Algorithm 8.3 is applied to minimize the auxiliary cluster function (8.1) for different $u \in U$ and to find the starting point for the lth cluster center. In its turn, Step 2 of Algorithm 8.3 is time-consuming as in this step, clusters are computed for each data point $\boldsymbol{a} \in \tilde{S}_2^{u_t} \cap A$. This requires the partial computation of the affinity matrix. In addition, centers of those clusters and values of the function $\bar{f}_l^u$ at these centers need to be computed. Since for each data point only one center is obtained the complexity of the computation of the function $\bar{f}_l^u$ is the same as the complexity of the computation of the affinity matrix.

In [29], two different approaches are introduced to reduce the computational complexity of Step 2 in Algorithm 8.3. Both approaches exploit the incremental nature of the algorithm. In these approaches a matrix, consisting of the distances between data points and cluster centers is used instead of the affinity matrix. Since the number of clusters is significantly less than the number of data points the former matrix is much smaller than the latter one. More precisely, in these approaches data points which are close to cluster centers from the $(l-1)$th partition are excluded. Therefore, these data points are removed from the list of points which can attract large clusters and also from the list of points which can be attracted by non-excluded data points.

Let $\boldsymbol{x}_1, \ldots, \boldsymbol{x}_{l-1},\ l \geq 2$ be known cluster centers. Assume v_{iq} is the squared Euclidean distance between the data point $\boldsymbol{a}_i,\ i = 1, \ldots, m$ and the cluster center $\boldsymbol{x}_q,\ q = 1, \ldots, l-1$, that is

$$v_{iq} = d_2(\boldsymbol{a}_i, \boldsymbol{x}_q).$$

Let $\boldsymbol{v}_q = \big(v_{1q}, \ldots, v_{mq}\big) \in \mathbb{R}^m,\ q = 1, \ldots, l-1$. Consider a matrix V of the size $m \times (l-1)$, whose columns are vectors $\boldsymbol{v}_q,\ q = 1, \ldots, l-1$:

$$V = [\boldsymbol{v}_{iq}], \quad i = 1, \ldots, m,\ q = 1, \ldots, l-1.$$

Let also $\boldsymbol{r} = (r_{l-1}^1, \ldots, r_{l-1}^m)$ be a vector of m components where r_{l-1}^i is the squared Euclidean distance between the data point $\boldsymbol{a}_i$ and its cluster center in the $(l-1)$th partition (see (7.6)). Note that the matrix V and the vector $\boldsymbol{r}$ are available after the $(l-1)$th iteration of the incremental clustering algorithm.

Now, we are ready to describe the following two approaches to reduce the computational complexity of Step 2 of Algorithm 8.3.

1. *Reduction of the number of pairwise distance computations.* Let a data point $\boldsymbol{a}_j \in A$ be given and $\boldsymbol{x}_{q(j)}$ be its cluster center. Here $q(j) \in \{1, \ldots, l-1\}$. For a given $u \in U$ and the data point $\boldsymbol{a}_i$ if

$$\Big(1 + \frac{1}{\sqrt{u}}\Big)^2 r_{l-1}^j \leq v_{iq(j)},$$

then $\boldsymbol{a}_j \notin \tilde{B}_3^u(\boldsymbol{a}_i)$. Indeed, we have

$$\|\boldsymbol{a}_i - \boldsymbol{a}_j\| \geq \|\boldsymbol{a}_i - \boldsymbol{x}_{q(j)}\| - \|\boldsymbol{a}_j - \boldsymbol{x}_{q(j)}\| \geq (1/\sqrt{u})\|\boldsymbol{a}_j - \boldsymbol{x}_{q(j)}\|.$$

Thus, we get $ud_2(\boldsymbol{a}_i, \boldsymbol{a}_j) \geq r_{l-1}^j$, and therefore $\boldsymbol{a}_j \notin \tilde{B}_3^u(\boldsymbol{a}_i)$. This condition allows us to reduce the number of pairwise distance computations. This reduction becomes substantial as the number of clusters increases. Define the set

$$\tilde{R}^u(\boldsymbol{a}_i) = \Big\{\boldsymbol{a}_j \in A : \Big(1 + \frac{1}{\sqrt{u}}\Big)^2 r_{l-1}^j > v_{iq(j)}\Big\}.$$

It is clear that

$$\tilde{B}_3^u(\boldsymbol{a}_i) \subseteq \tilde{R}^u(\boldsymbol{a}_i).$$

The set $\tilde{R}^u(\boldsymbol{a}_i)$ can be used instead of the set A to compute the value of the function $\bar{f}_l^u$ in Step 2 of Algorithm 8.3. In this case, one may not get the exact value of this function; however, it gives a good approximation to the exact value. Furthermore, take

$$w \in \left(1, \left(1 + \frac{1}{\sqrt{u}}\right)^2\right],$$

and consider the set

$$\bar{R}^u_w(\boldsymbol{a}_i) = \left\{\boldsymbol{a}_j \in A : \; wr^j_{l-1} > d_2(\boldsymbol{a}_i, \boldsymbol{a}_j)\right\}.$$

Then replace A by $\bar{R}^u_w(\boldsymbol{a}_i)$ for the computation of the function $\bar{f}^u_l$. This will further reduce the amount of computations in Step 2 of Algorithm 8.3.

2. *Reduction of the number of starting cluster centers.* This approach is similar to that of considered in Algorithm 7.4. More specifically, data points which are very close to previous cluster centers are not considered for being starting points to minimize the auxiliary cluster function (8.1). At the $(l-1)$th iteration a squared averaged radius

$$\bar{d}^q_{av} = \frac{1}{|A^q|} \sum_{\boldsymbol{a} \in A^q} d_2(\boldsymbol{x}_q, \boldsymbol{a}),$$

and a squared maximum radius

$$\bar{d}^q_{\max} = \max_{\boldsymbol{a} \in A^q} d_2(\boldsymbol{x}_q, \boldsymbol{a})$$

of each cluster $A^q, \; q = 1, \ldots, l-1$ are computed. Consider the numbers

$$\alpha_q = \frac{\bar{d}^q_{\max}}{\bar{d}^q_{av}} \ge 1 \quad \text{and} \quad \beta_q = \varepsilon(\alpha_q - 1),$$

where $\varepsilon > 0$ is a sufficiently small number. Let

$$\gamma_{ql} = 1 + \beta_q(l-1), \quad q = 1, \ldots, l-1.$$

It is clear that $\gamma_{ql} \ge 1, \; q = 1, \ldots, l-1$. Define the following subset of the cluster A^q:

$$\bar{A}^q = \left\{\boldsymbol{a} \in A^q : \; \gamma_{ql}\bar{d}^q_{av} \le d_2(\boldsymbol{x}_q, \boldsymbol{a})\right\}.$$

In other words, the set $\bar{A}^q$ is obtained from the cluster A^q by removing all points for which $d_2(\boldsymbol{x}_q, \boldsymbol{a}) < \gamma_q \bar{d}^q_{av}$. Since in the incremental approach the clusters are becoming more stable as the number l increases it follows that the numbers γ_{ql} are increased as l increases. Define the set

$$\bar{A} = \bigcup_{q=1}^{l-1} \bar{A}^q,$$

and consider only data points $\boldsymbol{a} \in \bar{A}$ as the candidates to be starting points for minimizing the auxiliary cluster function $\bar{f}_l$.

Summarizing, Steps 2 and 3 of Algorithm 8.3 can be rewritten as follows:

2': for each $\boldsymbol{a} \in \tilde{S}_2^{u_t} \cap \bar{A}$ compute the set $\tilde{B}_3^{u_t}(\boldsymbol{a})$, its center $\boldsymbol{c}$, and the value $\bar{f}_{l,\boldsymbol{a}}^{u_t} = \bar{f}_l^{u_t}(\boldsymbol{c})$ of the function $\bar{f}_l^{u_t}$ at the point $\boldsymbol{c}$ over the set $\tilde{R}_w^u(\boldsymbol{a})$.

3': compute

$$\bar{f}_{l,\min}^{u_t} = \min_{\boldsymbol{a} \in \tilde{S}_2^{u_t} \cap \bar{A}} \bar{f}_{l,\boldsymbol{a}}^{u_t} \quad \text{and} \quad \bar{\boldsymbol{a}} = \operatorname*{argmin}_{\boldsymbol{a} \in \tilde{S}_2^{u_t} \cap \bar{A}} \bar{f}_{l,\boldsymbol{a}}^{u_t},$$

and the corresponding center $\bar{\boldsymbol{c}}$.

The use of these two schemes allows us to significantly reduce the computational complexity of Algorithm 8.4 and accelerate its convergence.

8.4 Limited Memory Bundle Method for Clustering

In this section, we present the *limited memory bundle method for clustering* (LMB-CLUST) [171]. The LMB-CLUST has been developed specifically to solve clustering problems in large data sets. The algorithm combines two different approaches to solve the clustering problem when the squared Euclidian distance d_2 is used as a similarity measure. The MSINC-CLUST is used to solve the clustering problem globally and the LMBM is applied at each iteration of the algorithm to solve both the clustering problem (7.2) and the auxiliary clustering problem (7.4). The flowchart of the LMB-CLUST is given in Fig. 8.2.

The LMBM, given in Fig. 3.5, is originally developed for solving general large-scale nonconvex NSO problems. Here, this method is slightly modified to be better suited for solving the clustering and the auxiliary clustering problems. In particular, a nonmonotone line search is used to find step sizes t_L^h and t_R^h. In addition, different stopping tolerances are utilized for different problems. That is, the tolerance ε is set to be relatively large for the auxiliary clustering problem (7.4)—since this problem need not to be solved very accurately—and smaller for the clustering problem (7.2).

Next, we give the modified version of the LMBM in its step by step form. We use $\boldsymbol{x}_1$ for the starting point; $\varepsilon_c > 0$ for the stopping tolerance; ε_L and ε_R for line search parameters; γ for the distance measure parameter; $\hat{m}_c$ for the maximum number of stored correction vectors used to form limited memory matrix updates; $t_{\max}$ is an upper bound for serious steps; C is a control parameter for the length of the direction vector. We also use i_{type} to show the type of the problem, that is:

- $i_{\text{type}} = 0$: the auxiliary clustering problem (7.4); and
- $i_{\text{type}} = 1$: the clustering problem (7.2).

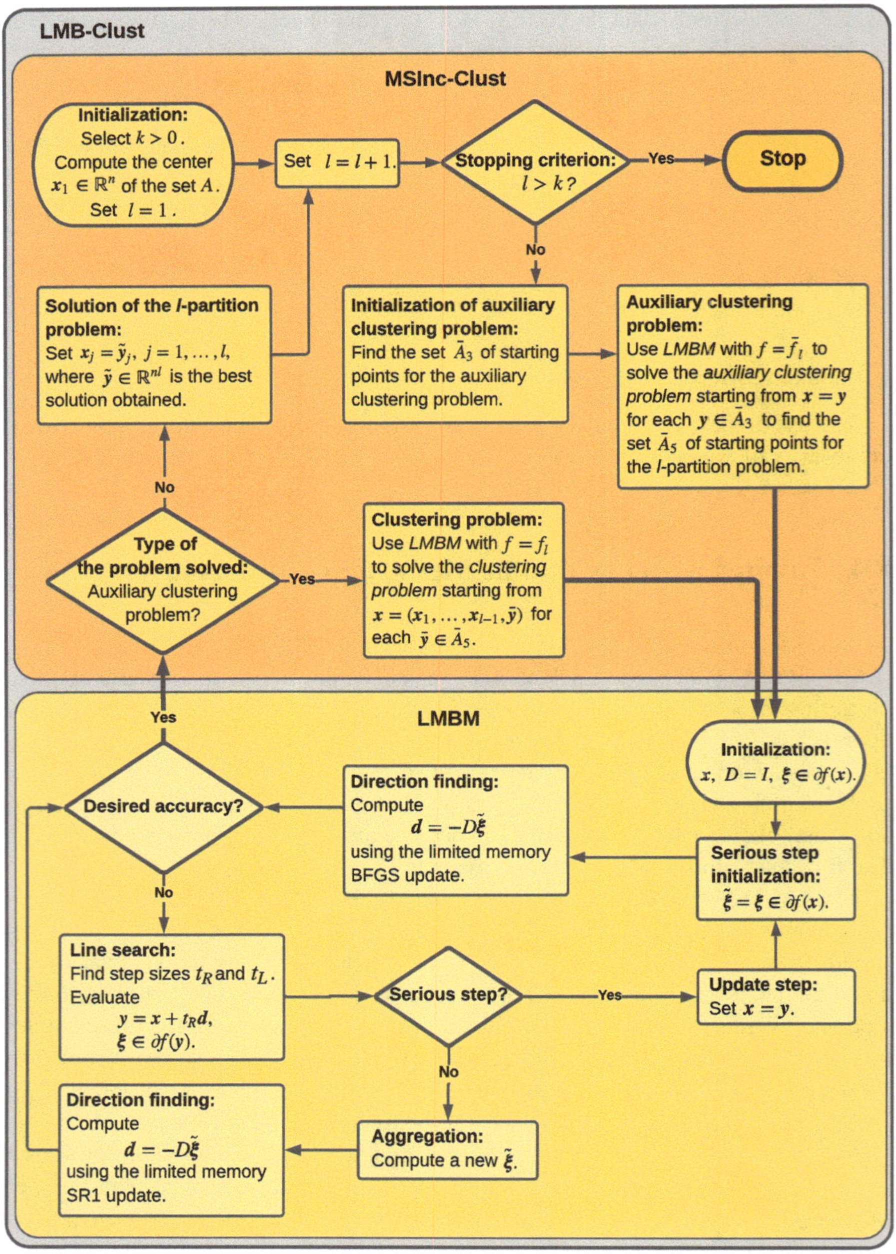

Fig. 8.2 Limited memory bundle method for clustering (LMB-CLUST)

In both cases, the objective function is denoted by f and the number of variables in the optimization problem is denoted by n. Hence $f = \bar{f}_l$ and $n = n$ for the auxiliary clustering problem and $f = f_l$ and $n = nl$ for the lth clustering problem.

Algorithm 8.5 Modified limited memory bundle algorithm

Input: $x_1 \in \mathbb{R}^n$, $\varepsilon_c > 0$, $\varepsilon_L \in (0, 1/2)$, $\varepsilon_R \in (\varepsilon_L, 1/2)$, $t_{\max} > 1$, $\gamma > 0$, $C > 0$, $\hat{m}_c \geq 3$ and $i_{\text{type}} \in \{0, 1\}$.
Output: Clarke stationary point x_h.

1: *(Initialization)* Set $y_1 = x_1$ and $\beta_1 = 0$. Compute $f_1 = f(x_1)$ and $\xi_1 \in \partial f(x_1)$. If $i_{\text{type}} = 1$, then set $\varepsilon = \varepsilon_c$. Otherwise, set $\varepsilon = 10^3\varepsilon_c$. Set $h = 1$.

2: *(Serious step initialization)* Set $\tilde{\xi}_h = \xi_h$, $\tilde{\beta}_h = 0$ and $m = h$.

3: *(Direction finding)* If $h = 1$, set $d_1 = -\xi_1$. Otherwise, compute

$$d_h = -D_h\tilde{\xi}_h$$

by the L-BFGS update if $m = h$ (use $\hat{m}_h = \min\{h-1, \hat{m}_c\}$ correction vectors in U_h and S_h) and by the L-SR1 update, otherwise.

4: *(Stopping criterion)* Compute $w_h = -\tilde{\xi}_h^T d_h + 2\tilde{\beta}_h$. If $w_h < \varepsilon$, then **stop** with x_h as the final solution.

5: *(Line search)* Set the scaling parameter θ_h for the length of the direction vector as $\theta_h = \min\{\, 1, C/\|d_h\|\,\}$. Use a nonmonotone line search to determine the step sizes $t_R^h \in (0, t_{\max}]$ and $t_L^h \in [0, t_R^h]$ and set the corresponding values

$$\begin{aligned} x_{h+1} &= x_h + t_L^h\theta_h d_h, & f_{h+1} &= f(x_{h+1}), \quad \text{and} \\ y_{h+1} &= x_h + t_R^h\theta_h d_h, & \xi_{h+1} &\in \partial f(y_{h+1}). \end{aligned}$$

Set $u_h = \xi_{h+1} - \xi_m$ and $s_h = y_{h+1} - x_h = t_R^h\theta_h d_h$ and append these values to U_h and S_h. If the modified serious step condition

$$t_R^h = t_L^h > 0 \quad \text{and} \quad f(y_{h+1}) \leq \max_{i\in M} f(x_i) - \varepsilon_L t_R^h w_h,$$

where $M \subseteq \{l : x_{l+1} = x_l + t_R^l\theta_l d_l\}$ such that M contains at most the ten greatest indices l, is satisfied, then set $\beta_{h+1} = 0$, $h = h+1$ and go to Step 2. Otherwise, calculate the locality measure β_{h+1} by

$$\beta_{h+1} = \max\left\{|f(x_h) - f(y_{h+1}) + \xi_{h+1}^T(y_{h+1} - x_h)|, \gamma\|y_{h+1} - x_h\|^2\right\}.$$

6: *(Aggregation)* Determine multipliers $\lambda_i^h \geq 0$ for all $i \in \{1, 2, 3\}$, $\sum_{i=1}^3 \lambda_i^h = 1$ that minimize the function $\varphi(\lambda_1, \lambda_2, \lambda_3)$ given in (3.9), where D_h is calculated by the same updating formula as in Step 3. Compute $\tilde{\xi}_{h+1}$ and $\tilde{\beta}_{h+1}$ as

$$\tilde{\xi}_{h+1} = \lambda_1^h\xi_m + \lambda_2^h\xi_{h+1} + \lambda_3^h\tilde{\xi}_h \quad \text{and} \quad \tilde{\beta}_{h+1} = \lambda_2^h\beta_{h+1} + \lambda_3^h\tilde{\beta}_h.$$

Set $h = h+1$ and go to Step 3.

The convergence properties of the LMBM are given in Sect. 3.4. Here, we recall the most important results in light of the clustering problem. Note that

Assumptions 3.1–3.3, needed to prove the global convergence of the LMBM, are trivially satisfied for both the clustering and the auxiliary clustering problems.

Proposition 8.2 *Assume that $\varepsilon_c = 0$. If the* LMBM *terminates after a finite number of iterations, say at the iteration h, then the point $\boldsymbol{x}_h$ is a Clarke stationary point of the (auxiliary) clustering problem.*

Proposition 8.3 *Assume that $\varepsilon_c = 0$. Every accumulation point $\bar{\boldsymbol{x}}$ generated by the* LMBM *is a Clarke stationary point of the (auxiliary) clustering problem.*

Remark 8.2 The LMBM terminates in a finite number of steps if we choose $\varepsilon_c > 0$.

Next, we describe the LMB-CLUST and give its step by step algorithm. Since the problem (7.2) is nonconvex it is important to select favorable starting points before applying a local search method like the LMBM to solve it. The LMB-CLUST uses the MSINC-CLUST for solving the clustering problem globally and the LMBM is applied at each iteration of the MSINC-CLUST to solve both the problems (7.2) and (7.4).

Algorithm 8.6 Limited memory bundle method for clustering (LMB-CLUST)

Input: Data set A and the number of clusters k to be computed.
Output: The l-partition of the set A with $l = 1, \dots, k$.

1: *(Initialization)* Compute the center $\boldsymbol{x}_1 \in \mathbb{R}^n$ of the set A. Set $l = 1$.

2: *(Stopping criterion)* Set $l = l + 1$. If $l > k$, then **stop**—the k-partition problem has been solved.

3: *(Computation of a set of starting points for the auxiliary clustering problem)* Apply Algorithm 7.2 to find the set $\bar{A}_3 \subset \mathbb{R}^n$ of starting points for the auxiliary clustering problem (7.4).

4: *(Computation of a set of starting points for the clustering problem)* For each $\boldsymbol{y} \in \bar{A}_3$ apply Algorithm 8.5 with $i_{\text{type}} = 0$ to solve the auxiliary clustering problem (7.4) and find $\bar{A}_5$, a set of starting points for the lth partition problem (7.2).

5: *(Computation of a set of cluster centers)* For each $\bar{\boldsymbol{y}} \in \bar{A}_5$ apply Algorithm 8.5 with $i_{\text{type}} = 1$ to solve the clustering problem (7.2) starting from the point $(\boldsymbol{x}_1, \dots, \boldsymbol{x}_{l-1}, \bar{\boldsymbol{y}})$ and find a solution $(\hat{\boldsymbol{y}}_1, \dots, \hat{\boldsymbol{y}}_l)$. Denote by $\bar{A}_6$ a set of all such solutions.

6: *(Computation of the best solution)* Compute

$$f_l^{\min} = \min\Big\{ f_l(\hat{\boldsymbol{y}}_1, \dots, \hat{\boldsymbol{y}}_l) \ : (\hat{\boldsymbol{y}}_1, \dots, \hat{\boldsymbol{y}}_l) \in \bar{A}_6 \Big\},$$

and the collection of cluster centers $(\tilde{\boldsymbol{y}}_1, \dots, \tilde{\boldsymbol{y}}_l)$ such that

$$f_l(\tilde{\boldsymbol{y}}_1, \dots, \tilde{\boldsymbol{y}}_l) = f_l^{\min}.$$

7: *(Solution to the lth partition problem)* Set $\boldsymbol{x}_j = \tilde{\boldsymbol{y}}_j$, $j = 1, \dots, l$ as a solution to the lth partition problem and go to Step 2.

8.5 Discrete Gradient Clustering Algorithm

In this section, we describe the *discrete gradient clustering algorithm* (DG-CLUST) to solve the clustering problem (7.2). As mentioned in Sect. 3.8, the underlying optimization solver DGM is a semi-derivative free method for solving nonconvex NSO problems. The DGM does not use subgradients or their approximations but only at the end of the solution process and thus, it can be used to solve problems which are not subdifferentially regular. Therefore, the clustering algorithm based on the DGM can be used to solve clustering problems with the similarity measures d_1 and d_∞, in addition to d_2 based clustering problems.

The flowchart of the DG-CLUST is given in Fig. 8.3. Similar to other optimization based clustering algorithms, the DG-CLUST uses the MSINC-CLUST for solving the clustering problem globally and the DGM is applied at each iteration of the MSINC-CLUST to solve both the problems (7.2) and (7.4).

The flowchart of the DGM with a more detailed description is given in Sect. 3.8. Here, we give this method in its step by step form. Note that we use $\boldsymbol{x}_1$ for the starting point; $\varepsilon > 0$ for the stopping tolerance; and ε_L and ε_R for line search parameters.

As before, we use the following notations: the objective function is denoted by f and n stands for the size of the optimization problem. That is, $f = \bar{f}_l$ and $n = n$ for the auxiliary clustering problem and $f = f_l$ and $n = nl$ for the l-partition problem.

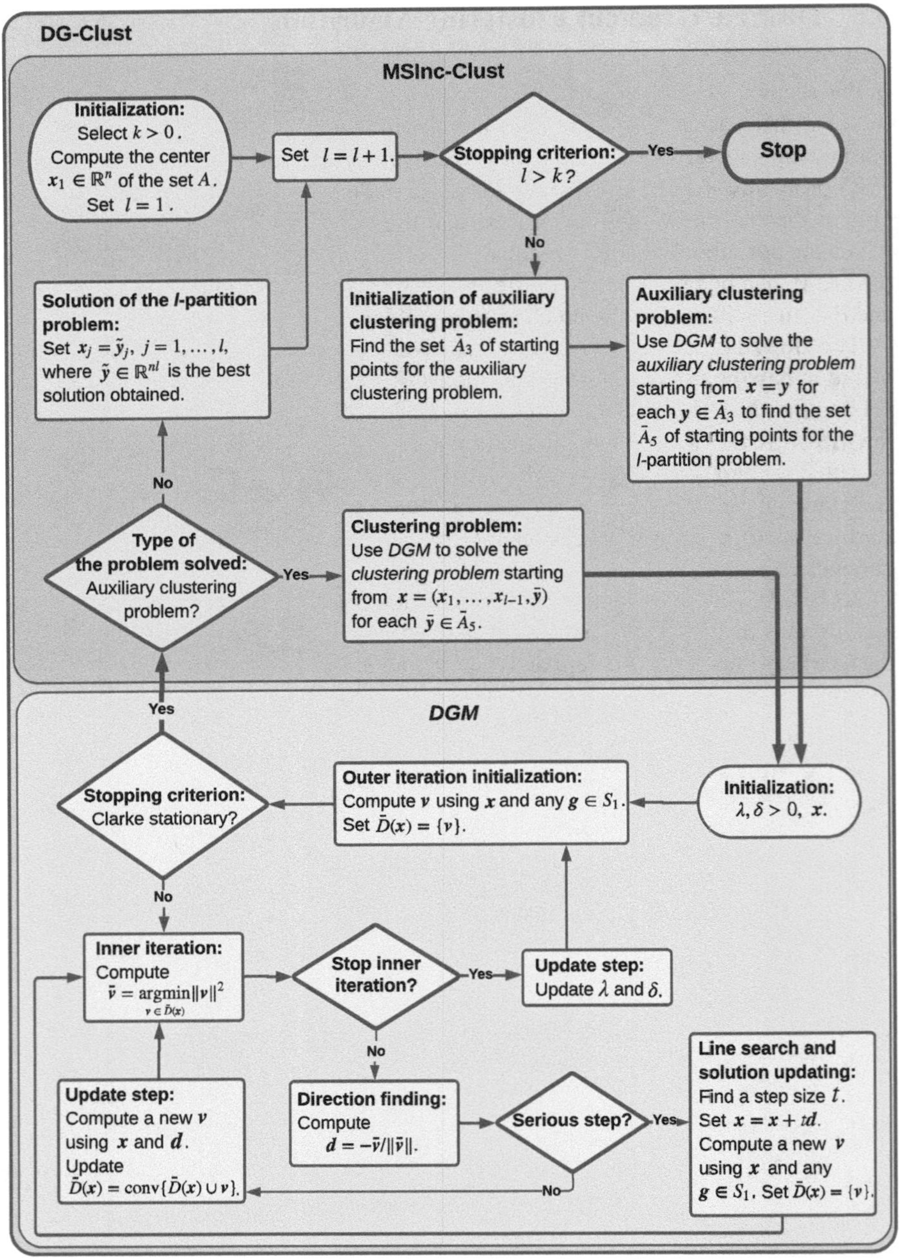

Fig. 8.3 Discrete gradient clustering algorithm (DG-CLUST)

Algorithm 8.7 Discrete gradient method

Input: $x_1 \in \mathbb{R}^n$, $\varepsilon > 0$, $\lambda_1 > 0$, $\delta_1 > 0$, $\alpha \in (0,1]$, $\varepsilon_L \in (0,1]$ and $\varepsilon_R \in (0, \varepsilon_L]$.
Output: Final solution x_h.

1: *(Outer iteration initialization)* Set $h = 1$.

2: *(Inner iteration initialization)* Set $s = 1$ and $x_{h_s} = x_h$. Choose any $g \in S_s$, $w \in G$. Compute a discrete gradient $v_{h_s} = \Gamma^i(x_{h_s}, g, w, \lambda_h, \alpha)$. Set $\bar{D}(x_{h_s}) = \{v_{h_s}\}$.

3: *(Stopping criterion)* If $\lambda_h < \varepsilon$ and $\delta_h < \varepsilon$, then **stop** with x_h as a final solution.

4: (Minimum norm). Compute the vector

$$\bar{v}_{h_s} = \operatorname*{argmin}_{v \in \bar{D}(x_{h_s})} \|v\|^2.$$

5: *(Inner iteration termination)* If $\|\bar{v}_{h_s}\| \leq \delta_h$, then update λ_{h+1} and δ_{h+1}. Set $x_{h+1} = x_{h_s}$, $h = h + 1$ and go to Step 2.

6: *(Search direction)* Compute the search direction

$$d_{h_s} = -\frac{\bar{v}_{h_s}}{\|\bar{v}_{h_s}\|}.$$

7: If $f(x_{h_s} + \lambda_h d_{h_s}) - f(x_{h_s}) > -\varepsilon_L \lambda_h \|\bar{v}_{h_s}\|$, then go to Step 9.

8: *(Serious step)* Construct $x_{h_{s+1}} = x_{h_s} + t_{h_s} d_{h_s}$, where the step size t_{h_s} is computed as

$$t_{h_s} = \operatorname{argmax} \left\{ t \geq 0 : f(x_{h_s} + t d_{h_s}) - f(x_{h_s}) \leq -\varepsilon_R t \|\bar{v}_{h_s}\| \right\}.$$

Compute a new discrete gradient $v_{h_{s+1}}$ using $x_{h_{s+1}}$ and any $g \in S_1$:

$$v_{h_{s+1}} = \Gamma^i(x_{h_{s+1}}, g, w, \lambda_h, \alpha).$$

Set $\bar{D}(x_{h_{s+1}}) = \{v_{h_{s+1}}\}$, $s = s + 1$ and go to Step 4.

9: *(Null step)* Compute a new discrete gradient $v_{h_{s+1}}$ using x_{h_s} and d_{h_s}:

$$v_{h_{s+1}} = \Gamma^i(x_{h_s}, d_{h_s}, w, \lambda_h, \alpha).$$

Update the set

$$\bar{D}(x_{h_{s+1}}) = \operatorname{conv} \left\{ \bar{D}(x_{h_s}) \cup \{v_{h_{s+1}}\} \right\}.$$

Set $x_{h_{s+1}} = x_{h_s}$, $s = s + 1$ and go to Step 4.

The global convergence of Algorithm 8.7 has been studied in Sect. 3.8. Note that assumptions needed to get its convergence are satisfied for both the cluster and the auxiliary cluster functions. Next, we present the step by step description of the DG-CLUST.

Algorithm 8.8 Discrete gradient clustering algorithm (DG-CLUST)

Input: Data set A and the number of clusters k to be computed.
Output: The l-partition of the set A with $l = 1, \dots, k$.

1: *(Initialization)* Compute the center $\boldsymbol{x}_1 \in \mathbb{R}^n$ of the set A. Set $l = 1$.

2: *(Stopping criterion)* Set $l = l + 1$. If $l > k$, then **stop**—the k-partition problem has been solved.

3: *(Computation of a set of starting points for the auxiliary clustering problem)* Apply Algorithm 7.2 to find the set $\bar{A}_3 \subset \mathbb{R}^n$ of starting points for solving the auxiliary clustering problem (7.4).

4: *(Computation of a set of starting points for the lth cluster center)* Apply Algorithm 8.7 to solve the auxiliary clustering problem (7.4) starting from each point $\boldsymbol{y} \in \bar{A}_3$. This algorithm generates a set $\bar{A}_5$ of starting points for the lth cluster center.

5: *(Computation of a set of cluster centers)* For each $\bar{\boldsymbol{y}} \in \bar{A}_5$ apply Algorithm 8.7 to solve the clustering problem (7.2) starting from the point $(\boldsymbol{x}_1, \dots, \boldsymbol{x}_{l-1}, \bar{\boldsymbol{y}})$ and find a solution $(\hat{\boldsymbol{y}}_1, \dots, \hat{\boldsymbol{y}}_l)$. Denote by $\bar{A}_6$ a set of all such solutions.

6: *(Computation of the best solution)* Compute

$$f_l^{\min} = \min\left\{ f_l(\hat{\boldsymbol{y}}_1, \dots, \hat{\boldsymbol{y}}_l) : (\hat{\boldsymbol{y}}_1, \dots, \hat{\boldsymbol{y}}_l) \in \bar{A}_6 \right\},$$

and the collection of cluster centers $(\tilde{\boldsymbol{y}}_1, \dots, \tilde{\boldsymbol{y}}_l)$ such that

$$f_l(\tilde{\boldsymbol{y}}_1, \dots, \tilde{\boldsymbol{y}}_l) = f_l^{\min}.$$

7: *(Solution to the lth partition problem)* Set $\boldsymbol{x}_j = \tilde{\boldsymbol{y}}_j, \ j = 1, \dots, l$ as a solution to the l-partition problem and go to Step 2.

Note that the DGM uses only discrete gradients to find an approximate solution to both the clustering and the auxiliary clustering problems. The calculation of discrete gradients can be simplified using a special structure of a problem such as the piecewise separability or piecewise partial separability of the objective functions (see Sect. 2.6.3).

It is proved in Propositions 4.5 and 4.12 that both the cluster function (7.3) and the auxiliary cluster function (7.5) are piecewise separable with the similarity measures d_1, d_2, and d_∞. Therefore, we can simplify the calculations of discrete gradients for both the cluster and the auxiliary cluster functions.

First, we consider the computation of discrete gradients of the cluster function f_k. This function is the special case of the function f, defined in (2.21) as

$$f(\boldsymbol{x}) = \sum_{i=1}^{m} \max_{h \in \mathcal{H}_i} \min_{j \in \mathcal{J}_h} f_{ihj}(\boldsymbol{x}).$$

The cluster function f_k does not depend on the index h and the sets $\mathcal{H}_i$, $i = 1, \dots, m$ are all singletons. Therefore, for all $h \in \mathcal{H}_i$ we have $\mathcal{J}_h = \{1, \dots, k\}$ and

$$f_k(\boldsymbol{x}) = \frac{1}{m} \sum_{i=1}^{m} \min_{l=1,\dots,k} f_{il}(\boldsymbol{x}_l),$$

where $f_{il}(\boldsymbol{x}_l) = d_p(\boldsymbol{x}_l, \boldsymbol{a}_i),\ l = 1, \ldots, k.$

Then the term functions are

$$(x_{lt} - a_{it})^2 \quad \text{for} \quad p = 2, \quad \text{and}$$
$$|x_{lt} - a_{it}| \quad \text{for} \quad p = 1, \infty.$$

Here, $t = 1, \ldots, n,\ l = 1, \ldots, k,\ i = 1, \ldots, m$, and therefore, the total number of such term functions is mnk. Since the function f_k has nk number of variables one needs $nk + 1$ evaluations of this function to compute its one discrete gradient. Then the total number of evaluations of term functions to compute one discrete gradient of f_k is $N_t = mnk(nk + 1)$.

According to the definition of the discrete gradients for a given $i \in \{1, \ldots, nk\}$ we compute values of the function f_k at the following $nk + 1$ points:

$$\boldsymbol{x}, \boldsymbol{x}^0, \boldsymbol{x}^1, \ldots, \boldsymbol{x}^{i-1}, \boldsymbol{x}^{i+1}, \ldots, \boldsymbol{x}^{nk}.$$

We need the full evaluation of the function f_k only at two points: at $\boldsymbol{x}$ and $\boldsymbol{x}^0$ which requires $2mnk$ calculations of the term functions. Other points from this sequence are obtained from the previous point by changing only one coordinate which is the coordinate of only one cluster center. This means that we need to update only m term functions at points $\boldsymbol{x}^1, \ldots, \boldsymbol{x}^{i-1}, \boldsymbol{x}^{i+1}, \ldots, \boldsymbol{x}^{nk}$ and the number of evaluations of the term functions at these points is $m(nk - 1)$. Therefore, the total number of evaluations of term functions for computation of one discrete gradient is $\bar{N}_t = m(3nk - 1)$.

Thus, in order to calculate one discrete gradient of the function f_k at the point $\boldsymbol{x}$ the following simplified scheme can be used. We compute the values of the function f_k at the points $\boldsymbol{x}$ and $\boldsymbol{x}^0$. Then we store values of all term functions calculated at $\boldsymbol{x}^0$. In order to calculate the value of f_k at $\boldsymbol{x}^1$ we update only those term functions which contain the first coordinate and keep all other term functions as they are. We repeat this scheme for all other coordinates. Note that we compute the function f_k at the point $\boldsymbol{x}$ when we compute the first discrete gradient at this point. The use of this scheme allows us to reduce the number of term functions evaluations for computation of the first discrete gradient

$$\frac{N_t}{\bar{N}_t} = \frac{mnk(nk + 1)}{m(3nk - 1)} \approx \frac{nk + 1}{3}$$

times and approximately $(nk + 1)/2$ times for the computation of all other discrete gradients at $\boldsymbol{x}$. This reduction becomes very significant as the number of clusters k increases.

The similar scheme can be designed to compute discrete gradients of the auxiliary cluster function $\bar{f}_k$. Here, the total number of term functions is mn. The function $\bar{f}_k$ has n variables and therefore, one needs $n+1$ evaluations of this function to compute its one discrete gradient. This means that the total number of evaluations of term functions to compute one discrete gradient of $\bar{f}_k$ is $N_t = mn(n + 1)$.

For a given $i \in \{1, \ldots, n\}$, we compute values of the function $\bar{f}_k$ at the following $n+1$ points:

$$\boldsymbol{x}, \boldsymbol{x}^0, \boldsymbol{x}^1, \ldots, \boldsymbol{x}^{i-1}, \boldsymbol{x}^{i+1}, \ldots, \boldsymbol{x}^n.$$

The full evaluation of the function $\bar{f}_k$ at points $\boldsymbol{x}$ and $\boldsymbol{x}^0$ requires $2mn$ calculations of the term functions. Other points from this sequence are obtained from the previous point by changing only one coordinate. This means that we need to update only m term functions at points $\boldsymbol{x}^1, \ldots, \boldsymbol{x}^{i-1}, \boldsymbol{x}^{i+1}, \ldots, \boldsymbol{x}^n$ and therefore, the total number of evaluations of the term functions for calculating of $\bar{f}_k$ at these points is $m(n-1)$. The total number of evaluations of term functions for computation of one discrete gradient is $\bar{N}_t = m(3n-1)$.

Therefore, we can apply the following simplified scheme to compute one discrete gradient of the $\bar{f}_k$ at the point $\boldsymbol{x}$. We compute the function $\bar{f}_k$ at the points $\boldsymbol{x}$ and $\boldsymbol{x}^0$ and store the values of all term functions calculated at $\boldsymbol{x}^0$. In order to calculate the value of $\bar{f}_k$ at $\boldsymbol{x}^1$ for each data point we update only the first term function and keep all other term functions as they are. This scheme is repeated for all other coordinates. Applying this scheme leads to the reduction of the number of term functions evaluations to compute the first discrete gradient

$$\frac{N_t}{\bar{N}_t} = \frac{mn(n+1)}{m(3n-1)} \approx \frac{n+1}{3}$$

times and approximately $(n+1)/2$ times for the computation of all other discrete gradients at $\boldsymbol{x}$.

8.6 Smooth Incremental Clustering Algorithm

In this section, we describe the *smooth incremental clustering algorithm* (IS-CLUST) where the objective functions in both the clustering and the auxiliary clustering problems are approximated by smooth functions [33]. To approximate objective functions, we apply the HSM, described in Sect. 3.9. The hyperbolic smoothings of the cluster function f_k and the auxiliary cluster function $\bar{f}_k$ are given in Sects. 4.7.2 and 4.7.3, respectively. For convenience, we recall these smooth functions for any $l = 2, \ldots, k$:

$$\begin{aligned}\Phi_{l,\tau}(\boldsymbol{x}, \boldsymbol{t}) &= -\frac{1}{m}\sum_{i=1}^{m}\left(t_i + \sum_{j=1}^{l}\frac{-d_p(\boldsymbol{x}_j, \boldsymbol{a}_i) - t_i + \sqrt{(d_p(\boldsymbol{x}_j, \boldsymbol{a}_i) + t_i)^2 + \tau^2}}{2}\right)\\ &= \frac{1}{m}\sum_{i=1}^{m}\left(-t_i + \sum_{j=1}^{l}\frac{t_i + d_p(\boldsymbol{x}_j, \boldsymbol{a}_i) - \sqrt{(d_p(\boldsymbol{x}_j, \boldsymbol{a}_i) + t_i)^2 + \tau^2}}{2}\right),\end{aligned}$$

and

$$\bar{\Phi}_{l,\tau}(\boldsymbol{y}) = \frac{1}{m}\sum_{i=1}^{m} r_{l-1}^{i} - \frac{1}{m}\sum_{i=1}^{m} \frac{r_{l-1}^{i} - d_p(\boldsymbol{y}, \boldsymbol{a}_i) + \sqrt{(r_{l-1}^{i} - d_p(\boldsymbol{y}, \boldsymbol{a}_i))^2 + \tau^2}}{2}$$
$$= \frac{1}{m}\sum_{i=1}^{m} \frac{r_{l-1}^{i} + d_p(\boldsymbol{y}, \boldsymbol{a}_i) - \sqrt{(r_{l-1}^{i} - d_p(\boldsymbol{y}, \boldsymbol{a}_i))^2 + \tau^2}}{2},$$

where $\boldsymbol{x} = (\boldsymbol{x}_1, \ldots, \boldsymbol{x}_l) \in \mathbb{R}^{nl}$, $\boldsymbol{y} \in \mathbb{R}^n$ and $\boldsymbol{t} = (t_1, \ldots, t_m)$, such that

$$t_i = -\min_{j=1,\ldots,l} d_p(\boldsymbol{x}_j, \boldsymbol{a}_i), \quad i = 1, \ldots, m.$$

As mentioned before, if the function d_p is defined using the squared Euclidean norm, then the functions $\Phi_{l,\tau}$ and $\bar{\Phi}_{l,\tau}$ are both smooth since d_2 is differentiable. However, the other two functions d_1 and d_∞ are nonsmooth, and we need to reapply the hyperbolic smoothing technique to these functions to approximate them with the smooth functions. These results are presented in Sect. 4.7.

Take any sequence $\{\tau_h\}$ such that $\tau_h \downarrow 0$ as $h \to \infty$, then the clustering and the auxiliary clustering problems (7.2) and (7.4) can be replaced by the sequence of the following smooth problems, respectively:

$$\begin{cases} \text{minimize} & \Phi_{l,\tau_h}(\boldsymbol{x}, \boldsymbol{t}) \\ \text{subject to} & \boldsymbol{x} = (\boldsymbol{x}_1, \ldots, \boldsymbol{x}_l) \in \mathbb{R}^{nl}, \end{cases} \tag{8.2}$$

and

$$\begin{cases} \text{minimize} & \bar{\Phi}_{l,\tau_h}(\boldsymbol{y}) \\ \text{subject to} & \boldsymbol{y} \in \mathbb{R}^n. \end{cases} \tag{8.3}$$

The IS-CLUST solves the clustering problem by combining the MSINC-CLUST and an optimization method. The IS-CLUST applies the MSINC-CLUST to solve the clustering problem globally. Since the clustering and the auxiliary clustering problems (8.2) and (8.3) are smooth problems the IS-CLUST can utilize any smooth optimization method to solve them. The flowchart of the IS-CLUST is presented in Fig. 8.4.

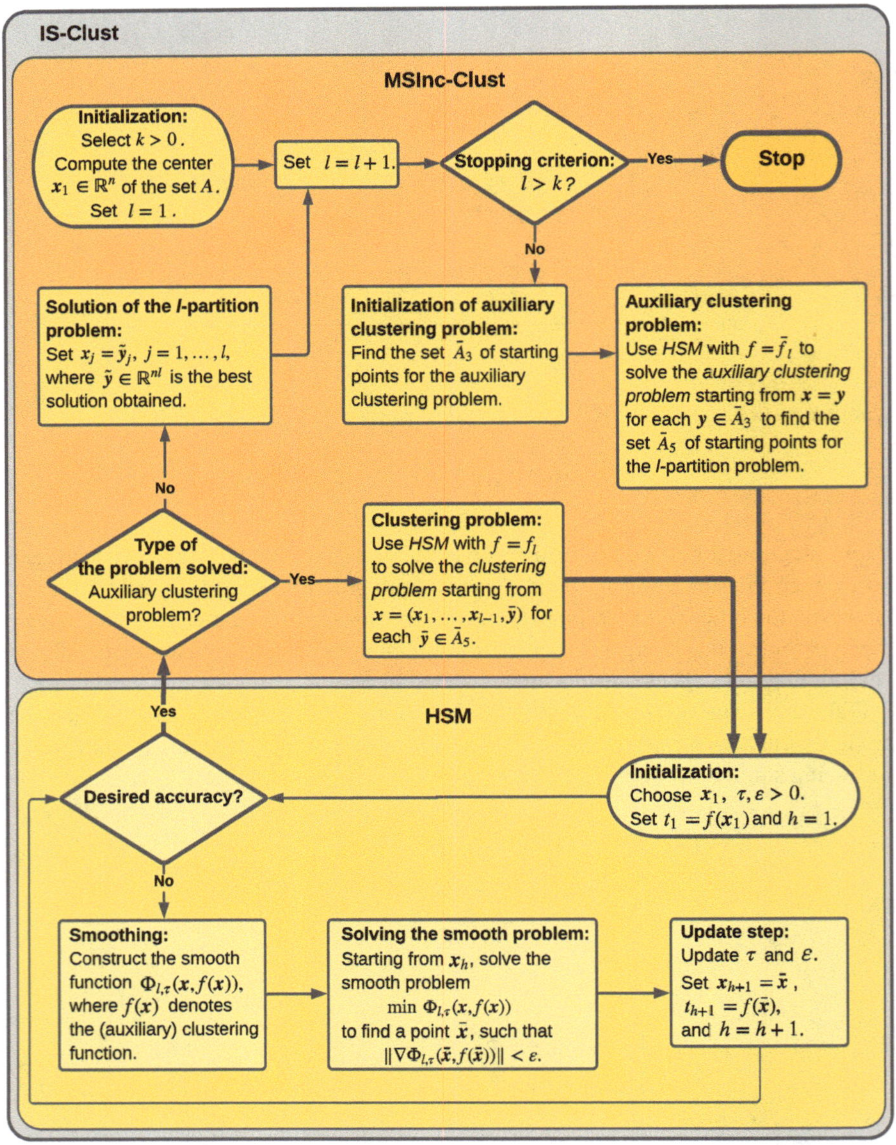

Fig. 8.4 Smooth incremental clustering algorithm (IS-CLUST)

The IS-CLUST is given in its step by step description as follows.

Algorithm 8.9 Smooth incremental clustering algorithm (IS-CLUST)

Input: Data set A and the number of clusters k to be computed.
Output: The l-partition of the set A with $l = 1, \dots, k$.

1: *(Initialization)* Compute the center $\boldsymbol{x}_1 \in \mathbb{R}^n$ of the set A. Set $l = 1$.

2: *(Stopping criterion)* Set $l = l + 1$. If $l > k$, then **stop**—the k-partition problem has been solved.

3: *(Computation of a set of starting points for the next cluster center)* Apply Algorithm 7.2 and by solving the smooth auxiliary clustering problem (8.3), compute the set $\bar{A}_5$.

4: *(Computation of a set of cluster centers)* For each $\bar{\boldsymbol{y}} \in \bar{A}_5$, take $(\boldsymbol{x}_1, \dots, \boldsymbol{x}_{l-1}, \bar{\boldsymbol{y}})$ as a starting point and solve the smooth clustering problem (8.2) and find a solution $(\hat{\boldsymbol{y}}_1, \dots, \hat{\boldsymbol{y}}_l)$. Denote by $\bar{A}_6$ a set of all such solutions.

5: *(Computation of the best solution)* Compute

$$f_l^{\min} = \min\Big\{ f_l(\hat{\boldsymbol{y}}_1, \dots, \hat{\boldsymbol{y}}_l) : (\hat{\boldsymbol{y}}_1, \dots, \hat{\boldsymbol{y}}_l) \in \bar{A}_6 \Big\},$$

and the collection of cluster centers $(\tilde{\boldsymbol{y}}_1, \dots, \tilde{\boldsymbol{y}}_l)$ such that

$$f_l(\tilde{\boldsymbol{y}}_1, \dots, \tilde{\boldsymbol{y}}_l) = f_l^{\min}.$$

6: *(Solution to the lth partition problem)* Set $\boldsymbol{x}_j = \tilde{\boldsymbol{y}}_j, \ j = 1, \dots, l$ as a solution to the lth partition problem and go to Step 2.

Note that in Step 3 of this algorithm when we apply Algorithm 7.2 to compute the set of starting points $\bar{A}_5$, the auxiliary cluster function $\bar{f}_l$ is approximated by the smooth function $\bar{\Phi}_{l,\tau}$.

Chapter 9
DC Optimization Based Clustering Algorithms

9.1 Introduction

This chapter presents the clustering algorithms based on the DC optimization approaches. In Chap. 4, the clustering problems are formulated using the DC representation of their objective functions. Using this representation we describe three different DC optimization algorithms.

For simplicity we use the following unconstrained DC programming problem to represent both the clustering and the auxiliary clustering problems (4.20) and (4.34):

$$\begin{cases} \text{minimize} & f(\mathbf{x}) = f_1(\mathbf{x}) - f_2(\mathbf{x}) \\ \text{subject to} & \mathbf{x} \in \mathbb{R}^n, \end{cases} \tag{9.1}$$

where both f_1 and f_2 are finite valued convex functions on $\mathbb{R}^n$. As mentioned before, if the squared Euclidean norm is used to define the similarity measure, then the function f_1 is smooth and the function f_2 is, in general, nonsmooth. However, with other two similarity measures d_1 and d_∞, both functions are nonsmooth. In this chapter, we only consider the first case and present three different algorithms to solve the clustering problem (9.1).

We start with the incremental nonsmooth DC clustering algorithm [36]. This algorithm combines the MSInc-Clust with the algorithm for finding inf-stationary points given in Fig. 3.7. The latter algorithm, in its turn, applies the NDCM presented in Fig. 3.8.

Then we present the DC diagonal bundle clustering algorithm [170]. Similar to the incremental DC clustering algorithm, the DC diagonal bundle clustering algorithm is a combination of the MSInc-Clust and the NSO methods. However, here we apply the DCD-Bundle given in Fig. 3.6 instead of the NDCM.

A. M. Bagirov et al., *Partitional Clustering via Nonsmooth Optimization*,
Unsupervised and Semi-Supervised Learning,
https://doi.org/10.1007/978-3-030-37826-4_9

Finally, we describe the incremental DCA for clustering [20]. The algorithm is a combination of the DCA (see Fig. 3.9) and the MSINC-CLUST.

9.2 Incremental Nonsmooth DC Clustering Algorithm

The *incremental nonsmooth DC clustering algorithm* (NDC-CLUST) is a combination of three different algorithms. The MSINC-CLUST is used to solve the clustering problem globally. At each iteration of this algorithm the algorithm for finding inf-stationary points is applied to solve both the clustering and the auxiliary clustering problems. In its turn, the later algorithm uses the NDCM to find Clarke stationary points of these problems. The flowchart of NDC-CLUST is given in Fig. 9.1.

Next, we present a detailed description of the NDC-CLUST. For a given point $\boldsymbol{x} \in \mathbb{R}^n$ and a number $\lambda > 0$, consider the set

$$Q_1(\boldsymbol{x}, \lambda) = \text{conv}\left\{\nabla f_1(\boldsymbol{x} + \lambda \boldsymbol{g}) : \boldsymbol{g} \in S_1\right\},$$

where S_1 is the sphere of the unit ball. It is obvious that the set $Q_1(\boldsymbol{x}, \lambda)$ is convex and since the function f_1 is smooth it is also compact for any $\boldsymbol{x} \in \mathbb{R}^n$ and $\lambda > 0$.

Recall that a point $\boldsymbol{x}^* \in \mathbb{R}^n$ is called (λ, δ)-inf-stationary of the problem (9.1) if and only if

$$\partial f_2(\boldsymbol{x}^*) \subset Q_1(\boldsymbol{x}^*, \lambda) + B(\boldsymbol{0}; \delta),$$

and (λ, δ)-stationary if there exists $\boldsymbol{\xi}_2 \in \partial f_2(\boldsymbol{x}^*)$ such that

$$\boldsymbol{\xi}_2 \in Q_1(\boldsymbol{x}^*, \lambda) + B(\boldsymbol{0}; \delta).$$

If a point $\boldsymbol{x} \in \mathbb{R}^n$ is not a (λ, δ)-stationary point, then $\|\boldsymbol{\xi}_2 - \boldsymbol{z}\| \geq \delta$ for all $\boldsymbol{\xi}_2 \in \partial f_2(\boldsymbol{x})$ and $\boldsymbol{z} \in Q_1(\boldsymbol{x}, \lambda)$. Take any $\boldsymbol{\xi}_2 \in \partial f_2(\boldsymbol{x})$ and construct the set

$$\widetilde{Q}(\boldsymbol{x}, \lambda, \boldsymbol{\xi}_2) = Q_1(\boldsymbol{x}, \lambda) - \boldsymbol{\xi}_2,$$

then we have

$$f(\boldsymbol{x} + \lambda \boldsymbol{u}) - f(\boldsymbol{x}) \leq \lambda \max_{\boldsymbol{z} \in \widetilde{Q}(\boldsymbol{x}, \lambda, \boldsymbol{\xi}_2)} \boldsymbol{z}^T \boldsymbol{u} \quad \text{for all} \quad \boldsymbol{u} \in \mathbb{R}^n.$$

It is shown in Proposition 3.9 that if the point $\boldsymbol{x}$ is not a (λ, δ)-stationary, then the set $\widetilde{Q}(\boldsymbol{x}, \lambda, \boldsymbol{\xi}_2)$ can be used to find a direction of sufficient decrease of the function f at $\boldsymbol{x}$. However, the computation of this set is not always possible. Next, we give a step by step algorithm which uses a finite number of elements from $\widetilde{Q}(\boldsymbol{x}, \lambda, \boldsymbol{\xi}_2)$ to compute descent directions, (λ, δ)-stationary points, and eventually Clarke stationary points of the problem (9.1). The flowchart and the more detailed

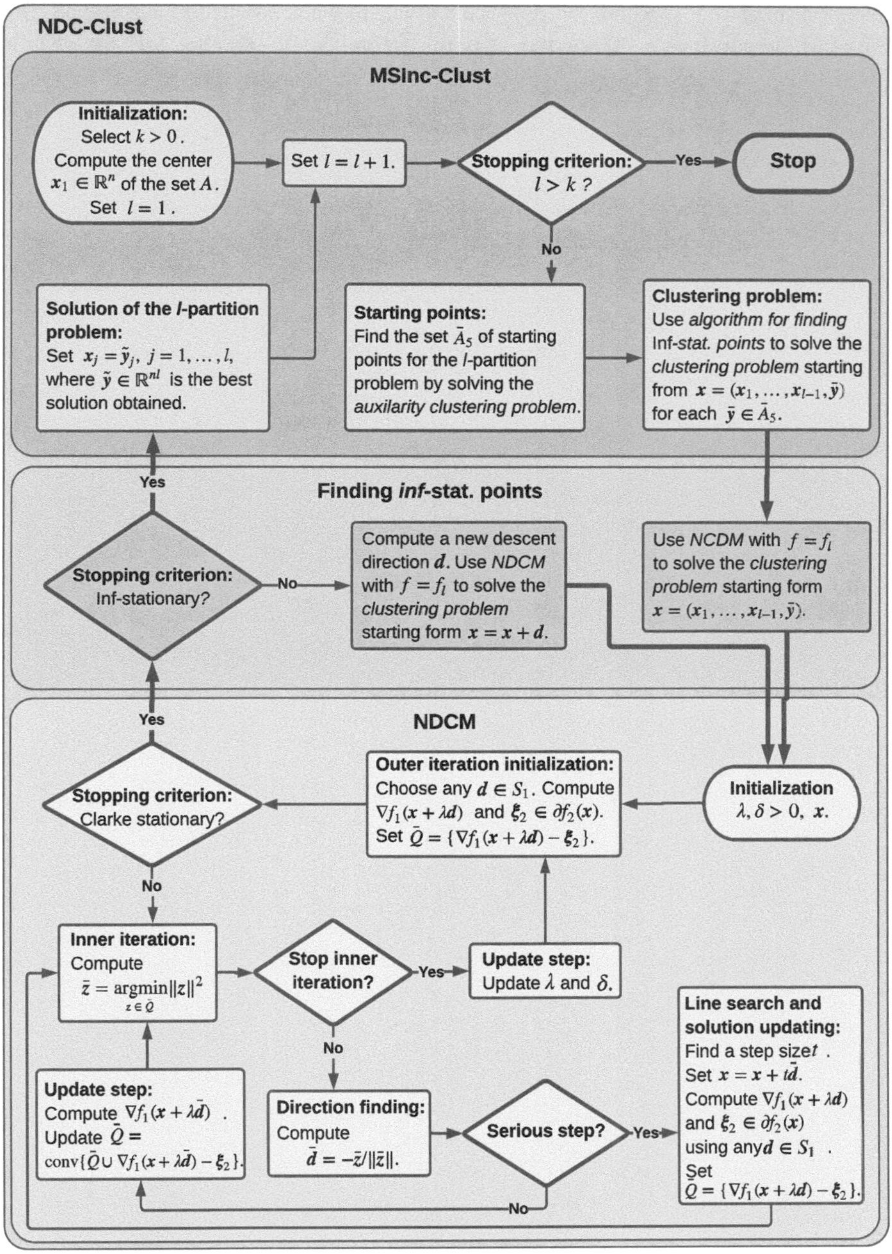

Fig. 9.1 Incremental nonsmooth DC clustering algorithm (NDC-CLUST)

description of this method (NDCM) are given in Sect. 3.6. Here, we use $\boldsymbol{x}_1$ for the starting point; $\varepsilon > 0$ for the stopping tolerance; ε_L and ε_R for line search parameters.

The convergence results for Algorithm 9.1 are given in Sect. 3.6. The next two propositions recall the most important results in light of the clustering problem.

Proposition 9.1 *Algorithm 9.1 finds* (λ, δ)*-stationary points of the clustering and the auxiliary clustering problems in at most* $h_{\max}$ *iterations where*

$$h_{\max} = \left\lceil \frac{f(\boldsymbol{x}_1)}{\lambda \delta \varepsilon_R} \right\rceil.$$

Proof The proof follows from Proposition 3.10 and the fact that $f^* = \inf\{f(\boldsymbol{x}),\ \boldsymbol{x} \in \mathbb{R}^n\} > 0$ for both the clustering and the auxiliary clustering problems. □

Proposition 9.2 *Assume that* $\varepsilon = 0$. *Then all limit points of the sequence* $\{\boldsymbol{x}_h\}$ *generated by Algorithm 9.1 are Clarke stationary points of the clustering or the auxiliary clustering problems.*

An algorithm for finding inf-stationary points of the problem (9.1) is presented next (see also Fig. 3.7). Assume that $\boldsymbol{x}^*$ is a Clarke stationary point found by Algorithm 9.1. If the subdifferential $\partial f_2(\boldsymbol{x}^*)$ is a singleton, then according to Proposition 3.7 the point is also an inf-stationary point.

Algorithm 9.1 Nonsmooth DC algorithm

Input: $x_1 \in \mathbb{R}^n$, $\varepsilon > 0$, $\lambda_1 > 0$, $\delta_1 > 0$, $\varepsilon_L \in (0, 1)$ and $\varepsilon_R \in (0, \varepsilon_L]$.
Output: Approximate Clarke stationary point x_h.

1: *(Outer iteration initialization)* Set $h = 1$.

2: *(Inner iteration initialization)* Set $s = 1$ and $x_{h_s} = x_h$. Choose any $g \in S_1$ and compute $\nabla f_1(x_{h_s} + \lambda_h g)$ and $\xi_{2,h_s} \in \partial f_2(x_{h_s})$. Set

$$\bar{Q}_h^s = \{\nabla f_1(x_{h_s} + \lambda_h g) - \xi_{2,h_s}\}.$$

3: *(Stopping criterion)* If $\lambda_h < \varepsilon$ and $\delta_h < \varepsilon$, then **stop** with x_h as a final solution.

4: *(Minimum norm)* Compute

$$z_{h_s} = \underset{z \in \bar{Q}_h^s}{\operatorname{argmin}} \|z\|^2.$$

5: *(Inner iteration termination)* If $\|z_{h_s}\| \le \delta_h$, then update λ_{h+1} and δ_{h+1}. Set $x_{h+1} = x_{h_s}$, $h = h + 1$ and go to Step 2.

6: *(Search direction)* Compute the search direction

$$d_{h_s} = -\frac{z_{h_s}}{\|z_{h_s}\|}.$$

7: If $f(x_{h_s} + \lambda_h d_{h_s}) - f(x_{h_s}) > -\varepsilon_L \lambda_h \|z_{h_s}\|$, then go to Step 9.

8: *(Serious step)* Construct $x_{h_{s+1}} = x_{h_s} + t_{h_s} d_{h_s}$, where the step size t_{h_s} is computed as

$$t_{h_s} = \operatorname{argmax}\Big\{t \ge 0 : f(x_{h_s} + t d_{h_s}) - f(x_{h_s}) \le -\varepsilon_R t \|z_{h_s}\|\Big\}.$$

Choose any $g \in S_1$ and compute $\nabla f_1(x_{h_{s+1}} + \lambda_h g)$ and $\xi_{2,h_{s+1}} \in \partial f_2(x_{h_{s+1}})$. Set

$$\bar{Q}_h^{s+1} = \Big\{\nabla f_1(x_{h_{s+1}} + \lambda_h g) - \xi_{2,h_{s+1}}\Big\},$$

$s = s + 1$ and go to Step 4.

9: *(Null step)* Compute $\nabla f_1(x_{h_s} + \lambda_h d_{h_s})$. Update the set

$$\bar{Q}_h^{s+1} = \operatorname{conv}\Big\{\bar{Q}_h^s \cup \{\nabla f_1(x_{h_s} + \lambda_h d_{h_s}) - \xi_{2,h_s}\}\Big\}.$$

Set $x_{h_{s+1}} = x_{h_s}$, $s = s + 1$ and go to Step 4.

If the subdifferential $\partial f_2(x^*)$ is not a singleton, Corollary 3.3 implies that the point x^* is not inf-stationary. Then according to Proposition 3.6 a descent direction from this point can be computed which in turn allows us to find a new starting point for Algorithm 9.1.

Algorithm 9.2 Finding inf-stationary points of clustering problems

Input: $x_1 \in \mathbb{R}^n$, $\varepsilon_A > 0$ and $\varepsilon_T \in (0, 1/2]$.
Output: Approximate inf-stationary point x_j.

1: *(Initialization)* Set $j = 1$.
2: *(Clarke stationary point)* Apply Algorithm 9.1 starting from the point x_j to find Clarke stationary point x^* with the optimality tolerance ε_A.
3: *(Stopping criterion)* If

$$\partial f_2(x^*) \subset \{\nabla f_1(x^*)\} + B(\mathbf{0}; \varepsilon_A),$$

then **stop**: x^* is an approximate inf-stationary point.
4: *(Descent direction)* Compute subgradients $\xi_2^1, \xi_2^2 \in \partial f_2(x^*)$ such that

$$r = \max_{i=1,2} \|\xi_2^i - \nabla f_1(x^*)\| \geq \varepsilon_A,$$

and the direction $u_j = -v/\|v\|$ at x^*, where

$$v = \operatorname*{argmax}_{i=1,2} \Big\{\|\nabla f_1(x^*) - \xi_2^i\|\Big\}.$$

5: *(Step size)* Compute $x_{j+1} = x^* + t_j u_j$ where

$$t_j = \operatorname{argmax} \Big\{t \geq 0 : \ f(x^* + tu_j) - f(x^*) \leq -\varepsilon_T tr\Big\}.$$

Set $j = j + 1$ and go to Step 2.

Note that if the subdifferential $\partial f_2(x)$ is not singleton, then the two subgradients $\xi_2^1, \xi_2^2 \in \partial f_2(x)$, such that $\xi_2^1 \neq \xi_2^2$ can be computed as described in Remarks 4.2 and 4.6. In addition, the following Lemmas show that the gradients of functions $\bar{f}_{k1}$ and f_{k1}, given respectively in (4.33) and (4.19), satisfy Lipschitz condition.

Lemma 9.1 *The gradient of the function $\bar{f}_{k1}$ satisfies Lipschitz condition on $\mathbb{R}^n$ with the constant $L = 2$.*

Proof Recall that the gradient of the function $\bar{f}_{k1}$ at a point $y \in \mathbb{R}^n$ is

$$\nabla \bar{f}_{k1}(y) = \frac{2}{m} \sum_{a \in A} (y - a).$$

Then for any $y_1, y_2 \in \mathbb{R}^n$ we get

$$\nabla \bar{f}_{k1}(y_1) - \nabla \bar{f}_{k1}(y_2) = 2(y_1 - y_2).$$

Therefore,

$$\|\nabla \bar{f}_{k1}(y_1) - \nabla \bar{f}_{k1}(y_2)\| = 2\|y_1 - y_2\|,$$

that is the gradient $\nabla \bar{f}_{k1}$ satisfies the Lipschitz condition on $\mathbb{R}^n$ with the constant $L = 2$. □

Lemma 9.2 *The gradient of the function f_{k1} satisfies Lipschitz condition on $\mathbb{R}^{nk}$ with the constant $L = 2$.*

Proof The proof is similar to that of Lemma 9.1. □

Considering clustering problems we can now get the following result.

Proposition 9.3 *Algorithm 9.2 terminates after the finite number of iterations at an approximate* inf*-stationary point of the (auxiliary) clustering problem.*

Proof The proof follows directly from Proposition 3.8 and Lemmas 9.1 and 9.2. □

Now we are ready to give the NDC-CLUST for solving the problem (9.1). The NDC-CLUST first uses Algorithm 7.2 to generate a set of promising starting points for the auxiliary clustering problem. In addition, Algorithm 9.2 is utilized to solve both the clustering and the auxiliary clustering problems. This algorithm, in its turn, applies Algorithm 9.1 to find Clarke stationary points of the clustering problems. The NDC-CLUST is described in Algorithm 9.3.

Algorithm 9.3 Incremental nonsmooth DC clustering algorithm (NDC-CLUST)

Input: Data set A and the number of clusters k to be computed.
Output: The l-partition of the set A with $l = 1, \ldots, k$.

1: *(Initialization)* Compute the center $\boldsymbol{x}_1 \in \mathbb{R}^n$ of the set A. Set $l = 1$.

2: *(Stopping criterion)* Set $l = l+1$. If $l > k$, then **stop**. The k-partition problem has been solved.

3: *(Computation of a set of starting points for the auxiliary clustering problem)* Apply Algorithm 7.2 to find the set $\bar{A}_3 \subset \mathbb{R}^n$ of starting points for solving the auxiliary clustering problem (4.34).

4: *(Computation of a set of starting points for the lth cluster center)* Apply Algorithm 9.2 to solve the auxiliary clustering problem (4.34) starting from each point $\boldsymbol{y} \in \bar{A}_3$. This algorithm generates a set $\bar{A}_5$ of starting points for the lth cluster center.

5: *(Computation of a set of cluster centers)* For each $\bar{\boldsymbol{y}} \in \bar{A}_5$ apply Algorithm 9.2 to solve the clustering problem (4.20) starting from the point $(\boldsymbol{x}_1, \ldots, \boldsymbol{x}_{l-1}, \bar{\boldsymbol{y}})$ and find a solution $(\hat{\boldsymbol{y}}_1, \ldots, \hat{\boldsymbol{y}}_l)$. Denote by $\bar{A}_6$ a set of all such solutions.

6: *(Computation of the best solution)* Compute

$$f_l^{\min} = \min\left\{ f_l(\hat{\boldsymbol{y}}_1, \ldots, \hat{\boldsymbol{y}}_l) : (\hat{\boldsymbol{y}}_1, \ldots, \hat{\boldsymbol{y}}_l) \in \bar{A}_6 \right\},$$

and the collection of cluster centers $(\tilde{\boldsymbol{y}}_1, \ldots, \tilde{\boldsymbol{y}}_l)$ such that

$$f_l(\tilde{\boldsymbol{y}}_1, \ldots, \tilde{\boldsymbol{y}}_l) = f_l^{\min}.$$

7: *(Solution to the lth partition problem)* Set $\boldsymbol{x}_j = \tilde{\boldsymbol{y}}_j,\ j = 1, \ldots, l$ as a solution to the lth partition problem and go to Step 2.

Remark 9.1 Algorithm 9.3 can be used to solve clustering problems with the distance functions d_1 and d_∞ if we apply the partial smoothing to the functions f_k and $\bar{f}_k$, described in Sects. 4.7.4 and 4.7.5, respectively (see [23]). More specifically, if we approximate the first component of the (auxiliary) cluster function by applying a smoothing technique then Algorithm 9.3 becomes applicable to solve clustering problems with the distance functions d_1 and d_∞.

9.3 DC Diagonal Bundle Clustering Algorithm

In this section, we describe the *DC diagonal bundle clustering algorithm* (DCDB-CLUST) for solving the problem (9.1) in large data sets [170]. The algorithm is a combination of three different algorithms. The MSINC-CLUST is used to solve the clustering problem globally. At each iteration of this algorithm a modified version of the algorithm for finding inf-stationary points (Algorithm 9.2) is applied to solve both the clustering and the auxiliary clustering problems. The later algorithm uses the DCD-BUNDLE to find Clarke stationary points of these problems. The flowchart of DCDB-CLUST is given in Fig. 9.2.

The DCD-BUNDLE is developed specifically to solve the clustering problems that are formulated as the nonsmooth DC optimization problem. The flowchart and more details of this method are given in Sect. 3.5. Here, we give the algorithm in its step by step form. We use $\boldsymbol{x}_1$ for the starting point; $\varepsilon_c > 0$ for the stopping

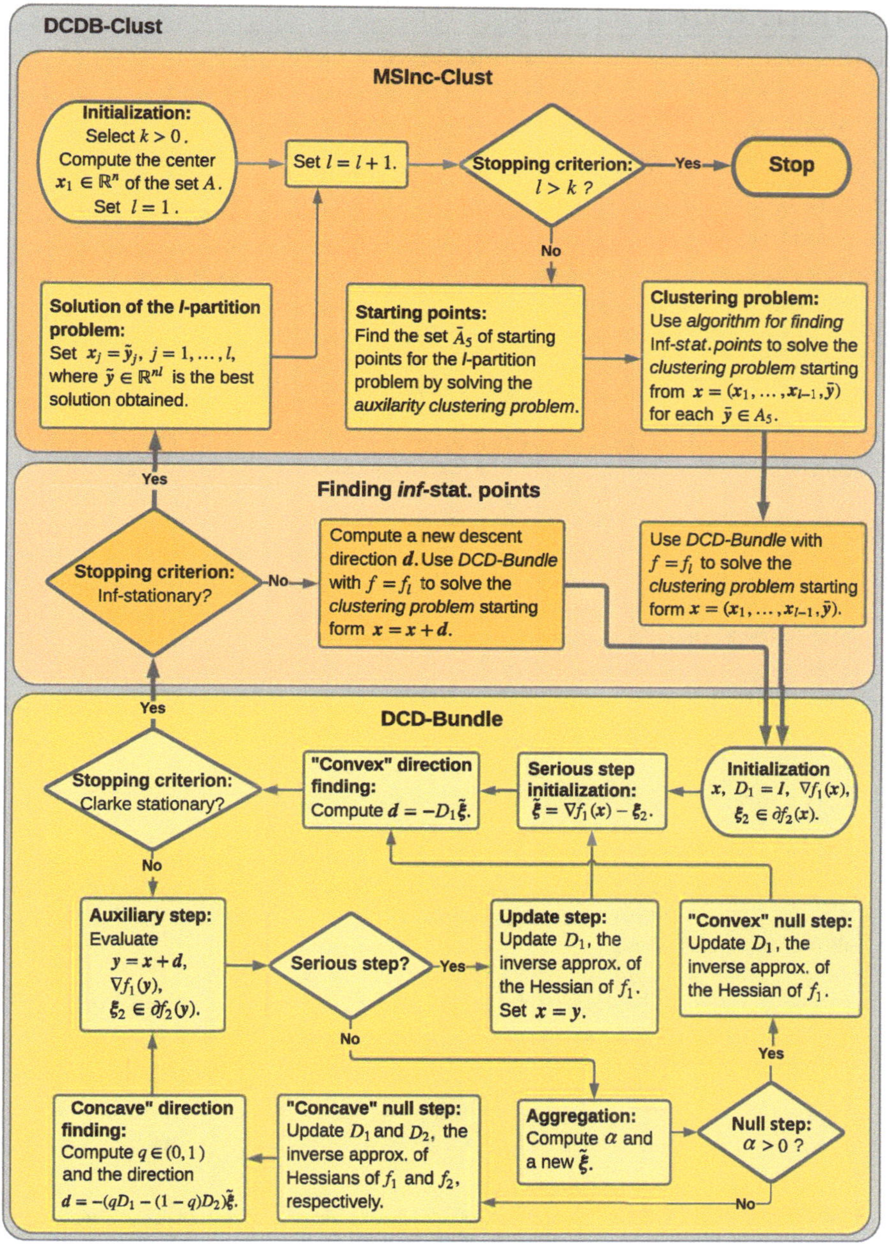

Fig. 9.2 DC diagonal bundle clustering algorithm (DCDB-CLUST)

Algorithm 9.4 DC diagonal bundle algorithm

Input: $x_1 \in \mathbb{R}^n$, $\varepsilon_c > 0$, $\varepsilon_L \in (0, 1/2)$, $\varepsilon_R \in (\varepsilon_L, 1)$, $\hat{m}_c \geq 1$ and $i_{\text{type}} \in \{0, 1\}$.
Output: Clarke stationary point x_h.

1: *(Initialization)* Set $D_{1,1} = I$. Compute $f(x_1)$, $\nabla f_{1,1} = \nabla f_1(x_1)$ and $\xi_{2,1} \in \partial f_2(x_1)$. If $i_{\text{type}} = 1$, set $\varepsilon = \varepsilon_c$. Otherwise, set $\varepsilon = 10^3\varepsilon_c$. Set $h = 1$.

2: *(Serious step initialization)* Set $\tilde{\xi}_h = \xi_h = \nabla f_{1,h} - \xi_{2,h}$ and $\tilde{\beta}_h = 0$. Set $m = h$.

3: *(Convex direction)* Compute $d_h = -D_{1,h}\tilde{\xi}_h$.

4: *(Stopping criterion)* Calculate $w_h = \tilde{\xi}_h^T D_{1,h}\tilde{\xi}_h + 2\tilde{\beta}_h$. If $w_h < \varepsilon$, then **stop** with x_h as a final solution.

5: *(Auxiliary step)* Evaluate

$$y_{h+1} = x_h + d_h, \quad \nabla f_{1,h+1} = \nabla f_1(y_{h+1}) \quad \text{and} \quad \xi_{2,h+1} \in \partial f_2(y_{h+1}).$$

Set $s_h = d_h$, $u_{1,h} = \nabla f_{1,h+1} - \nabla f_{1,m}$, $u_h = \xi_{2,h+1} - \xi_{2,m}$, and add these values to the correction matrices S_h, $U_{1,h}$, and $U_{2,h}$ (delete the earliest values if $|S_h| = |U_{1,h}| = |U_{2,h}| > \hat{m}_c$).

6: *(Serious step)* If

$$f(y_{h+1}) - f(x_h) \leq -\varepsilon_L w_h,$$

then compute $D_{1,h+1}$ using S_h and $U_{1,h}$. Set $x_{h+1} = y_{h+1}$, $f(x_{h+1}) = f(y_{h+1})$ and go to Step 2.

7: *(Aggregation)* Compute

$$\alpha_{h+1} = f(x_h) - f(y_{h+1}) + (\nabla f_{1,h+1} - \xi_{2,h+1})^T d_h,$$

and $t \in (0, 1]$ such that $\xi_{h+1}^t \in \partial f(x_h + t d_h)$ satisfies the condition

$$-\beta_{h+1} + (\xi_{h+1}^t)^T d_h \geq -\varepsilon_R w_h,$$

with β_{h+1} given in (3.19). Determine multipliers $\lambda_i^k \geq 0$ for all $i \in \{1, 2, 3\}$, $\sum_{i=1}^3 \lambda_i^h = 1$ that minimize the function

$$\begin{aligned}\varphi(\lambda_1, \lambda_2, \lambda_3) = (\lambda_1\xi_m + \lambda_2\xi_{h+1}^t + \lambda_3\tilde{\xi}_h)^T D_{1,h}(\lambda_1\xi_m + \lambda_2\xi_{h+1}^t + \lambda_3\tilde{\xi}_h) \\ + 2(\lambda_2\beta_{h+1} + \lambda_3\tilde{\beta}_h).\end{aligned}$$

Set $\tilde{\xi}_{h+1}^t = \lambda_1^h\xi_m + \lambda_2^h\xi_{h+1}^t + \lambda_3^h\tilde{\xi}_h$ and $\tilde{\beta}_{h+1} = \lambda_2^h\beta_{h+1} + \lambda_3^h\tilde{\beta}_h$.

8: *(Null step)* If $m = h$, then compute $D_{1,h+1}$ using S_h and $U_{1,h}$. Otherwise, set $D_{1,h+1} = D_{1,h}$. Two cases can occur.

(i) *(Convex Null Step)* If $\alpha_{h+1} \geq 0$, then set $x_{h+1} = x_h$, $h = h + 1$ and go to Step 3.
(ii) *(Concave Null Step)* If $\alpha_{h+1} < 0$, then compute $D_{2,h+1}$ using S_h and $U_{2,h}$. Set $x_{h+1} = x_h$, $h = h + 1$.

9: *(Concave direction)* Compute the smallest $q \in (0, 1)$ such that the matrix $qD_{1,h} - (1-q)D_{2,h}$ remains positive semidefinite. Compute

$$d_h = -\left(qD_{1,h} - (1-q)D_{2,h}\right)\tilde{\xi}_h$$

and go to Step 4.

tolerance; ε_L and ε_R for line search parameters; γ for the distance measure parameter; $\hat{m}_c$ for the maximum number of stored correction vectors used to form diagonal updates. We also use i_{type} to show the type of the problem, that is:

- $i_{\text{type}} = 0$: the auxiliary clustering problem (7.4);
- $i_{\text{type}} = 1$: the clustering problem (7.2).

The convergence properties of the DCD-BUNDLE are studied in Sect. 3.5. Here, we recall the most important results for clustering problems. Note that Assumptions 3.5–3.6 are trivially satisfied for both the cluster and the auxiliary cluster functions.

Proposition 9.4 *Assume $\varepsilon_c = 0$. If Algorithm 9.4 terminates at the hth iteration, then the point $\boldsymbol{x}_h$ is a Clarke stationary point of the (auxiliary) clustering problem.*

Proposition 9.5 *Assume $\varepsilon_c = 0$. Every accumulation point of the sequence $\{\boldsymbol{x}_h\}$ generated by Algorithm 9.4 is a Clarke stationary of the (auxiliary) clustering problem.*

If the function f_2 in the problem (9.1) is smooth, then the point found by Algorithm 9.4 is also inf-stationary. Otherwise, a slight modification of Algorithm 9.2 is applied to find an inf-stationary point of the problem. This modification is given in Algorithm 9.5.

Algorithm 9.5 Finding inf-stationary points of clustering problems

Input: $\boldsymbol{x}_1 \in \mathbb{R}^n$, $\varepsilon_A > 0$ and $\varepsilon_T \in (0, 1/2]$.
Output: Approximate inf-stationary point $\boldsymbol{x}^*$.

1: *(Initialization)* Set $j = 1$.

2: *(Clarke stationary point)* Apply Algorithm 9.4 starting from the point $\boldsymbol{x}_j$ to find the Clarke stationary point $\boldsymbol{x}^*$ with the optimality tolerance ε_A.

3: *(Stopping criterion)* If

$$\partial f_2(\boldsymbol{x}^*) \subset \left\{\nabla f_1(\boldsymbol{x}^*)\right\} + B(\boldsymbol{0}; \varepsilon_A),$$

then **stop**: $\boldsymbol{x}^*$ is an approximate inf-stationary point.

4: *(Descent direction)* Compute subgradients $\boldsymbol{\xi}_2^1, \boldsymbol{\xi}_2^2 \in \partial f_2(\boldsymbol{x}^*)$ such that

$$r = \max_{i=1,2} \|\boldsymbol{\xi}_2^i - \nabla f_1(\boldsymbol{x}^*)\| \geq \varepsilon_A,$$

and the direction $\boldsymbol{u}_j = -\boldsymbol{v}/\|\boldsymbol{v}\|$ at $\boldsymbol{x}^*$, where

$$\boldsymbol{v} = \operatorname*{argmax}_{i=1,2} \|\nabla f_1(\boldsymbol{x}^*) - \boldsymbol{\xi}_2^i\|.$$

5: *(Step size)* Compute $\boldsymbol{x}_{j+1} = \boldsymbol{x}^* + t_j \boldsymbol{u}_j$ where

$$t_j = \operatorname{argmax}\left\{t \geq 0 : \; f(\boldsymbol{x}^* + t\boldsymbol{u}_j) - f(\boldsymbol{x}^*) \leq -\varepsilon_T t r\right\}.$$

Set $j = j + 1$ and go to Step 2.

If the subdifferential $\partial f_2(\boldsymbol{x})$ is not a singleton, then we can compute two different subgradients $\boldsymbol{\xi}_2^1, \boldsymbol{\xi}_2^2 \in \partial f_2(\boldsymbol{x})$ in Step 4 of Algorithm 9.5 (see Remarks 4.2 and 4.6). In addition, in Lemmas 9.1 and 9.2, we proved that the gradients of functions $\bar{f}_{k1}$ and f_{k1} (see (4.20) and (4.34)) satisfy the Lipschitz condition. Then we get the following convergence result for clustering problems.

Proposition 9.6 *Algorithm 9.5 terminates after finite number of iterations at an approximate* inf*-stationary point of the (auxiliary) clustering problem.*

Proof The proof follows directly from Proposition 3.8 and Lemmas 9.1 and 9.2. □

Next, we give the step by step description of the DCDB-CLUST.

Algorithm 9.6 DC diagonal bundle clustering algorithm (DCDB-CLUST)

Input: Data set A and the number of clusters k to be computed.
Output: The l-partition of the set A with $l = 1, \ldots, k$.

1: *(Initialization)* Compute the center $\boldsymbol{x}_1 \in \mathbb{R}^n$ of the set A. Set $l = 1$.

2: *(Stopping criterion)* Set $l = l + 1$. If $l > k$, then **stop**—the k-partition problem has been solved.

3: *(Computation of a set of starting points for the auxiliary clustering problem)* Apply Algorithm 7.2 to find the set $\bar{A}_3 \subset \mathbb{R}^n$ of starting points for the auxiliary clustering problem (4.34).

4: *(Computation of a set of starting points for the clustering problem)* For each $\boldsymbol{y} \in \bar{A}_3$ apply Algorithm 9.5 to solve the auxiliary clustering problem (4.34) and find $\bar{A}_5$, a set of starting points for the lth cluster center in the lth clustering problem (4.20).

5: *(Computation of a set of cluster centers)* For each $\bar{\boldsymbol{y}} \in \bar{A}_5$ apply Algorithm 9.5 to solve the clustering problem (4.20) starting from the point $(\boldsymbol{x}_1, \ldots, \boldsymbol{x}_{l-1}, \bar{\boldsymbol{y}})$ and find a solution $(\hat{\boldsymbol{y}}_1, \ldots, \hat{\boldsymbol{y}}_l)$. Denote by $\bar{A}_6$ a set of all such solutions.

6: *(Computation of the best solution)* Compute

$$f_l^{\min} = \min\left\{ f_l(\hat{\boldsymbol{y}}_1, \ldots, \hat{\boldsymbol{y}}_l) \ : (\hat{\boldsymbol{y}}_1, \ldots, \hat{\boldsymbol{y}}_l) \in \bar{A}_6 \right\},$$

and the collection of cluster centers $(\tilde{\boldsymbol{y}}_1, \ldots, \tilde{\boldsymbol{y}}_l)$ such that

$$f_l(\tilde{\boldsymbol{y}}_1, \ldots, \tilde{\boldsymbol{y}}_l) = f_l^{\min}.$$

7: *(Solution to the lth partition problem)* Set $\boldsymbol{x}_j = \tilde{\boldsymbol{y}}_j, \ j = 1, \ldots, l$ as a solution to the lth partition problem and go to Step 2.

Remark 9.2 Similar to Algorithm 9.3, Algorithm 9.6 can be applied to solve clustering problems with the distance functions d_1 and d_∞ if we apply the partial smoothing to the cluster function f_k and the auxiliary cluster function $\bar{f}_k$.

9.4 Incremental DCA for Clustering

In this section, we describe an *incremental DCA for clustering* (IDCA-CLUST) to solve the clustering problem (9.1) [20]. The IDCA-CLUST is based on the MSINC-CLUST and the DCA, where the latter algorithm is utilized at each iteration of the MSINC-CLUST to solve the clustering and the auxiliary clustering problems. Figure 9.3 illustrates the flowchart of the IDCA-CLUST.

First, we recall the DCA for solving the unconstrained DC programming problem (9.1) when the first DC component f_1 is continuously differentiable.

Algorithm 9.7 DC algorithm

Input: Starting point $\boldsymbol{x}_1 \in \mathbb{R}^n$.
Output: Critical point $\boldsymbol{x}_h$.

1: *(Initialization)* Set $h = 1$.
2: Compute $\boldsymbol{\xi}_{2,h} \in \partial f_2(\boldsymbol{x}_h)$.
3: *(Stopping criterion)* If $\boldsymbol{\xi}_{2,h} = \nabla f_1(\boldsymbol{x}_h)$, then **stop**.
4: Find the solution $\boldsymbol{x}_{h+1}$ to the convex optimization problem

$$\begin{cases} \text{minimize} & f_1(\boldsymbol{x}) - \boldsymbol{\xi}_{2,h}^T(\boldsymbol{x} - \boldsymbol{x}_h) \\ \text{subject to} & \boldsymbol{x} \in \mathbb{R}^n. \end{cases} \tag{9.2}$$

5: Set $h = h + 1$ and go to Step 2.

Next, we explain how this algorithm can be applied to solve the clustering and the auxiliary clustering problems (9.1). We start with the clustering problem. Let $\boldsymbol{x}_h = (\boldsymbol{x}_{h,1}, \dots, \boldsymbol{x}_{h,k}) \in \mathbb{R}^{nk}$ be a vector of cluster centers at the iteration h and $A^1, \dots, A^k$ be the cluster partition of the data set A provided by these centers.

We discussed the subdifferentials of the functions f_1 and f_2 in Sect. 4.4. Here, we recall them when the similarity measure d_2 is used in these functions. In this case, the function f_1 is continuously differentiable and we have

$$\nabla f_{k1}(\boldsymbol{x}) = 2(\boldsymbol{x} - \tilde{\boldsymbol{a}}), \quad \boldsymbol{x} \in \mathbb{R}^{nk},$$

where $\tilde{\boldsymbol{a}} = (\bar{\boldsymbol{a}}, \dots, \bar{\boldsymbol{a}})$ and $\bar{\boldsymbol{a}} = \frac{1}{m}\sum_{i=1}^m \boldsymbol{a}_i$.

For the subdifferential of the function f_2, recall the function $\varphi_{\boldsymbol{a}}(\boldsymbol{x})$ and the set $\widetilde{\mathcal{R}}_{\boldsymbol{a}}(\boldsymbol{x}),\ \boldsymbol{x} \in \mathbb{R}^{nk}$, defined in (4.22) and (4.23), respectively:

$$\varphi_{\boldsymbol{a}}(\boldsymbol{x}) = \max_{j=1,\dots,k} \sum_{s=1, s\neq j}^{k} d_2(\boldsymbol{x}_s, \boldsymbol{a}),$$

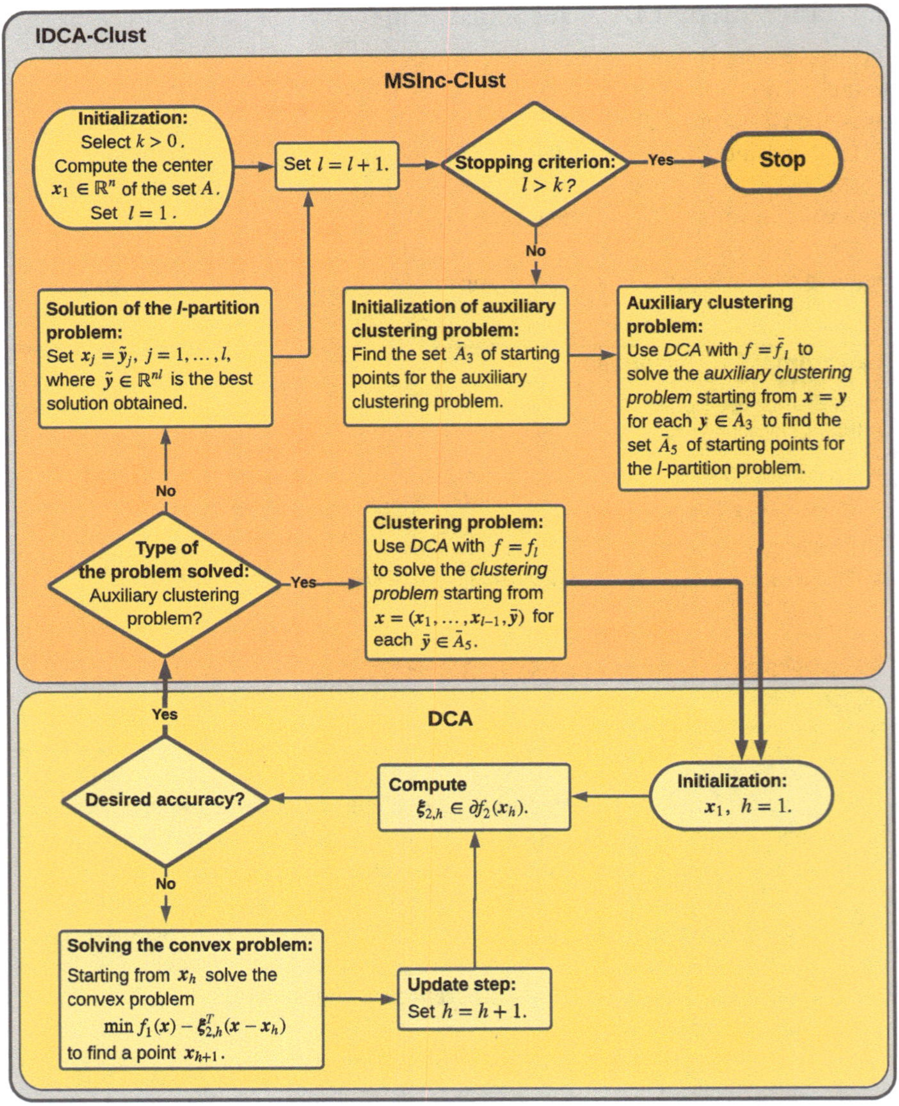

Fig. 9.3 Incremental DCA for clustering (IDCA-CLUST)

and

$$\widetilde{\mathcal{R}}_{\boldsymbol{a}}(\boldsymbol{x}) = \Big\{ j \in \{1, \ldots, k\} \ : \ \sum_{s=1, s \neq j}^{k} d_2(\boldsymbol{x}_s, \boldsymbol{a}) = \varphi_{\boldsymbol{a}}(\boldsymbol{x}) \Big\}.$$

Then we have

$$\partial \varphi_{\boldsymbol{a}}(\boldsymbol{x}) = \operatorname{conv}\Big\{ 2\big(\boldsymbol{x}_1 - \boldsymbol{a}, \ldots, \boldsymbol{x}_{j-1} - \boldsymbol{a}, \boldsymbol{0}, \boldsymbol{x}_{j+1} - \boldsymbol{a}, \ldots, \boldsymbol{x}_k - \boldsymbol{a}\big), \\ j \in \widetilde{\mathcal{R}}_{\boldsymbol{a}}(\boldsymbol{x}) \Big\},$$

and

$$\partial f_{k2}(\boldsymbol{x}) = \frac{1}{m} \sum_{\boldsymbol{a} \in A} \partial \varphi_{\boldsymbol{a}}(\boldsymbol{x}).$$

Applying these formulas for subdifferentials, the subgradient $\boldsymbol{\xi}_{2,h} \in \partial f_2(\boldsymbol{x}_h)$ in Step 2 of Algorithm 9.7 is

$$\begin{aligned} \boldsymbol{\xi}_{2,h} &= \frac{2}{m} \Big(\sum_{\boldsymbol{a} \in A \setminus A^1} (\boldsymbol{x}_{h,1} - \boldsymbol{a}), \ldots, \sum_{\boldsymbol{a} \in A \setminus A^k} (\boldsymbol{x}_{h,k} - \boldsymbol{a}) \Big) \\ &= \frac{2}{m} \Big((m - |A^1|)\boldsymbol{x}_{h,1} - (m\bar{\boldsymbol{a}} - |A^1|\bar{\boldsymbol{a}}_1), \ldots, \\ &\qquad (m - |A^k|)\boldsymbol{x}_{h,k} - (m\bar{\boldsymbol{a}} - |A^k|\bar{\boldsymbol{a}}_k) \Big), \end{aligned}$$

where $\bar{\boldsymbol{a}}_l$ is the center of the cluster A^l, $l = 1, \ldots, k$ and $\bar{\boldsymbol{a}}$ is the center of the whole set A. In addition, the solution $\boldsymbol{x}_{h+1} = (\boldsymbol{x}_{h+1,1}, \ldots, \boldsymbol{x}_{h+1,k})$ to the problem (9.2) in Step 4 of Algorithm 9.7 is

$$\boldsymbol{x}_{h+1,t} = \Big(1 - \frac{|A^t|}{m}\Big)\boldsymbol{x}_{h,t} + \frac{|A^t|}{m}\bar{\boldsymbol{a}}_t, \quad t = 1, \ldots, k,$$

and the stopping criterion in Step 3 of this algorithm can be given as

$$\boldsymbol{x}_{h,t} = \Big(1 - \frac{|A^t|}{m}\Big)\boldsymbol{x}_{h,t} + \frac{|A^t|}{m}\bar{\boldsymbol{a}}_t, \quad t = 1, \ldots, k.$$

In order to apply Algorithm 9.7 for solving the auxiliary clustering problem, recall the sets $B_i(\boldsymbol{y}),\ i = 1, 2, 3$, defined in (4.30) for $p = 2$ and $\boldsymbol{y} = \boldsymbol{x}_h \in \mathbb{R}^n$:

$$\begin{aligned} B_1(\boldsymbol{x}_h) &= \big\{ \boldsymbol{a} \in A : \ r_{l-1}^{\boldsymbol{a}} < d_2(\boldsymbol{x}_h, \boldsymbol{a}) \big\}, \\ B_2(\boldsymbol{x}_h) &= \big\{ \boldsymbol{a} \in A : \ r_{l-1}^{\boldsymbol{a}} = d_2(\boldsymbol{x}_h, \boldsymbol{a}) \big\}, \quad \text{and} \end{aligned}$$

$$B_3(\boldsymbol{x}_h) = \left\{\boldsymbol{a} \in A : r_{l-1}^{\boldsymbol{a}} > d_2(\boldsymbol{x}_h, \boldsymbol{a})\right\}.$$

Then the subgradient $\boldsymbol{\xi}_{2,h} \in \partial f_2(\boldsymbol{x}_h)$ in Step 2 of Algorithm 9.7 is computed as

$$\boldsymbol{\xi}_{2,h} = \frac{2}{m} \sum_{\boldsymbol{a} \in B_1(\boldsymbol{x}_h)} (\boldsymbol{x}_h - \boldsymbol{a}), \quad \boldsymbol{x}_h \in \mathbb{R}^n.$$

Furthermore, the solution $\boldsymbol{x}_{h+1}$ to the problem (9.2) in Step 4 is

$$\boldsymbol{x}_{h+1} = \frac{1}{m} \left(|B_1(\boldsymbol{x}_h)| \boldsymbol{x}_h + \sum_{\boldsymbol{a} \in B_2(\boldsymbol{x}_h) \cup B_3(\boldsymbol{x}_h)} \boldsymbol{a} \right).$$

Finally, the stopping criterion in Step 3 of Algorithm 9.7 can be given by

$$\sum_{\boldsymbol{a} \in B_2(\boldsymbol{x}_h) \cup B_3(\boldsymbol{x}_h)} (\boldsymbol{x}_h - \boldsymbol{a}) = 0.$$

These results demonstrate that there is no need to apply any optimization algorithm to solve the problem (9.2) for both the DC clustering and the DC auxiliary clustering problems. In both cases solutions can be expressed explicitly.

Proposition 9.7 *All accumulation points of the sequence* $\{\boldsymbol{x}_h\}$ *generated by Algorithm 9.7 are Clarke stationary points of the problem* (9.1) *when* d_2 *is used as a similarity measure.*

Proof Since the function f_1 in the problem (9.1) with the similarity measure d_2 is smooth the sets of critical points and Clarke stationary points of this problem coincide (see Theorem 2.27 and Fig. 2.9). □

Now, we are ready to design an IDCA-CLUST. This algorithm is based on the MSINC-CLUST and the DCA. The IDCA-CLUST applies the MSINC-CLUST for solving the clustering problem globally and the DCA is utilized at each iteration of the MSINC-CLUST to solve both the clustering and the auxiliary clustering problems. The step by step description of the IDCA-CLUST is given in Algorithm 9.8.

Remark 9.3 Similar to Algorithms 9.3 and 9.6, we can apply Algorithm 9.2 in Steps 4 and 5 of Algorithm 9.8. Then we obtain inf-stationary points of both the clustering and the auxiliary clustering problems.

Algorithm 9.8 Incremental DCA for clustering (IDCA-CLUST)

Input: Data set A and the number of clusters k to be computed.
Output: The l-partition of the set A with $l = 1, \ldots, k$.

1: *(Initialization)* Compute the center $\boldsymbol{x}_1 \in \mathbb{R}^n$ of the set A. Set $l = 1$.

2: *(Stopping criterion)* Set $l = l + 1$. If $l > k$, then **stop**—the k-partition problem has been solved.

3: *(Computation of a set of starting points for the auxiliary clustering problem)* Apply Algorithm 7.2 to find the set $\bar{A}_3$ of starting points for solving the auxiliary clustering problem (4.34).

4: *(Computation of a set of starting points for the lth cluster center)* Apply Algorithm 9.7 to solve the auxiliary clustering problem (4.34) starting from each point $\boldsymbol{y} \in \bar{A}_3$. This algorithm generates a set $\bar{A}_5$ of starting points for the lth cluster center.

5: *(Computation of a set of cluster centers)* For each $\bar{\boldsymbol{y}} \in \bar{A}_5$ apply Algorithm 9.7 to solve the clustering problem (4.20) starting from the point $(\boldsymbol{x}_1, \ldots, \boldsymbol{x}_{l-1}, \bar{\boldsymbol{y}})$ and find a solution $(\hat{\boldsymbol{y}}_1, \ldots, \hat{\boldsymbol{y}}_l)$. Denote by $\bar{A}_6$ a set of all such solutions.

6: *(Computation of the best solution)* Compute

$$f_l^{\min} = \min\left\{f_l(\hat{\boldsymbol{y}}_1, \ldots, \hat{\boldsymbol{y}}_l) : (\hat{\boldsymbol{y}}_1, \ldots, \hat{\boldsymbol{y}}_l) \in \bar{A}_6\right\},$$

and the collection of cluster centers $(\tilde{\boldsymbol{y}}_1, \ldots, \tilde{\boldsymbol{y}}_l)$ such that

$$f_l(\tilde{\boldsymbol{y}}_1, \ldots, \tilde{\boldsymbol{y}}_l) = f_l^{\min}.$$

7: *(Solution to the lth partition problem)* Set $\boldsymbol{x}_j = \tilde{\boldsymbol{y}}_j,\ j = 1, \ldots, l$ as a solution to the lth partition problem and go to Step 2.

Part III
Implementations and Evaluations of Clustering Algorithms

Chapter 10
Performance and Evaluation Measures

10.1 Introduction

In cluster analysis it is important to apply special evaluation and performance measures to assess the quality of clustering solutions and to compare the performance of different clustering algorithms. Here, we differentiate evaluation and performance measures. Evaluation measures are predominantly used to judge the quality of clustering solutions whereas performance measures are applied to compare the efficiency of the algorithms using the computational time and/or the number of the objective and constraint functions evaluations.

A good clustering algorithm is able to find a true number of clusters and to compute well-separated clusters which are compact and connected. Nevertheless, quantifying and measuring these objectives are not a trivial task. In some applications when a data set contains a small number of instances and a very few attributes, it might be possible to intuitively evaluate clustering results. However, such an evaluation is not possible in large and medium sized data sets or even in small data sets with several attributes.

The objectives of clustering—separability, connectivity, and compactness—are defined as follows:

- *separability* of clusters means that they are pairwise separable: that is, for each pair of clusters there exists a hyperplane separating them in the n-dimensional space. Roughly speaking in this case, data points from different clusters are away from each other;
- *connectivity* is the degree to which neighboring data points are placed in the same cluster [137]. This degree is defined by a neighborhood algorithm. The most commonly used neighborhood construction algorithms are the k-nearest

A. M. Bagirov et al., *Partitional Clustering via Nonsmooth Optimization*,
Unsupervised and Semi-Supervised Learning,
https://doi.org/10.1007/978-3-030-37826-4_10

neighbors, the ε-neighborhood [99], and the NC algorithm [151]. In general, the connectivity decreases when the number of clusters increases;
- *compactness* of clusters is characterized by the degree of density of data points around cluster centers. The compactness improves when the number of clusters increases.

The quality of clustering results can be evaluated by the *cluster validity indices*. In general, validation criteria can be divided into two groups: the *internal validation*—that is based on the information intrinsic to data, and the *external validation*—that is based on the previous knowledge of data. For instance, such knowledge can be a class distribution of a data set. In this case, the known class distributions in data sets can be used to compare the quality of solutions obtained by clustering algorithms. More precisely, cluster distributions are matched with class distributions and for example, the notion of *purity* is used to compare clustering algorithms.

In this chapter, we describe some evaluation measures including various cluster validity indices, silhouette coefficients, Rand index, purity, normalized mutual information, and F-score to mention but a few. Among these measures, the last three are external criteria. The Rand index and its modification can be used both as an internal and an external criteria. All other evaluation measures considered in this chapter are internal criteria.

Clustering algorithms can be compared using their accuracy, the required computational time, and the number of distance function evaluations. Using these three measures, we introduce performance profiles for clustering and apply them to compare the clustering algorithms.

10.2 Optimal Number of Clusters

Determining the optimal number of clusters is among the most challenging problems in cluster analysis. Various cluster validity indices can be used to find the optimal number. The values of these indices are calculated for different number of clusters and a curve of the index values with respect to the number of clusters is drawn. With most cluster validity indices, the optimal number corresponds to the global (or local) minimum or maximum of a cluster validity index. The following algorithm describes the necessary steps to find the optimal number of clusters with respect to a given validity index.

Algorithm 10.1 Finding optimal number of clusters

Input: Minimum and maximum number of clusters, $k_{\min}$ and $k_{\max}$, respectively.
Output: An optimal number of clusters $k \in [k_{\min}, k_{\max}]$.

1: *(Initialization)* Set $j = k_{\min}$.
2: Run a clustering algorithm with j clusters.
3: Compute the cluster distribution $\bar{A} = \{A^1, \ldots, A^j\}$ and the corresponding cluster centers $\boldsymbol{x} = (\boldsymbol{x}_1, \ldots, \boldsymbol{x}_j) \in \mathbb{R}^{nj}$. Calculate the index value for the jth partition.
4: *(Stopping criterion)* If $j < k_{\max}$, then set $j = j + 1$ and go to Step 2.
5: Select $j \in [k_{\min}, k_{\max}]$ for which the partition provides the best result according to some criteria (minimum or maximum) as the optimal number of clusters. Set $k = j$ and **stop**.

Remark 10.1 The value of $k_{\min}$ is usually chosen to be 2.

Remark 10.2 In the case of the incremental clustering algorithms, it is sufficient to select only $k_{\max}$ and there is no need to apply this algorithm repeatedly as it calculates clusters incrementally.

As mentioned above, the optimal number of clusters usually corresponds to optimizers of cluster validity indices. However, it is not always the case as some indices may monotonously decrease (or increase) depending on the number of clusters or indices may have several local maximum or minimum values if $k_{\max}$ is very large. Therefore, it may be more appropriate to use "*knee*" points to find the optimal number of clusters. In the univariate case knee is a point where the index curve is best approximated by a pair of lines. Although the knee point on the index curve may indicate the optimal number of clusters, locating it is not an easy task.

One way to find the knee point is to use the difference between successive values of indices. If the difference is significant, then the previous value of the index can be accepted as the knee point. This type of detection uses only local information and does not reflect the global trend of the index curve. Another way is to apply the *L*-method that examines the boundary between the pair of straight lines in which they most closely fit the index curve in hierarchical/segmentation clustering (see [256] for more details).

10.3 Cluster Validity Indices

Finding the optimal number of clusters usually relies on the notion of the cluster validity index as mentioned in the previous section. In addition, cluster validity indices can be applied to evaluate the quality of cluster solutions and also can be used as objective functions in clustering problems. Cluster validity indices have been widely studied and applied in cluster analysis [134, 135, 155, 214, 303, 314]. Most of them assume certain geometrical shapes for clusters. When these assumptions are not met, then such indices may fail. Therefore, there is no universal cluster validity index applicable to all data sets and different indices should be tried in order to decide about the true cluster structure of a data set.

Let $k \geq 2$ and $\bar{A} = \{A^1, \ldots, A^k\}$ be the cluster distribution of the data set A and $\boldsymbol{x}_1, \ldots, \boldsymbol{x}_k$ be its cluster centers. Let also $m_j = |A^j|$ be the number of points in the cluster A^j, $j = 1, \ldots, k$. Two important numbers are used in most cluster validity indices. The first one is the sum of squares within the clusters *(intra-cluster)* defined as

$$W_k = \sum_{j=1}^{k} \sum_{\boldsymbol{a} \in A^j} d_2(\boldsymbol{x}_j, \boldsymbol{a}). \tag{10.1}$$

Note that W_k is the value of the clustering objective function multiplied by the number of points in the set A. It can be rewritten as

$$W_k = \sum_{\boldsymbol{a} \in A} \min_{j=1,\ldots,k} d_2(\boldsymbol{x}_j, \boldsymbol{a}). \tag{10.2}$$

The second number is the sum of squares between each cluster center and the center of the entire set A—called the *(inter-cluster)*—defined as

$$B_k = \sum_{j=1}^{k} m_j d_2(\boldsymbol{x}_j, \bar{\boldsymbol{x}}), \tag{10.3}$$

where $\bar{\boldsymbol{x}} \in \mathbb{R}^n$ is the center of the set A.

Note that W_k is used to measure the compactness of clusters whereas B_k is considered as a measure of separation of clusters.

10.3.1 Optimal Value of Objective Function

An optimal value of the objective function in the clustering problem can be used to find the optimal number of clusters. There are different optimization models of clustering problems, and the aim in these models is to minimize the objective function by attaining a high intra-cluster and a low inter-cluster similarity. This means that the clustering objective function is an internal criterion for the quality of clustering and we can get compact and in some cases well-separated clusters by minimizing the objective. However, this aim may not be achieved always. The main reason is that clustering is a global optimization problem—it has many local solutions and only global or nearly global ones provide a good quality cluster solutions. Nevertheless, most clustering algorithms are local search methods. They start from any initial solution and find the closest local solution which might be far away from the global one.

The optimal value of the clustering function found by a clustering algorithm decreases usually as the number of clusters increases. However, this is not always the case. For example, optimal values may not decrease when the k-means or the k-medians algorithms are applied. In these cases, the use of such values may lead to an erroneous decision on the number of clusters as these values may be generated by unexpected local minimizers.

The use of the incremental approach helps us to avoid such undesirable situations. The optimal value of the clustering function found by an incremental clustering algorithm monotonously decreases as the number of clusters increases. Monotonicity of these values with respect to the number of clusters means that there are no (local) minimizers and the knee points should be used to estimate the optimal number of clusters. The following approach can be used to estimate the optimal number of clusters. Let $\varepsilon > 0$ be a given tolerance. For the clustering problem, if

$$\frac{W_k - W_{k+1}}{W_1} \leq \varepsilon,$$

then k is accepted as the optimal number of clusters (W_k is defined in (10.1)). The number W_1 is used to define the relative error as this number is a characteristic of the whole data set.

Note that the optimal value of the cluster function alone may not provide an accurate optimal number of clusters and additional measures are needed. In what follows, we introduce cluster validity indices using the similarity measure d_2; however, most of them can also be given for the similarity measures d_1 and d_∞.

10.3.2 Davies–Bouldin Index

Davies–Bouldin (DB) index [75] is a function of the ratio of the sum of within-cluster scatter to between-cluster separation. For the k-partition problem consider

$$S(A^j) = \frac{1}{m_j} \sum_{\boldsymbol{a} \in A^j} d_2(\boldsymbol{x}_j, \boldsymbol{a}),$$

which is the average squared Euclidean distance of all data points from the cluster A^j to their cluster center $\boldsymbol{x}_j,\ j = 1, \ldots, k$. Let $d_2(\boldsymbol{x}_l, \boldsymbol{x}_j)$ be the squared Euclidean distance between cluster centers $\boldsymbol{x}_l$ and $\boldsymbol{x}_j,\ l, j = 1, \ldots, k,\ j \neq l$. Introduce the following two numbers:

$$R_{lj} = \frac{S(A^l) + S(A^j)}{d_2(\boldsymbol{x}_l, \boldsymbol{x}_j)}, \quad \text{and}$$

$$\bar{R}_j = \max_{l=1,\ldots,k} R_{lj}, \quad j = 1, \ldots, k.$$

The small value of R_{lj} means that the lth and jth clusters are separated, and the small value of $\bar{R}_j$ indicates that the jth cluster is separated from all other clusters. The DB index is defined as

$$DB_k = \frac{1}{k}\sum_{j=1}^{k} \bar{R}_j.$$

Note that the DB index is small if the clusters are compact and well-separated. Consequently, the DB index will have a small value for a good clustering. More specifically, the optimal number of clusters can be identified using local minimizers of the DB index.

10.3.3 Dunn Index

Dunn (Dn) index was introduced in [93, 94]. For $k \geq 2$, consider the k-partition problem. Let

$$\text{dist}(A^j, A^l) = \min_{\boldsymbol{a}\in A^j,\ \boldsymbol{b}\in A^l} d_2(\boldsymbol{a}, \boldsymbol{b})$$

be a squared Euclidean distance between the lth and jth clusters, $j, l = 1, \ldots, k,\ j \neq l$. The squared diameter of the cluster A^j is given by

$$\text{diam}(A^j) = \max_{\boldsymbol{a},\boldsymbol{b}\in A^j} d_2(\boldsymbol{a}, \boldsymbol{b}).$$

Let

$$\Delta_k = \max_{j=1,\ldots,k} \text{diam}(A^j).$$

Then the Dn index is defined as

$$Dn_k = \frac{1}{\Delta_k} \min_{j=1,\ldots,k} \ \min_{l=1,\ldots,k,\ l\neq j} \text{dist}(A^j, A^l).$$

The Dn index maximizes the inter-cluster distances and minimizes the intra-cluster distances. This index measures the minimum separation to maximum compactness ratio, so the higher the Dn index value is the better clustering is. Therefore, the number of clusters that maximizes Dn_k can be chosen as the optimal number of clusters. Some generalizations of the Dn index have been proposed, for instance, in [255].

It should be noted that the squared distance between clusters can be defined in many different ways. Here, we define it by calculating pairwise squared distances between points from clusters. However, cluster centers can be used to define this distance. Note that the similarity measures d_1 and d_∞ can also be utilized to define the Dn index.

10.3.4 Hartigan Index

Hartigan (H) index is one of the first cluster validity indices introduced in [143]. The H index is defined as

$$H_k = \Big(\frac{W_k}{W_{k+1}} - 1\Big)(m - k - 1). \tag{10.4}$$

Another expression for this index is

$$H_k = \log \frac{B_k}{W_k},$$

where W_k and B_k are defined in (10.1) and (10.3), respectively.

Let us consider the first definition of the H index. We assume that $W_{k+1} \leq W_k$ for all $k \geq 1$. However, the difference between two successive values of W_k becomes smaller and smaller as the number k of clusters increases. This means that the first term in (10.4) is nonnegative and approaches to 0 as the number of clusters increases. The second term decreases as the number of clusters increases. Therefore, the maximum value of the H index may correspond to the optimal number of clusters.

The second expression for the H index is based on the compactness and the separation of clusters. For a good clustering, the value of B_k is expected to be as large as possible and the value of W_k to be as small as possible. This means that the (local) maximum of the H index corresponds to a good clustering distribution. Note that different similarity measures can be used to define the H index.

10.3.5 Krzanowski–Lai Index

Krzanowski–Lai (KL) index was introduced in [187] and is defined as

$$KL_k = \frac{|v_k|}{|v_{k+1}|}, \quad k > 1.$$

Here

$$v_k = \big(k-1\big)^{\frac{2}{n}} W_{k-1} - k^{\frac{2}{n}} W_k,$$

and n is the number of attributes in the data set A.

If the set A contains a large number of attributes, then $v_k \approx W_{k-1} - W_k$ and the optimal number of clusters may coincide with a local maximizer of KL_k or with its knee point. On the other hand, when the number of attributes is very small (say for example, less than six), then the optimal number of clusters can be identified using knee points of the KL index curve.

10.3.6 Ball & Hall Index

Ball & Hall (BH) index was introduced in [40]. This index is very simple and can be easily calculated as

$$BH_k = \frac{W_k}{k}.$$

Since one can expect that $W_{k+1} \leq W_k$ for all $k > 1$ it follows that the BH index decreases as k increases. Therefore, in general, it is not expected that BH_k has a local minimizer or maximizer. This is always true for incremental clustering algorithms for which BH_k decreases monotonously. In this case, we can define a tolerance $\varepsilon > 0$. If

$$BH_k - BH_{k+1} \leq \varepsilon \quad \text{for some} \quad k \geq 2,$$

then k can be accepted as the optimal number of clusters. Alternatively, the optimal number of clusters can be determined using knee points on the BH index curve.

10.3.7 Bayesian Information Criterion

Bayesian information criterion (BIC) was introduced in [315]. The BIC is defined as

$$BIC_k = Lm - \frac{1}{2}k(n+1)\sum_{j=1}^{k} \log\big(m_j\big),$$

where $L > 0$ is a log-likelihood in the BIC, m_j is the number of data points in the jth cluster, $j = 1, \ldots, k$, and n is the number of attributes in the data set A. Note that knee points of the BIC can be considered as a possible optimal number of clusters.

10.3.8 WB Index

The sum-of-squares based index, called the WB index was introduced in [316] and its modifications are studied in [314]. The WB index is defined as

$$WB_k = \frac{kW_k}{B_k},$$

where W_k and B_k are defined in (10.1) and (10.3), respectively. The name of this index originates from notations used for these two numbers.

As mentioned before, the smaller the value of W_k is the better compactness of clusters and the larger value of B_k is the better separated clusters. Therefore, the minimum of WB_k corresponds to the optimal number of clusters.

10.3.9 Xu Index

Xu index [304] is defined as

$$Xu_k = \frac{n}{2}\log\left(\frac{W_k}{nm^2}\right) + \log(k).$$

Since for a good clustering the values of W_k are expected to be as small as possible it follows that the optimal number of clusters can be identified using local minimizers of the Xu index.

10.3.10 Xie-Beni Index

Xie-Beni (XB) index [303] is applicable to fuzzy clustering problems. Nevertheless, it can also be applied to hard clustering problems with some slight modifications.

Let w_{ij} be a membership degree of the ith data point $\boldsymbol{a}_i$ to the cluster A^j, $i = 1, \ldots, m$, $j = 1, \ldots, k$. Introduce the following numbers:

$$U_k = \sum_{i=1}^{m}\sum_{j=1}^{k} w_{ij}^2 d_2(\boldsymbol{x}_j, \boldsymbol{a}_i), \quad \text{and}$$

$$\tilde{U}_k = \min_{j=1,\ldots,k} \; \min_{t=j+1,\ldots,k} d_2(\boldsymbol{x}_j, \boldsymbol{x}_t).$$

Here, U_k is the clustering objective function value on the fuzzy clustering problem multiplied by the number of data points. The value of $\tilde{U}_k$ indicates how far cluster centers lie from each other. Then the XB index is defined as

$$XB_k = \frac{U_k}{m\tilde{U}_k}.$$

It is clear that the smaller value of U_k means more compact clusters. The larger value of $\tilde{U}_k$ indicates well-separated clusters. Therefore, the minimum value of the XB index corresponds to the optimal number of clusters.

To make the XB index applicable to hard clustering problems, U_k needs to be modified as follows:

$$U_k = \sum_{j=1}^{k} \sum_{\boldsymbol{a} \in A^j} d_2(\boldsymbol{x}_j, \boldsymbol{a}), \quad \text{or}$$

$$U_k = \sum_{i=1}^{m} \min_{j=1,\ldots,k} d_2(\boldsymbol{x}_j, \boldsymbol{a}_i).$$

In this case, U_k coincides with W_k defined in (10.1). The similarity measures d_1 and d_∞ can also be used to define the XB index.

10.3.11 Sym Index

Sym (Sm) index [42] is used to measure the overall average symmetry with respect to cluster centers. Take any cluster $\hat{A}$ from the k-partition $\bar{A} = \{A^1, \ldots, A^k\}$ and denote its center by $\hat{\boldsymbol{x}}$. Let $\bar{\boldsymbol{y}} \in \hat{A}$. Compute $2\hat{\boldsymbol{x}} - \bar{\boldsymbol{y}}$ which reflects the point $\bar{\boldsymbol{y}}$ with respect to the center $\hat{\boldsymbol{x}}$. Denote the reflected point by $\hat{\boldsymbol{y}}$. Assume that k_N nearest neighbors $\boldsymbol{y}_i,\ i = 1, \ldots, k_N$ of $\hat{\boldsymbol{y}}$ are at the squared Euclidean distances $d_2(\hat{\boldsymbol{y}}, \boldsymbol{y}_i)$. Compute the symmetry measure d_s of $\bar{\boldsymbol{y}}$ with respect to $\hat{\boldsymbol{x}}$ as

$$d_s(\bar{\boldsymbol{y}}, \hat{\boldsymbol{x}}) = \frac{\sum_{i=1}^{k_N} d_2(\hat{\boldsymbol{y}}, \boldsymbol{y}_i)}{k_N}.$$

Then the point symmetry based distance $d_{ps}(\bar{\boldsymbol{y}}, \hat{\boldsymbol{x}})$ between $\bar{\boldsymbol{y}}$ and $\hat{\boldsymbol{x}}$ is computed as

$$d_{ps}(\bar{\boldsymbol{y}}, \hat{\boldsymbol{x}}) = d_s(\bar{\boldsymbol{y}}, \hat{\boldsymbol{x}}) \cdot d_2(\bar{\boldsymbol{y}}, \hat{\boldsymbol{x}}).$$

Note that the number of neighbor points k_N cannot be 1. Otherwise, the point $\hat{\boldsymbol{y}}$ is in a data set and $d_{ps}(\bar{\boldsymbol{y}}, \hat{\boldsymbol{x}}) = 0$ and therefore, the impact of the Euclidean distance is ignored. On the other hand, large values of k_N may reduce the symmetry property of a point with respect to a particular cluster center. In practice, k_N is a user defined number and can be chosen as 2.

The maximum separation between a pair of clusters over all possible pairs of clusters is defined by

$$D_k = \max_{i=1,\ldots,k} \max_{j=1,\ldots,k} \|\hat{\boldsymbol{x}}_i - \hat{\boldsymbol{x}}_j\|. \tag{10.5}$$

For each cluster $A^j,\ j = 1, \ldots, k$ compute

$$E_j = \sum_{\boldsymbol{a} \in A^j} d_{ps}(\boldsymbol{a}, \hat{\boldsymbol{x}}_j), \quad \text{and}$$

$$\mathcal{E}_k = \sum_{j=1}^{k} E_j.$$

Then the Sm index is defined as

$$Sm_k = \frac{D_k}{k\mathcal{E}_k}.$$

The Sm index should be maximized in order to obtain the optimal number of clusters. This index can also be extended to use the similarity measures d_1 and d_∞.

10.3.12 *I Index*

I index [8, 208] is defined as

$$I_k = \Big(\frac{1}{k} \times \frac{F_1}{F_k} \times D_k\Big)^q,$$

where D_k is given in (10.5) and q is any positive integer—usually $q = 2$. The number F_k is defined as the value of clustering function multiplied by the number of data points. That is

$$F_k = \sum_{\boldsymbol{a} \in A} \min_{j=1,\ldots,k} d_2(\boldsymbol{a}, \boldsymbol{x}_j).$$

It is obvious that F_1 corresponds to F_k when $k = 1$.

There are three factors in the definition of the I index: the first one is the reciprocal of the number of clusters and it decreases as the number of clusters increases; the second factor is the ratio of F_1 and F_k. Here, F_1 is a constant for a given data set. Since, in general, with an increase of k the value of F_k will be decreased this factor ensures the formation of more compact clusters; the third factor, D_k generally increases as k increases. Due to the complementary nature of these three factors, it is guaranteed that the I index is able to determine the optimal partitioning. It is clear that the I index is maximized to obtain the optimal number of clusters.

10.3.13 Calinski–Harabasz Index

Calinski–Harabasz (CH) index [57] is defined using the sum of squares within the clusters and the sum of squares between the clusters. The CH index for the k-partition problem with $k \geq 2$ is given by

$$CH_k = \frac{(m-k)B_k}{(k-1)W_k}, \tag{10.6}$$

where W_k and B_k are defined in (10.1) and (10.3), respectively.

The CH index can be expressed with a different formulation. Let d_A be the general mean of all squared distances between points $\boldsymbol{a}_i, \boldsymbol{a}_j \in A$:

$$d_A = \frac{2\sum_{i=1}^{m}\sum_{j=i+1}^{m} d_2(\boldsymbol{a}_i, \boldsymbol{a}_j)}{m(m-1)},$$

and d_j be the mean value for each cluster $A^j,\ j = 1, \ldots, k$:

$$d_j = \frac{2\sum_{\boldsymbol{a}\in A^j}\sum_{\boldsymbol{b}\in A^j} d_2(\boldsymbol{a}, \boldsymbol{b})}{m_j(m_j-1)}.$$

Then the sum of squares within the clusters can be alternatively computed as (cf. (10.1))

$$W_k = \frac{1}{2}\Big(\sum_{j=1}^{k}(m_j-1)d_j\Big),$$

and the sum of squares between the clusters is alternatively defined as (cf. (10.3))

$$B_k = \frac{1}{2}\Big((k-1)d_A + (m-k)Q_k\Big).$$

Here, Q_k is a weighted mean of the differences between the general and the within-cluster mean squared distances, that is

$$Q_k = \frac{1}{m-k}\sum_{j=1}^{k}(m_j-1)(d_A-d_j).$$

Then the CH index, given in (10.6), can be reformulated as

$$CH_k = \frac{d_A + \left(\frac{m-k}{k-1}\right)Q_k}{d_A - Q_k}.$$

If the distances between all pairs of points are equal, then $Q_k = 0$ and we get $CH_k = 1$. Since d_A is a constant for a data set A it follows that the minimum value of W_k maximizes Q_k for a given k.

The parameter Q_k can also be used to compare partitions obtained for different number of clusters: the difference $Q_k - Q_{k-1}$ indicates an average gain in the compactness of clusters resulting from the change from $k-1$ to k clusters. Hence, the behavior of Q_k depending on k may be sensitive to the existence of such clusters. Let

$$q_k = \frac{Q_k}{d_A}.$$

It is clear that $q_k \in [0, 1]$. The case $q_k = 0$ means that all distances between pairs of data points are equal while $q_k = 1$ implies $k = m$.

Then the CH index can be rewritten as

$$CH_k = \frac{1 + \left(\frac{m-k}{k-1}\right) q_k}{1 - q_k}.$$

If data points are grouped into k clusters with a small within-cluster variation, then the change from $k-1$ to k causes a considerable increase in q_k. This in turn leads to a rapid increase of the CH index. Therefore, this index can be applied to identify the optimal number of clusters. It is suggested in [57] to choose the value of k for which the CH index has a (local) maximum or at least a comparatively rapid increase. The latter point can be considered as a knee point on the CH index curve. If there are several such local maxima or knee points, then the smallest corresponding value of k can be chosen. In practice, this means that the computation can be stopped when the first local maximum or the knee point is found.

10.4 Silhouette Coefficients and Plots

The aim of the *silhouette plot* is to identify compact and well-separated clusters [251] (see, also [174]). More precisely, the silhouette plot is used to interpret and validate consistency within clusters. It provides a concise graphical representation of how well each data point lies within its cluster.

Silhouettes are constructed using the k-partition $\bar{A} = \{A^1, \ldots, A^k\}$, its cluster centers $\boldsymbol{x}_1, \ldots, \boldsymbol{x}_k$, and the collection of all proximities between data points. Take any point $\boldsymbol{a} \in A$ and denote by $A^j,\ j \in \{1, \ldots, k\}$ the cluster to which this point belongs. Assuming that this cluster contains other data points apart from $\boldsymbol{a}$, compute

$$\bar{d}_{\boldsymbol{a}} = \frac{1}{m_j - 1} \sum_{\boldsymbol{b} \in A^j, \boldsymbol{b} \neq \boldsymbol{a}} d_2(\boldsymbol{a}, \boldsymbol{b}).$$

Here, $\bar{d}_{\boldsymbol{a}}$ is the average dissimilarity of the point $\boldsymbol{a}$ to all other points of the cluster A^j. Choose any $t \in \{1, \dots, k\},\ t \neq j$ and compute

$$\bar{d}_{t,\boldsymbol{a}} = \frac{1}{m_t} \sum_{\boldsymbol{b} \in A^t} d_2(\boldsymbol{a}, \boldsymbol{b}),$$

which is the average dissimilarity of the point $\boldsymbol{a}$ to points from the cluster A^t. Calculate the minimum average dissimilarity

$$\hat{d}_{\boldsymbol{a}} = \min_{t=1,\dots,k,\ t \neq j} \bar{d}_{t,\boldsymbol{a}}.$$

Let $\bar{d}_{t_0,\boldsymbol{a}} = \hat{d}_{\boldsymbol{a}},\ t_0 \in \{1, \dots, k\},\ t_0 \neq j$. The cluster A^{t_0} is called the *neighbor cluster* of the data point $\boldsymbol{a}$. For each point $\boldsymbol{a} \in A$, calculate the *silhouette coefficient* $s(\boldsymbol{a})$ using the following formula:

$$s(\boldsymbol{a}) = \begin{cases} 1 - \bar{d}_{\boldsymbol{a}}/\hat{d}_{\boldsymbol{a}}, & \text{if } \bar{d}_{\boldsymbol{a}} < \hat{d}_{\boldsymbol{a}}, \\ 0, & \text{if } \bar{d}_{\boldsymbol{a}} = \hat{d}_{\boldsymbol{a}}, \\ \hat{d}_{\boldsymbol{a}}/\bar{d}_{\boldsymbol{a}} - 1, & \text{if } \bar{d}_{\boldsymbol{a}} > \hat{d}_{\boldsymbol{a}}. \end{cases}$$

This formula can be rewritten as

$$s(\boldsymbol{a}) = \frac{\hat{d}_{\boldsymbol{a}} - \bar{d}_{\boldsymbol{a}}}{\max\{\bar{d}_{\boldsymbol{a}},\ \hat{d}_{\boldsymbol{a}}\}}.$$

It is clear that $s(\boldsymbol{a}) \in [-1, 1]$ for all $\boldsymbol{a} \in A$. When $s(\boldsymbol{a})$ is close to 1, the *within dissimilarity* $\bar{d}_{\boldsymbol{a}}$ is much smaller than the smallest *between dissimilarity* $\hat{d}_{\boldsymbol{a}}$. Therefore, we can say that $\boldsymbol{a}$ is *well-clustered* as it seems that $\boldsymbol{a}$ is assigned to an appropriate cluster and the second best choice A^{t_0} is not nearly as close as the actual choice A^j. If $s(\boldsymbol{a}) = 0$ or it is close to 0, then $\bar{d}_{\boldsymbol{a}}$ and $\hat{d}_{\boldsymbol{a}}$ are approximately equal. This means that the point $\boldsymbol{a}$ lies equally far away from A^j and A^{t_0}, and it is not clear which cluster it should be assigned. In this case, the data point can be considered as an *intermediate case*. When $s(\boldsymbol{a})$ is close to -1, $\bar{d}_{\boldsymbol{a}}$ is much larger than $\hat{d}_{\boldsymbol{a}}$. This means that on the average $\boldsymbol{a}$ is much closer to A^{t_0} than A^j. Hence, it would be more natural to assign $\boldsymbol{a}$ to A^{t_0} than to A^j and we can conclude that $\boldsymbol{a}$ is *misclassified*. To conclude, $s(\boldsymbol{a})$ measures how well the data point $\boldsymbol{a}$ matches with the clustering at hand, that is, how well the point has been assigned.

The numbers $s(\boldsymbol{a})$ can be used to draw the silhouette plot. The silhouette of the cluster A^j is a plot of $s(\boldsymbol{a})$ ranked in a decreasing order for all points $\boldsymbol{a}$ in A^j. The silhouette plot shows which data points lie well within the cluster and which ones are merely somewhere in between clusters. A wide silhouette indicates large $s(\boldsymbol{a})$ values and hence a pronounced cluster. The other dimension of the silhouette plot is its height which simply equals the number of objects in A^j.

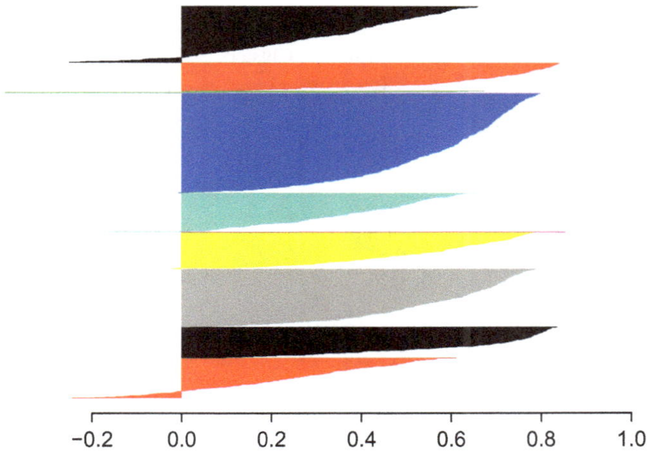

Fig. 10.1 Illustration of silhouette plot for ten clusters

Combining the silhouette of different clusters we can draw a single silhouette plot for the whole data set A. This allows us to distinguish *clear-cut* clusters from the *weak* (not well-separated) ones. In addition, the number of objects in each cluster A^j can be detected by using the height of the silhouette of the cluster A^j.

To illustrate silhouette plots we consider the 10-partition of image segmentation data set obtained using the MGKM. The silhouette plot of the 10-partition is given in Fig. 10.1. It is depicted using the R package available from https://cran.r-project.org/web/packages/clues/clues.pdf (see [290], for details). In this figure, clusters are shown using different colors. The height and area of the cluster depend on the number of its data points. The very narrow "pink" cluster is the smallest and the "blue" cluster is the largest one. If any cluster has data points with negative silhouettes, then this cluster is not well-separated. The absence of such points means that the cluster is considered as well-separated. The top "black," "green," "light blue," "yellow" and the bottom "red" clusters are not well-separated from other clusters. The "blue" cluster contains a very few "misclassified" points. The remaining four clusters are considered as separable from other clusters. The right hand side of clusters reflects the number of data points well lying inside their own clusters. The top "red," "blue," and the bottom "black" clusters contain more such points than any other cluster.

10.5 Rand Index

Rand (Rn) index [243] measures similarity between two different cluster distributions of the same data set (see, also [215, 257]). Let $\bar{A}_1$ and $\bar{A}_2$ be two cluster distributions of the data set A:

Table 10.1 Contingency table for comparing partitions $\bar{A}_1$ and $\bar{A}_2$

		$\bar{A}_2$				
Partition	Clusters	A_2^1	A_2^2	...	$A_2^{k_2}$	Total
$\bar{A}_1$	A_1^1	n_{11}	n_{12}	...	n_{1k_2}	s_1
	A_1^2	n_{21}	n_{22}	...	n_{2k_2}	s_2
	$\vdots$	$\vdots$	$\vdots$	$\ddots$	$\vdots$	$\vdots$
	$A_1^{k_1}$	n_{k_11}	n_{k_12}	...	$n_{k_1k_2}$	s_{k_1}
Total		r_1	r_2	...	r_{k_2}	m

$$\bar{A}_1 = \{A_1^1, \dots, A_1^{k_1}\}, \quad k_1 > 1, \quad \text{and} \tag{10.7}$$

$$\bar{A}_2 = \{A_2^1, \dots, A_2^{k_2}\}, \quad k_2 > 1. \tag{10.8}$$

Denote the elements of the $k_1 \times k_2$ matrix by n_{ij}, where n_{ij} represents the number of data points belonging to the ith cluster A_1^i of the cluster partition $\bar{A}_1$ and the jth cluster of the cluster partition $\bar{A}_2$, $i = 1, \dots, k_1, \ j = 1, \dots, k_2$. Then the contingency table for distributions $\bar{A}_1$ and $\bar{A}_2$ can be formed as in Table 10.1. In this table, the entry s_i indicates the number of data points in the ith cluster of the cluster partition $\bar{A}_1$ and r_j shows the number of data points in the jth cluster of the cluster partition $\bar{A}_2$.

Let $C(m, 2)$ be the number of 2-combinations from m data points. Among all possible combinations of pairs $C(m, 2)$, there are the following different types of pairs:

- data points in a pair belong to the same cluster in $\bar{A}_1$ and to the same cluster in $\bar{A}_2$. Denote the number of such pairs by N_1;
- data points in a pair belong to the same cluster in $\bar{A}_1$ and to different clusters in $\bar{A}_2$. Denote the number of such pairs by N_2;
- data points in a pair belong to the same cluster in $\bar{A}_2$ and to different clusters in $\bar{A}_1$. Denote the number of such pairs by N_3; and
- data points in a pair belong to different clusters in $\bar{A}_1$ and to different clusters in $\bar{A}_2$. Denote the number of such pairs by N_4.

The values of N_1, N_2, N_3, and N_4 can be calculated using entries from Table 10.1:

$$N_1 = \sum_{i=1}^{k_1} \sum_{j=1}^{k_2} C(n_{ij}, 2) = \frac{1}{2} \left(\sum_{i=1}^{k_1} \sum_{j=1}^{k_2} n_{ij}^2 - m \right),$$

$$N_2 = \sum_{i=1}^{k_1} C(s_i, 2) - N_1 = \frac{1}{2} \left(\sum_{i=1}^{k_1} s_i^2 - \sum_{i=1}^{k_1} \sum_{j=1}^{k_2} n_{ij}^2 \right),$$

$$N_3 = \sum_{j=1}^{k_2} C(r_j, 2) - N_1 = \frac{1}{2} \left(\sum_{j=1}^{k_2} r_j^2 - \sum_{i=1}^{k_1} \sum_{j=1}^{k_2} n_{ij}^2 \right), \quad \text{and}$$

$$N_4 = C(m,2) - N_1 - N_2 - N_3 = \frac{1}{2}\left(\sum_{i=1}^{k_1}\sum_{j=1}^{k_2} n_{ij}^2 + m^2 - \sum_{i=1}^{k_1} s_i^2 - \sum_{j=1}^{k_2} r_j^2\right).$$

Then the Rn index can be calculated as

$$Rn = \frac{N_1 + N_4}{N_1 + N_2 + N_3 + N_4}.$$

Note that the Rn index is ranging from 0 to 1. If $Rn = 0$, then two cluster distributions have no similarity, and if $Rn = 1$, then they are identical.

The Rn index can also be used to measure the similarity between a cluster distribution and a class distribution in data sets with class outputs. In this case, we can replace one of cluster distributions above by the class distribution. Therefore, the Rn index is both an internal and an external index.

10.6 Adjusted Rand Index

The Rn index of two random cluster distributions may not have a constant value (say zero) or it may approach its upper limit of unity as the number of clusters increases. To overcome these limitations several other measures have been introduced, for example, the Fowlkes–Mallows index proposed in [109]. Another example of such measures is the *Adjusted Rand* (ARn) index which is an improvement of the Rn index [150]. This index is considered as one of the successful cluster validation indices [215].

Recall the cluster distributions $\bar{A}_1$ and $\bar{A}_2$, given in (10.7) and (10.8), for the data set A. Then the ARn index is

$$ARn = \frac{C(m,2)(N_1 + N_4) - \Big((N_1 + N_2)(N_1 + N_3) + (N_2 + N_4)(N_3 + N_4)\Big)}{\Big(C(m,2)\Big)^2 - \Big((N_1 + N_2)(N_1 + N_3) + (N_2 + N_4)(N_3 + N_4)\Big)},$$

or

$$ARn = \frac{C(m,2)\sum_i \sum_j C(n_{ij},2) - \Big(\sum_i C(s_i,2)\sum_j C(r_j,2)\Big)}{\frac{1}{2}C(m,2)\Big(\sum_i C(s_i,2) + \sum_j C(r_j,2)\Big) - \Big(\sum_i C(s_i,2)\sum_j C(r_j,2)\Big)},$$

where, n_{ij}, s_i, r_j are values from Table 10.1.

The ARn index can be used both as an internal and an external index. In the latter case it is assumed that the data set A has class outputs and its cluster distribution is compared with the class distribution.

10.7 Purity

Purity measures the quality of clustering in data sets with class outputs [83]. More precisely, it shows how well the cluster distribution obtained by a certain clustering algorithm reflects the existing class structure of a data set. Therefore, the purity is applicable when a data set has a class label, that is, the purity is an external criterion. The purity can be computed for each cluster $A^j,\ j = 1, \ldots, k$ and the whole data set A.

Let $\bar{A} = \{A^1, \ldots, A^k\}$ be the cluster distribution of the set A obtained by a clustering algorithm. It is assumed that $k \geq 2$ and $A^j \neq \emptyset,\ j = 1, \ldots, k$. Let also $\bar{C} = \{C_1, \ldots, C_l\}$ be the set of true classes of the set A. The cluster A^j may contain different number of points from classes $C_t,\ t = 1, \ldots, l$. Denote by n_{tj} the number of points from the tth class belonging to the jth cluster. For this cluster compute

$$\bar{m}_j = \max_{t=1,\ldots,l} n_{tj}, \quad j = 1, \ldots, k.$$

Then the purity of the cluster A^j is defined as

$$Pu(A^j) = \frac{\bar{m}_j}{m_j},$$

where m_j is the number of points in the cluster A^j. Usually the purity is expressed in percentage, hence this formula can be rewritten as

$$Pu(A^j) = \frac{\bar{m}_j}{m_j} \times 100\%.$$

The purity for the cluster distribution $\bar{A}$ is

$$Pu(\bar{A}) = \frac{\sum_{j=1}^{k} \bar{m}_j}{m} \times 100\%.$$

The purity is 1 (100%) if each cluster contains data points only from one class. Note that increasing the number of clusters, in general, will increase the purity. In particular, if each cluster has only one data point, then the purity is equal to 1 (100%). This implies that the purity cannot be used to evaluate the trade-off between the quality and the number of clusters.

10.8 Normalized Mutual Information

Normalized mutual information (NMI) is a combination of the mutual information and the entropy [189]. The mutual information is used to measure how well the computed clusters and the true classes predict one another. The entropy is used to measure the amount of information inherent in both the cluster distribution and the true classes. The entropy is also used to normalize the mutual information which allows us to evaluate the trade-off between the quality and the number of clusters. The NMI is an external criterion to evaluate the quality of clusters.

Recall that $\bar{A}$ is the cluster distribution obtained by a clustering algorithm for the data set A and $\bar{C}$ is the set of its true classes. Let n_{tj} be the number of points from the tth class belonging to the jth cluster, n_t be the number of points in the class C_t, and m_j be the number of points in the cluster A^j.

The estimation $\bar{P}_j$ of probability for a data point being in the jth cluster is $\bar{P}_j = m_j/m,\ j = 1, \ldots, k$, the estimation $\hat{P}_t$ of probability for a data point being in the tth class is $\hat{P}_t = n_t/m$ and finally, the estimation $\bar{P}_{tj}$ of probability for a data point being in the intersection $A^j \cap C_t$ is $\bar{P}_{tj} = n_{tj}/m$. Then the mutual information $I(\bar{A}, \bar{C})$ between the cluster distribution $\bar{A}$ and the class distribution $\bar{C}$ is defined as

$$
\begin{aligned}
I(\bar{A}, \bar{C}) &= \sum_{j=1}^{k} \sum_{t=1}^{l} \bar{P}_{tj} \log\left(\frac{\bar{P}_{tj}}{\bar{P}_j \hat{P}_t}\right) \\
&= \sum_{j=1}^{k} \sum_{t=1}^{l} \frac{n_{tj}}{m} \log\left(\frac{m n_{tj}}{m_j n_t}\right).
\end{aligned}
$$

The estimation $H(\bar{A})$ for the entropy of the cluster distribution $\bar{A}$ is computed as

$$
\begin{aligned}
H(\bar{A}) &= -\sum_{j=1}^{k} \bar{P}_j \log \bar{P}_j \\
&= -\sum_{j=1}^{k} \frac{m_j}{m} \log\left(\frac{m_j}{m}\right)
\end{aligned}
$$

and the estimation $H(\bar{C})$ for the entropy of the class distribution $\bar{C}$ is

$$
\begin{aligned}
H(\bar{C}) &= -\sum_{t=1}^{l} \hat{P}_t \log \hat{P}_t \\
&= -\sum_{t=1}^{l} \frac{n_t}{m} \log\left(\frac{n_t}{m}\right).
\end{aligned}
$$

Then the NMI is calculated by

$$NMI(\bar{A}, \bar{C}) = \frac{2I(\bar{A}, \bar{C})}{H(\bar{A}) + H(\bar{C})}.$$

It is clear that $NMI(\bar{A}, \bar{C}) \in [0, 1]$. $NMI(\bar{A}, \bar{C}) = 1$ means that the cluster distribution $\bar{A}$ and the class distribution $\bar{C}$ of the data set A are identical, that is the cluster distribution obtained by a clustering algorithm has a high quality.

10.9 F-Score

F-score (known also as *F-measure*) is an external validity index to evaluate the quality of clustering solutions [191]. The introduction of this index is inspired by the information retrieval metric known as the F-measure. We describe the F-score using notations from the previous section.

The F-score combines the concepts of *precision* (Pr) and *recall* (Re). For the cluster $A^j,\ j = 1, \ldots, k$ and the class $C_t,\ t = 1, \ldots, l$, Pr and Re are defined as

$$Pr(A^j, C_t) = \frac{n_{tj}}{n_t}, \quad \text{and}$$

$$Re(A^j, C_t) = \frac{n_{tj}}{m_j}.$$

Then the F-score of the cluster A^j and the class C_t is given by

$$F(A^j, C_t) = \frac{2Pr(A^j, C_t) \times Re(A^j, C_t)}{Pr(A^j, C_t) + Re(A^j, C_t)}.$$

Note that $F(A^j, C_t) \in [0, 1],\ t = 1, \ldots, l,\ j = 1, \ldots, k$. The total F-score F_t is computed as

$$F_t = \frac{1}{m} \sum_{j=1}^{k} m_j \max_{t=1,\ldots,l} F(A^j, C_t).$$

It is obvious that $F_t \in [0, 1]$. The higher the F-score is, the better the clustering solution is. This measure has an advantage over the purity since it measures both the homogeneity and the completeness of a clustering solution. The *homogeneity* of a clustering solution means that all its clusters contain only data points which are members of a single class. The *completeness* of a clustering solution means that data points that are members of a given class are elements of the same cluster.

10.10 Performance Profiles in Cluster Analysis

Performance profiles, introduced in [86], are widely used to compare optimization algorithms. Such profiles have been introduced for comparison of (sub)gradient-based and derivative free optimization algorithms. The number of function (and gradient) evaluations and the computational time are usually used to compute these profiles.

Clustering is a global optimization problem and the ability of a clustering algorithm to find global or nearly global solutions is important as such solutions provide the best cluster structure of a data set with the least number of clusters. Therefore, here we introduce performance profiles for comparison of clustering algorithms. The profiles are defined using three parameters: the accuracy, the number of distance function evaluations, and the computational time.

10.10.1 Accuracy

Accuracy of clustering algorithms can be determined using the known global solutions or the best known solutions of clustering problems. Consider the k-partition problem in a data set A. Assume that $f_k^* > 0$ is the best known value of the objective function in the k-partition problem and $\bar{f}_k$ is the lowest value of this function obtained by a clustering algorithm Υ. Then the accuracy (or error) E_Υ of the algorithm Υ for solving the k-partition problem is defined as

$$E_\Upsilon^k = \frac{\bar{f}_k - f_k^*}{f_k^*}. \tag{10.9}$$

In some cases, it is convenient to present the error (or accuracy) in %, therefore the error can be defined as

$$E_\Upsilon^k = \frac{\bar{f}_k - f_k^*}{f_k^*} \times 100\%. \tag{10.10}$$

We describe performance profiles using the percentage representation of the error. Assume that the algorithm Υ is applied to solve the l-clustering problems for $l = 2, \ldots, k$ in t data sets. Then the total number of clustering problems is $t(k-1)$. Denote by $E_\Upsilon^{l,q}$ the error of the solution obtained by the algorithm Υ for solving the l-clustering problem in the qth data set, where $l \in \{2, \ldots, k\}$ and $q \in \{1, \ldots, t\}$. Let $\tau \geq 0$ be any given number. For the algorithm Υ, define the set

$$\sigma_\Upsilon(\tau) = \{(l, q) : E_\Upsilon^{l,q} \leq \tau\}.$$

It is clear that the set $\sigma_\Upsilon(0)$ contains only those indices (l, q) in which the algorithm Υ finds the best known solutions. Furthermore, for sufficiently large τ we have

$$\sigma_\Upsilon(\tau) = \big\{(l, q) : l \in \{2, \ldots, k\},\ q \in \{1, \ldots, t\}\big\}.$$

Then the probability that the algorithm Υ solves the collection of clustering problems with the accuracy $\tau \geq 0$ is given as

$$P_\Upsilon(\tau) = \frac{|\sigma_\Upsilon(\tau)|}{t(k-1)}.$$

Define a threshold $\tau_0 > 0$. An algorithm Υ is unsuccessful in solving the l-clustering problem in the qth data set if $E_\Upsilon^{l,q} > \tau_0$ $(l \in \{2, \ldots, k\},\ q \in \{1, \ldots, t\})$. This allows us to draw the graph of the function $P_\Upsilon(\tau)$ in the interval $[0, \tau_0]$.

Note that the higher the graph on the left hand side is more accurate the algorithm is in finding the best known solutions than other algorithms. The higher on the right hand side means that the algorithm is more robust in finding the nearly best known solutions than the other algorithms.

10.10.2 Number of Distance Function Evaluations

Most optimization based clustering algorithms use values and subgradients (or gradients) of the cluster function to solve clustering problems. To compute them, we need to calculate distance functions and also apply minimum operations for each data point. Recall that the data set A has m data points and n attributes. It is expected that, in average, the number of distance function evaluations in a clustering algorithm depends linearly or almost linearly on the number of clusters. At each iteration, the number of such evaluations is $O(mk)$, where k is the number of clusters.

Assume that a clustering algorithm uses M iterations to solve the k-clustering problem. Then the expected value for the number of distance function evaluations required by the algorithm is

$$N(m, k) = cMmk. \tag{10.11}$$

Here, $c > 0$ is a given constant. The number N can be used as a benchmark to evaluate the performance of clustering algorithms. Note that unlike the performance profiles usually used in optimization [86], this benchmark does not depend on any solver. Therefore, we can use it to rank clustering algorithms both in the sense of efficiency and robustness.

First, we define when one can consider the performance of an algorithm as success or failure. Since clustering is a global optimization problem we consider any run of a clustering algorithm Υ as success if it finds either global (or best known) or nearly global (or nearly best known) solution. Thus, we can define some threshold $\tau > 0$ (in %). If the error $E_{\Upsilon}^{k} \leq \tau$, then the algorithm Υ is considered to be successful otherwise, it fails to solve the clustering problem.

Assume that v clustering solvers $\Upsilon_1, \ldots, \Upsilon_v$ are applied to solve the l-clustering problems for $l = 2, \ldots, k$ in t data sets. Let $N(l, q)$ be a benchmark number computed using (10.11) for the qth data set with l clusters. For each solver Υ_s, $s = 1, \ldots, v$, denote by Q_s the set of clustering problems successfully solved applying this solver:

$$Q_s \subseteq \big\{(l, q) : q \in \{1, \ldots, t\};\ l \in \{2, \ldots, k\}\big\}. \tag{10.12}$$

Let

$$V = \big\{s \in \{1, \ldots, v\} : Q_s \neq \emptyset\big\}, \tag{10.13}$$

and take any $s \in V$. Define the number

$$\sigma_s(l, q) = \frac{N_{lq}^{s}}{N(l, q)},$$

where N_{lq}^{s} is the number of distance function evaluations used by the solver Υ_s for solving the l-clustering problem in the qth data set.

Compute the number

$$\tau_{\max} = \max_{s \in V} \max_{(l,q) \in Q_s} \sigma_s(l, q).$$

For any $s \in V$ and a number $\tau \in (0, \tau_{\max}]$ consider the set

$$X_s(\tau) = \big\{(l, q) \in Q_s : \sigma_s(l, q) \leq \tau\big\}.$$

Then the performance profiles for the solver Υ_s, $s \in V$ can be defined as

$$\rho_s(\tau) = \frac{|X_s(\tau)|}{t(k-1)}.$$

For other solvers Υ_s, $s \notin V$ we set $\rho_s(\tau) = 0$ for all $\tau \geq 0$.

10.10.3 Computational Time

Performance profiles using the computational time can be obtained in a similar way to those using the distance function evaluations. The number of operations for one evaluation of the squared Euclidean distance function is $3n - 1$, where n is the number of attributes in the data set A. Consider the k-clustering problem. The number of distance function evaluations for one iteration of an algorithm and the total number of operations are estimated by $O(mk)$ and $O(nmk)$, respectively.

Assume that a clustering algorithm uses about M iterations to solve the k-clustering problem. Then we get a number

$$T(m, k) = \bar{c}Mnmk \tag{10.14}$$

as an expected value for the number of operations, where $\bar{c} > 0$ is a given constant. Dividing T by 10^9 (this number depends on characteristics of a computer), we get the expected value $\overline{T}$ for the computational time. This number is used as a benchmark to evaluate the performance of clustering algorithms based on the computational time.

As in the case of the number of distance function evaluations, we introduce a threshold $\tau > 0$ (in %) to define whether an algorithm fails or succeeds to solve the clustering problem. Assume that v clustering solvers $\Upsilon_1, \ldots, \Upsilon_v$ are applied to solve the l-clustering problems for $l = 2, \ldots, k$ in t data sets. For each solver $\Upsilon_s,\ s = 1, \ldots, v$ define the set Q_s using (10.12) and the set V by applying (10.13).

Take any $s \in V$ and compute

$$\beta_s(l, q) = \frac{T_{lq}^s}{\overline{T}(l, q)},$$

where T_{lq}^s is the computational time used by the solver Υ_s for solving the l-clustering problem in the qth data set. Calculate the number

$$\tau_{\max} = \max_{s \in V} \max_{(l,q) \in Q_s} \beta_s(l, q).$$

For any $s \in V$ and a number $\tau \in (0, \tau_{\max}]$, consider the set

$$X_s(\tau) = \big\{(l, q) \in Q_s : \beta_s(l, q) \le \tau\big\}.$$

For each solver $\Upsilon_s,\ s \in V$, we can define the performance profiles as follows:

$$\rho_s(\tau) = \frac{|X_s(\tau)|}{t(k-1)}.$$

For other solvers $\Upsilon_s,\ s \notin V$ we set $\rho_s(\tau) = 0$ for all $\tau \ge 0$.

Chapter 11
Implementations and Data Sets

11.1 Introduction

We discuss the implementations of various incremental clustering algorithms and provide some recommendations on the choice of their parameters. These algorithms are designed by combining the multi-start incremental algorithm with either NSO methods or variants of the k-means algorithm. All the NSO based incremental clustering algorithms involve the algorithm for finding initial cluster centers. Using numerical results, we discuss the choice of parameters for the latter algorithm. In addition, we describe some real-world data sets that are used to evaluate the clustering algorithms.

We start this chapter by introducing the algorithms applied in our numerical experiments and by providing details of their implementations. Then, in Sect. 11.3 we present data sets used in our experiments. Finally, in Sect. 11.4 we discuss the parameters selection for the algorithm that is used to find initial cluster centers (i.e., Algorithm 7.2 given in Sect. 7.4).

11.2 Implementations of Clustering Algorithms

We use the following algorithms in our numerical experiments:

- MS-KM—multi-start k-means algorithm;
- GKM—global k-means algorithm (Sect. 5.2.3);
- MGKM—modified global k-means algorithm (Sect. 8.2);
- LMB-CLUST—limited memory bundle method for clustering (Sect. 8.4);
- DG-CLUST—discrete gradient clustering algorithm (Sect. 8.5);
- IS-CLUST—smooth incremental clustering algorithm (Sect. 8.6);
- NDC-CLUST—incremental nonsmooth DC clustering algorithm (Sect. 9.2);

A. M. Bagirov et al., *Partitional Clustering via Nonsmooth Optimization*,
Unsupervised and Semi-Supervised Learning,
https://doi.org/10.1007/978-3-030-37826-4_11

- DCDB-CLUST—DC diagonal bundle clustering algorithm (Sect. 9.3);
- IDCA-CLUST—incremental DCA for clustering (Sect. 9.4).

The source codes (Fortran77 or Fortran95) of these algorithms are available at https://github.com/SnTa2019/Clustering-via-Nonsmooth-Optimization and also at http://napsu.karmitsa.fi/clustering/.

In the first three clustering algorithms, MS-KM, GKM, and MGKM, the underlying solver is the k-means algorithm (see Sect. 5.2). We use the standard implementation of the k-means algorithm in our experiments. The stopping criterion in this algorithm is formulated using the maximum number of data points which are allowed to change their clusters. More specifically, for some $\delta \geq 0$ the number $m_c = \delta m$ is defined, where m is the number of points in a data set. If the number of points changing their clusters is less than or equal to m_c, then the k-means algorithm terminates. In extra small, small, and medium sized data sets we set $\delta = 0$ and for large and very large data sets we use $\delta = 10^{-3}$.

Unlike other clustering algorithms listed above, the MS-KM is not an incremental algorithm and starting points for the clustering problem are generated using a simple multi-start randomized scheme. Note that, this algorithm does not provide any intermediate solutions for $1 < l < k$ and therefore, several runs need to be performed with different number of clusters. We denote by `MS-KM` the implementation of the MS-KM and set the maximum number of starting points in `MS-KM` to 1000.

`GKM` is an implementation of the incremental algorithm with the k-means algorithm, where each data point is used as a starting point for the kth cluster center and no auxiliary clustering problem is utilized. The original `GKM` utilizes the mixed integer programming formulation (4.2) of the clustering problem, however, the NSO formulation (4.3) of this problem can also be used. In its original design, `GKM` is only suitable for clustering relatively small data sets. In the implementation, we use only one starting point for the kth cluster center. More specifically, a data point which provides the largest decrease of the cluster function is utilized. This allows us to apply `GKM` to large data sets. Furthermore, we consider two different implementations of the GKM as follows:

- the affinity (or distance) matrix of a data set is computed before the application of the GKM. This implementation is applicable to small and medium sized data sets as the affinity matrix for large data sets cannot be stored in the memory of a computer and
- the affinity matrix is computed at each iteration. We can apply this implementation to large data sets.

Similar to other incremental clustering algorithms, `GKM` solves all the intermediate l-partition problems where $l = 1, \ldots, k$.

`MGKM` is an implementation of the MGKM. This is an incremental algorithm which utilizes the auxiliary clustering problem and the algorithm for finding starting cluster centers. For the parameters of the latter one, we refer to Sect. 11.4. At each iteration of the MGKM, the clustering problem is solved by applying the k-means algorithm and the auxiliary clustering problem is solved by utilizing the partial k-means algorithm where all cluster centers are fixed but one. We use the stopping tolerance $\varepsilon = 0$ (see Algorithm 8.2 in Sect. 8.2) and therefore, `MGKM` always computes up to the maximum number of clusters provided by the user.

`LMB-Clust` is an implementation of the LMB-CLUST specifically developed for solving clustering problems in large and very large data sets. In our experiments, we use the stopping tolerance $\varepsilon_c = 10^{-3}$. For small and medium sized data sets, the stopping tolerance should be set smaller than 10^{-3}. In addition, we apply the modified version of the underlying optimization solver LMBM with the initial number of stored correction pairs (used to form the variable metric update) equal to 7 and the maximum number of stored correction pairs equal to 15. Otherwise, the default values of parameters of the LMBM are used.

`DG-Clust` is an implementation of the DG-CLUST. As mentioned before, the DG-CLUST exploits piecewise separability of the cluster and auxiliary cluster functions. It does not utilize the subgradients of the objective cluster functions, instead it approximates them using values of the objective functions. Thus `DG-Clust` is suitable also for solving clustering problems with the similarity measures d_1 and d_∞. The simplified scheme, described in Sect. 8.5, is applied to compute discrete gradients of the cluster and the auxiliary cluster functions which allows us to significantly reduce the number of distance function evaluations. The most important parameter in `DG-Clust` is $\lambda > 0$ for approximation of the subgradients. Its initial value is chosen $\lambda_1 = 1$ and at the hth ($h > 1$) iteration it is updated as $\lambda_h = \mu\lambda_{h-1}$ where $\mu = 0.2$. Other parameters are chosen as $\delta_h = 10^{-7}$ for all h, $\varepsilon = 10^{-6}$ and $\alpha = 1$.

`IS-Clust` is an implementation of the IS-CLUST. The smoothing (or precision) parameter is chosen as $\tau_h = \mu\tau_{h-1},\ h > 1$ where $\tau_1 = 1$ and $\mu = 0.1$. To solve the smooth optimization problems, `IS-Clust` applies the Quasi-Newton method with the BFGS update.

`NDC-Clust` is an implementation of the NDC-CLUST. This algorithm takes into account the DC representation of the objective cluster function. The main parameter in `NDC-Clust` is the step $\lambda > 0$ used to approximate the first DC component. This parameter is chosen as $\lambda_h = \mu\lambda_{h-1},\ h > 1$, where $\mu = 0.1$ and $\lambda_1 = 1$. Otherwise, the default parameters of the underlying optimization solver are used. In particular, $\delta_h = 10^{-7}$ for all h, and $\varepsilon = 10^{-6}$.

`DCDB-Clust` is an implementation of the DCDB-CLUST, where the DC structure of the clustering and the auxiliary clustering problems are utilized. This method is designed for solving clustering problems in large data sets. In our experiments, we use the stopping tolerance $\varepsilon_c = 10^{-3}$. Similar to `LMB-Clust`, the stopping tolerance should be set smaller for small and medium sized data sets. The underlying optimization solver DCD-BUNDLE uses seven correction pairs to form the diagonal updates. In addition, the default values of other parameters of the optimization solver are used.

`IDCA-Clust` is an implementation of the IDCA-CLUST. Since the objective functions in the subproblem (9.2) of the IDCA-CLUST are convex quadratic functions with the simple structure their unique minima can be calculated explicitly without involving any optimization procedure or tuneable parameters.

All computational experiments were carried out in a PC with the CPU Intel(R) Core(TM) i5-8250U 1.60 GHz and RAM 8 GB working under Windows 10.

11.3 Data Sets

Data sets can be classified based on different parameters. Such parameters include the number of data points, the number of attributes, the number of classes if they are available (binary or multi-class), types of attributes (numeric, categorical, or mixture of both), completeness or incompleteness of data, in particular, absence of values of some attributes in some data points (missing values). We will use data sets with numeric attributes and no missing values.

The definition of the complexity of a data set is a problem and a model dependent. For instance, the complexity of a data set may be different from the perspective of the supervised data classification and clustering. Different models of a clustering problem contain different types and numbers of variables. Some of these models have constraints. Therefore, the complexity of a data set with respect to each of these models might be different. It is also not an easy task to determine how the complexity of a data set affects the time complexity of an algorithm. If a data set has well-separated clusters, then most clustering algorithms may require significantly less iterations than in data sets with not well-separated clusters.

In most optimization based clustering algorithms, considered in this book, optimization algorithms require the calculation of the value of the cluster functions and their one subgradient at each point. In the case of the MSSC problems, the objective is piecewise quadratic and quadratic functions can be represented as a sum of the squared Euclidean distances. In the case of the similarity measures d_1 and d_∞, the objective cluster function is piecewise linear. For such functions, the complexity of calculation of their values and subgradients depends on the number of points and attributes in a data set. This means that the complexity depends on the number of entries in a data set. Therefore, we use the number of entries to classify data sets as extra small, small, medium sized, large, and very large data sets.

11.3.1 Extra Small Data Sets

We group those data sets that contain no more than 3000 entries as extra small. The brief description of these data sets, used in our numerical experiments, and their references are given in Table 11.1.

Table 11.1 Brief description of extra small data sets

Data sets	Number of instances	Number of attributes	Number of classes	Number of entries	Refs.
German towns	59	2	–	116	[272]
Bavaria postal 1	84	3	–	252	[272]
Bavaria postal 2	84	4	–	336	[272]
Iris plant	150	3	3	450	[91]
TSPLIB1060	1060	2	–	2120	[246]
Liver disorder	356	6	2	2136	[91]

Table 11.2 Brief description of small data sets

Data sets	Number of instances	Number of attributes	Number of classes	Number of entries	Refs.
Heart disease	297	13	2	3861	[91]
TSPLIB3038	3038	2	–	6076	[246]
Pima Indians diabetes	768	8	2	6144	[91]
Breast cancer Wisconsin	683	9	2	6147	[91]
Ionosphere	351	34	2	11,934	[91]
Vehicle silhouettes	846	18	4	15,228	[91]

Table 11.3 Brief description of medium sized data sets

Data sets	Number of instances	Number of attributes	Number of classes	Number of entries	Refs.
D15112	15,112	2	–	30,224	[246]
Image segmentation	2310	19	7	43,890	[91]
Page blocks	5473	10	5	54,730	[91]
Pla85900	85,900	2	–	171,800	[246]
Pen-based recognition of handwritten digits	10,992	16	10	175,872	[91]

11.3.2 Small Data Sets

The data sets containing more than 3000 but less than 20,000 entries are classified as small. The brief description of such data sets, including their references, is given in Table 11.2.

11.3.3 Medium Sized Data Sets

Medium sized data sets are those containing more than 20,000 but less than 2.0×10^5 entries. These data sets are presented in Table 11.3 with their corresponding references.

11.3.4 Large Data Sets

We group the data sets containing more than 2×10^5 but less than 2×10^6 entries as large. The brief description of these data sets, including their corresponding references, is given in Table 11.4.

11.3.5 Very Large Data Sets

Very large data sets are those containing more than 2×10^6 entries. We give the brief description of these data sets and their corresponding references in Table 11.5.

Table 11.4 Brief description of large data sets

Data sets	Number of instances	Number of attributes	Number of classes	Number of entries	Refs.
EEG eye state	14,980	14	2	209,720	[91]
Landsat satellite	6435	36	6	231,660	[91]
Letter recognition	20,000	16	26	320,000	[91]
Optical recognition of handwritten digits	5620	64	10	359,680	[91] [91]
Person activity	164,860	3	11	494,580	[91]
Shuttle control	58,000	9	7	522,000	[91]
Skin segmentation	245,057	3	2	735,171	[91]
KEGG metabolic relation network	53,413	20	–	1,068,260	[91] [91]
3D road network	434,874	3	–	1,304,622	[91]
Gas sensor array drift	13,910	128	6	1,780,480	[91]

Table 11.5 Brief description of very large data sets

Data sets	Number of instances	Number of attributes	Number of classes	Number of entries	Refs.
Online news popularity	39,644	58	2	2,299,352	[91]
Sensorless drive diagnosis	58,509	48	11	2,866,941	[91]
ISOLET	7797	616	26	4,802,952	[91]
Covertype	581,012	10	7	5,810,012	[91]
MiniBooNE particle identification	130,064	50	2	6,503,200	[91] [91]
Gisette	13,500	5000	–	67,500,000	[91]

11.4 Parameters Selection in Finding Starting Cluster Centers

One of the most important components of the incremental clustering algorithms is the procedure for finding starting cluster centers. This procedure is given in Algorithm 7.2. It contains three parameters $\gamma_1, \gamma_2 \in [0, 1]$, and $\gamma_3 \in [1, \infty)$ whose optimal values depend on the size of a data set. More precisely, they depend more on the number of data points than on the number of attributes.

Assume that $l \geq 2$ and the solution to the $(l-1)$-clustering problem is known. Algorithm 7.2 consists of the following three main steps:

- first, each data point is considered as a candidate starting cluster center together with the solution of the previous $(l-1)$-clustering problem obtained by some incremental clustering algorithm. A data point which provides the largest decrease of the clustering function is determined. Then a threshold is computed by multiplying this maximum decrease by the parameter $\gamma_1 \in [0, 1]$. The data points which provide the decrease of the clustering function greater than this threshold are selected as potential starting cluster centers while the rest of the data points are removed;
- second, the selected data points are replaced by the centers of clusters around them, and the decrease of the clustering function is calculated using these centers. Similar to the previous step, a threshold is computed using the maximum decrease with the parameter $\gamma_2 \in [0, 1]$, and the centers providing the decrease greater than this threshold are kept as potential starting cluster centers; and
- in the last step, the auxiliary clustering problem is solved starting from each point left in the list of potential starting cluster centers, and the values of the auxiliary clustering function are computed at each (local) solution obtained. Then using the smallest value of this function, one more threshold parameter—$\gamma_3 \in [1, \infty)$—is defined. Solutions of the auxiliary clustering problem with the function values less than this threshold are included to the final list of starting cluster centers. Together with the solution of the previous $(l-1)$-clustering problem this list is used as a set of starting points for solving the l-clustering problem.

Note that there is no theoretical result which would help us to find exact values of the parameters γ_1, γ_2, and γ_3. We use numerical experiments on some data sets with different sizes to provide some recommendations on the values of these parameters. Let us denote by $\bar{A}_1(\gamma_1)$, the set $\bar{A}_1$ obtained from (7.12) using $\gamma_1 \in [0, 1]$; by $\bar{A}_3(\gamma_1, \gamma_2)$, the set $\bar{A}_3$ obtained from (7.14) using $\gamma_1, \gamma_2 \in [0, 1]$; and by $\bar{A}_5(\gamma_1, \gamma_2, \gamma_3)$, the set $\bar{A}_5$ obtained from (7.16) using $\gamma_1, \gamma_2 \in [0, 1]$, $\gamma_3 \in [1, \infty)$. It is clear that for all $\gamma_1 \leq \mu_1$, $\gamma_2 \leq \mu_2$, $\gamma_3 \geq \mu_3$ we have

$$\begin{aligned}
&\bar{A}_1(\mu_1) \subseteq \bar{A}_1(\gamma_1), \\
&\bar{A}_3(\mu_1, \mu_2) \subseteq \bar{A}_3(\gamma_1, \gamma_2), \quad \text{and} \\
&\bar{A}_5(\mu_1, \mu_2, \mu_3) \subseteq \bar{A}_5(\gamma_1, \gamma_2, \gamma_3).
\end{aligned} \tag{11.1}$$

Notice that the set $\bar{A}_1(0)$ contains all data points which are not cluster centers. This means that if $\gamma_1 = \gamma_2 = 0$ and γ_3 is sufficiently large, then the number of starting points in Step 5 of Algorithm 7.2 is the number of data points which are not cluster centers. This is the largest number of starting points for the clustering problem which can be obtained. The least number of starting points is obtained when $\gamma_1 = \gamma_2 = \gamma_3 = 1$.

The inclusions in (11.1) imply that an incremental clustering algorithm obtains its best solution when $\gamma_1 = \gamma_2 = 0$ and γ_3 is sufficiently large, and this solution cannot be improved using any other values of $\gamma_1, \gamma_2 \in [0, 1]$ and $\gamma_3 \in [1, \infty)$. Nevertheless, computational effort required by any incremental clustering algorithm reduces as parameters γ_1 and γ_2 increase and the parameter γ_3 decreases. Therefore, for a given data set we are interested in finding the largest values of γ_1 and γ_2, and the smallest value of γ_3 such that an incremental clustering algorithm can still obtain its best solution to the clustering problem and further increase of γ_1, γ_2 or decrease of γ_3 deteriorates the solution.

Let us denote by $\tilde{f}_k(\gamma_1, \gamma_2, \gamma_3)$ the value of the cluster function f_k, given in (4.4), obtained by an incremental clustering algorithm for the given values of $\gamma_1, \gamma_2, \gamma_3$, and k clusters. First, we set $\gamma_1 = \gamma_2 = 0$ and $\gamma_3 = 10$ (assuming that this value is sufficiently large) and find $\tilde{f}_k(0, 0, 10)$. This means that we compute the largest possible sets $\bar{A}_1$ and $\bar{A}_3$ and find the best possible solution by the incremental algorithm. Then we calculate $\tilde{f}_k(\gamma_1, \gamma_2, \gamma_3)$ for

$$\begin{aligned} \gamma_1, \gamma_2 &= 0.05i, & i &= 1, \dots, 20, \quad \text{and} \\ \gamma_3 &= 1 + 0.01(i-1), & i &= 1, \dots, 101. \end{aligned}$$

The largest values of γ_1, γ_2 and the smallest value of γ_3 satisfying the condition

$$\frac{\tilde{f}_k(\gamma_1, \gamma_2, \gamma_3) - \tilde{f}_k(0, 0, 10)}{\tilde{f}_k(0, 0, 10)} \leq \varepsilon \tag{11.2}$$

are accepted as an estimation of the optimal values of parameters $\gamma_j,\ j = 1, 2, 3$. Here, $\varepsilon \geq 0$ is a given tolerance.

Next, we demonstrate how to find estimations for the optimal values of $\gamma_j,\ j = 1, 2, 3$. For this aim, we apply `IS-Clust` on data sets Iris plant and the Breast cancer Wisconsin (see Tables 11.1 and 11.2 for details of these data sets). The results with different values of parameters are presented in Tables 11.6 and 11.7. We give first the results with $\gamma_1 = \gamma_2 = 0$ and $\gamma_3 = 10$, and then the results with the largest values of γ_1, γ_2 and the smallest value of γ_3 that satisfy condition (11.2). Finally, we present results with nondecreasing values of γ_1 and γ_2. In these tables, we use the following notations:

- k—the number of clusters;
- E—the error in % computed using (10.9); and
- α—the parameter defined by

$$\alpha = \frac{N_d(\gamma_1, \gamma_2, \gamma_3)}{N_d(0, 0, 10)},$$

where $N_d(\gamma_1, \gamma_2, \gamma_3)$ is the number of distance function evaluations by the incremental algorithm for given values of $\gamma_1, \gamma_2, \gamma_3$. It is clear that $N_d(0, 0, 10)$ is the number of distance function evaluations by the same algorithm when $\gamma_1 = \gamma_2 = 0$ and $\gamma_3 = 10$. The parameter α reflects the ratio of computational effort for given $\gamma_1, \gamma_2 \in [0, 1]$ and $\gamma_3 \geq 1$ with respect to that of for $\gamma_1 = \gamma_2 = 0$ and $\gamma_3 = 10$.

Since the data set Iris plant is extra small we take $\varepsilon = 0$ for it. From Table 11.6, we see that the largest values of γ_1, γ_2 and the smallest value of γ_3 satisfying the condition (11.2) are: $\gamma_1 = 0.50, \gamma_2 = 0.55, \gamma_3 = 1.10$. Results show that any increase of γ_1, γ_2 and any decrease of γ_3 deteriorate the best solution. It can also be observed that values of $\gamma_2 < \gamma_1$ does not improve the accuracy of the algorithm. Results given in columns at the bottom right corner of Table 11.6 confirm this claim. Results for the parameter α demonstrate that the selection of optimal values of $\gamma_j,\ j = 1, 2, 3$ allows us to significantly reduce the computational effort without deteriorating the clustering solution.

For the data set Breast cancer Wisconsin, we take $\varepsilon = 0.01$. The largest values of γ_1, γ_2 and the smallest value of γ_3 satisfying the condition (11.2) are: $\gamma_1 = 0.60, \gamma_2 = 0.80, \gamma_3 = 1.01$. Results in Table 11.7 show that any increase of γ_1, γ_2 or

Table 11.6 Results for different values of $(\gamma_1, \gamma_2, \gamma_3)$: Iris plant

k	E	α	E	α	E	α	E	α
	(0.00, 0.00, 10.00)		(0.50, 0.55, 1.10)		(0.50, 0.55, 1.05)		(0.55, 0.55, 1.10)	
2	0.00	1.00	0.00	0.11	0.00	0.11	0.00	0.09
3	0.00	1.00	0.00	0.34	0.00	0.34	0.00	0.30
4	0.00	1.00	0.00	0.32	0.00	0.31	0.00	0.28
5	0.00	1.00	0.00	0.28	0.00	0.25	0.00	0.23
6	0.00	1.00	0.00	0.36	0.00	0.33	0.00	0.31
7	0.00	1.00	0.00	0.28	0.01	0.26	0.01	0.24
8	0.00	1.00	0.00	0.20	0.25	0.19	0.25	0.17
9	0.00	1.00	0.00	0.17	0.27	0.17	0.27	0.15
10	0.00	1.00	0.00	0.15	0.00	0.16	0.00	0.14
	(0.55, 0.55, 2.00)		(0.50, 0.60, 1.10)		(0.50, 0.60, 2.00)		(0.55, 0.00, 2.00)	
2	0.00	0.38	0.00	0.11	0.00	0.48	0.00	0.38
3	0.00	0.39	0.00	0.34	0.00	0.45	0.00	0.39
4	0.00	0.33	0.00	0.32	0.00	0.39	0.00	0.33
5	0.00	0.26	0.00	0.27	0.00	0.32	0.00	0.26
6	0.00	0.26	0.00	0.26	0.00	0.29	0.00	0.26
7	0.01	0.25	0.01	0.25	0.01	0.28	0.01	0.25
8	0.25	0.21	0.25	0.22	0.25	0.24	0.25	0.21
9	0.27	0.21	0.27	0.23	0.27	0.24	0.27	0.21
10	0.00	0.23	0.00	0.25	0.00	0.26	0.00	0.23

Table 11.7 Results for different values of $(\gamma_1, \gamma_2, \gamma_3)$: Breast cancer Wisconsin

k	E	α	E	α	E	α	E	α
	(0.00, 0.00, 10.00)		(0.60, 0.80, 1.01)		(0.60, 0.80, 2.00)		(0.60, 0.80, 1.00)	
2	0.00	1.00	0.00	0.34	0.00	0.34	0.00	0.06
5	0.00	1.00	0.00	0.16	0.00	0.16	0.61	0.02
10	0.00	1.00	0.00	0.08	0.00	0.08	0.03	0.01
15	0.00	1.00	0.85	0.06	0.85	0.06	1.22	0.01
20	0.00	1.00	0.76	0.06	0.76	0.06	0.70	0.01
25	0.00	1.00	0.00	0.06	0.00	0.06	1.22	0.01
	(0.65, 0.80, 1.01)		(0.65, 0.80, 2.00)		(0.60, 0.85, 1.01)		(0.60, 0.85, 2.00)	
2	0.00	0.27	0.00	0.27	0.00	0.34	0.00	0.34
5	0.00	0.13	0.00	0.13	0.00	0.14	0.00	0.14
10	0.00	0.07	0.00	0.07	0.00	0.06	0.00	0.06
15	1.03	0.06	1.03	0.05	1.02	0.05	1.02	0.05
20	0.99	0.05	0.99	0.05	0.80	0.04	0.80	0.04
25	0.00	0.05	0.00	0.05	0.22	0.04	0.00	0.04

any decrease of γ_3 slightly deteriorates the best solution obtained by the incremental algorithm. Results for the parameter α demonstrate that the selection of estimations of the optimal values of $\gamma_j,\ j = 1, 2, 3$ allows us to significantly reduce the number of distance function evaluations in comparison with those required by the algorithm when $\gamma_1 = \gamma_2 = 0$ and $\gamma_3 = 10$. We can see that values of γ_1 and γ_2 are larger and γ_3 is smaller for this data set comparing to those for Iris plant.

Results presented in Tables 11.6 and 11.7, in general, lead to the following observations:

- the optimal values of $\gamma_1,\ \gamma_2 \in [0, 1]$ increase and the optimal value of $\gamma_3 \geq 1$ decreases as the size of a data set increases;
- we can select $\gamma_2 \in [\gamma_1, 1]$;
- the optimal value of γ_3 seems to be close to 1. This means that the solutions found by solving the auxiliary clustering problem are very close to the local minimizers of the clustering problem; and
- the values of α demonstrate that the use of the optimal values (or their estimations) of parameters leads to a significant decrease in computational effort. This decrease becomes even more significant as the number of clusters and the size of a data set increase.

To conclude, small values of $\gamma_1,\ \gamma_2$ and large values of γ_3 will increase computational time considerably in large data sets without any significant improvement in the quality of the solution to the clustering problem. Therefore, in the implementations of incremental clustering algorithms, we recommend to select the parameters $\gamma_j,\ j = 1, 2, 3$ as follows:

- for small and extra small data sets: γ_1 and γ_2 lie in the interval [0.4, 0.6] while $\gamma_3 \in [1.1, 2]$;

- for medium sized data sets: we can set $\gamma_1 \in [0.55, 0.65]$, $\gamma_2 \in [0.75, 0.85]$, and $\gamma_3 \in [1.01, 1.05]$;
- for large data sets: we set $\gamma_1 \in [0.90, 0.95]$, $\gamma_2 \in [0.925, 0.975]$, and $\gamma_3 \in [1.005, 1.0075]$; and
- for extra large data sets: we set $\gamma_1 \in [0.975, 0.995]$, $\gamma_2 \in [0.99, 0.999]$, and $\gamma_3 = 1.001$.

Note that these values are not optimal and can only be considered as recommended values.

Chapter 12
Numerical Experiments

12.1 Introduction

This chapter is primarily devoted to the study of the performance of optimization based incremental clustering algorithms. Since the procedure of finding starting cluster centers is an important part of all these algorithms we start the chapter with discussing on the impact of this procedure to the solution obtained by a clustering algorithm.

Then, we demonstrate the performance of the clustering algorithms using data sets with different number of data points and attributes described in Chap. 11: extra small, small, medium sized, large, and very large. The performance profiles are used to evaluate the accuracy of clustering solutions, the number of distance function evaluations and CPU time. In addition, we apply the DB index, the purity, the NMI index and silhouettes to compare different clustering algorithms. In all these algorithms, we consider the MSSC problem. To compare the performance of the incremental clustering algorithms when different similarity measures—d_1, d_2 and d_∞—are used in their objective functions, we apply `DG-Clust` on three real-world data sets given in Chap. 11: German towns, TSPLIB1060 and TSPLIB3038. We use the Voronoi diagrams for this purpose.

12.2 Importance of Procedure for Finding Starting Cluster Centers

In this section, we study the contribution of the procedure for generating starting cluster centers to the quality of the final clustering solutions and also to the overall performance of an incremental clustering algorithm. For this aim, we use three data sets with different number of entries: Ionosphere (small size), Image

A. M. Bagirov et al., *Partitional Clustering via Nonsmooth Optimization*, Unsupervised and Semi-Supervised Learning,
https://doi.org/10.1007/978-3-030-37826-4_12

segmentation (medium sized) and KEGG metabolic network (large scale). The number of attributes in these data sets ranges from 19 to 34. This choice of data sets allows us to clearly demonstrate the importance of this procedure.

In our experiments, we apply MGKM with and without the procedure for generating starting cluster centers. The performance of this solver with different procedures is compared using three performance measures: accuracy, the number of distance function calculations, and CPU time. The accuracy (or the error) is defined using the formula (10.10).

Consider the k-clustering problem with $k \geq 2$ and recall Algorithm 7.2—the procedure for finding starting cluster centers. This procedure has the following steps:

1: using radii of clusters from the previous iteration determine the list of data points which are candidates to be a starting point;

2: consider each point as the kth cluster center and compute the decrease of the cluster function f_k, defined in (4.4), in comparison with the optimal value f_{k-1}^* for the $(k-1)$-clustering problem;

3: remove some data points from the list using a threshold for the decrease;

4: for each point from the list compute the cluster around it and replace this data point with the corresponding cluster center;

5: solve the auxiliary clustering problem starting from each of these centers.

Steps 1, 2 and 3 use only data points where the preliminary list of candidate starting points is determined. Our aim is to demonstrate that Steps 4 and 5 are very important and make a significant contribution to the quality of the final solution in clustering. Therefore, we consider the following versions of MGKM:

- $V0$: MGKM0—the version where both Steps 4 and 5 are excluded;
- $V1$: MGKM4—the version where only Step 4 is used and Step 5 is excluded;
- $V2$: MGKM5—the version where Step 5 is used and Step 4 is excluded;
- $V3$: MGKM45—the full version with Steps 4 and 5.

The results are presented in Table 12.1, where as before k stands for the number of clusters. In our experiments, we use 5 hours time limit in all versions, that is if the algorithm exceeds this time limit, then its performance is considered as a *failure* (denoted by "fail" in the table). To avoid writing very large numbers for distance function evaluations (denoted by "distance function evals") in the table, we include a number after the name of each data set in brackets and to get the correct values, numbers in distance function evaluations columns should be multiplied by these numbers.

Results show that accuracies of all versions are comparable and differences between them are not significant. Furthermore, as the size of a data set increases the differences become even more insignificant. Regarding the number of distance function evaluations, we can see that the use of the full version $V3$ leads to a significant reduction of the number in all cases. Results for versions $V2$ and $V3$ imply that the use of the auxiliary clustering problem in the procedure allows

Table 12.1 Results with and without the procedure for finding starting points

	Accuracy				Distance function evals				CPU time			
k	V_0	V_1	V_2	V_3	V_0	V_1	V_2	V_3	V_0	V_1	V_2	V_3
Ionosphere ($\times 10^6$)												
2	0.00	0.00	0.00	0.00	1.28	0.55	0.71	0.44	0.33	0.11	0.11	0.06
3	0.03	0.03	0.03	0.03	2.77	1.15	1.24	0.81	0.67	0.22	0.22	0.11
5	0.06	0.07	0.11	0.11	6.28	2.12	2.22	1.51	1.38	0.36	0.36	0.20
7	0.05	0.00	0.10	0.52	10.69	3.40	3.47	2.28	2.14	0.55	0.56	0.30
10	0.27	0.30	0.53	0.28	17.84	5.27	4.91	3.42	3.28	0.81	0.80	0.45
15	0.84	0.65	0.95	2.10	26.07	8.05	6.65	5.05	4.50	1.17	1.06	0.64
20	1.31	1.22	1.54	3.32	34.07	10.34	8.39	6.72	5.70	1.45	1.33	0.86
22	1.50	1.31	1.21	2.80	36.73	11.30	9.05	7.32	6.11	1.59	1.44	0.94
25	1.58	2.00	1.38	3.03	42.81	12.89	9.96	8.42	7.03	1.81	1.56	1.08
Image segmentation ($\times 10^6$)												
2	0.00	0.00	0.00	0.00	0.29	0.22	0.14	0.19	0.27	0.08	0.06	0.05
3	0.00	0.00	0.00	0.00	0.77	0.60	0.36	0.42	0.56	0.22	0.13	0.08
5	0.00	0.00	0.00	0.00	5.38	2.51	2.10	1.28	2.45	0.83	0.48	0.17
7	0.00	2.31	2.30	2.30	12.20	3.92	3.40	2.17	4.41	1.09	0.72	0.30
10	0.00	1.75	1.75	1.75	21.10	7.86	4.82	3.54	6.05	1.81	0.92	0.44
15	0.49	0.49	0.48	1.90	42.58	14.50	8.79	6.61	8.91	2.59	1.45	0.73
20	0.61	0.76	0.79	0.62	92.14	24.05	14.71	11.16	14.94	3.61	2.16	1.14
22	0.64	0.83	0.85	0.66	118.05	29.38	17.93	13.03	17.94	4.16	2.52	1.30
25	0.43	0.86	0.84	0.65	161.32	36.15	23.14	16.95	22.72	4.83	3.14	1.64
KEGG metabolic network ($\times 10^7$)												
2	0.00	0.00	0.00	0.00	0.44	0.35	0.28	0.35	285.28	9.72	10.30	10.28
3	0.00	0.00	0.00	0.00	2.92	1.29	0.79	1.04	4097.22	398.09	201.55	51.69
5	fail	0.07	0.07	0.07	Fail	2.61	1.83	2.19	Fail	794.25	593.31	234.08
7	fail	0.18	0.18	0.18	Fail	3.41	2.55	2.78	Fail	1096.13	842.88	298.98
10	fail	0.01	0.01	0.01	Fail	5.03	4.07	4.10	Fail	1658.86	1311.27	583.70
15	fail	2.96	2.98	2.98	Fail	7.89	6.38	6.48	Fail	1996.03	1514.92	682.95
20	fail	1.26	1.42	1.18	Fail	11.99	10.43	10.48	Fail	2087.19	1677.55	760.66
22	fail	1.74	2.01	1.74	Fail	13.81	12.06	12.33	Fail	2113.00	1698.78	787.09
25	fail	0.21	0.04	1.74	Fail	17.79	15.99	16.10	Fail	2236.72	1936.06	867.17

us to significantly reduce the number of distance function evaluations without deteriorating the final solution. This is due to the fact that the solution obtained by solving the auxiliary clustering problem is close to the solutions of the clustering problem and the k-means algorithm requires a limited number of iterations to obtain them. Similar observations are true also for the required CPU time. Here, we can see that the use of the version $V3$ allows us to significantly decrease the CPU time even in comparison with the versions $V1$ and $V2$.

These results clearly show that regardless of the size of the data sets, the use of the procedure for finding starting cluster centers allows us to significantly reduce the computational effort while preserving almost the same accuracy for the obtained clustering solutions. Furthermore, the auxiliary clustering problem is an important component of the incremental clustering algorithms.

12.3 Performance Results of Incremental Clustering Algorithms

In this section, we present results on the performance of the incremental clustering algorithms as well as results obtained by the MS-KM. We include the latter algorithm for the comparison purpose. In `MS-KM`, the number of randomly generated starting points is limited by 1000, however, we also applied the time limit which is twice of the CPU time required by `MGKM`. We do not present results of `MS-KM` based on performance profiles using the CPU time and the number of distance function evaluations as the CPU time and also in some sense the number of distance function evaluations are fixed for this algorithm.

We present the results of our experiments for each class of data sets separately. The best known value of the cluster function for a given k is denoted by f_{best}. Note that, in all tables in order to find the true best values of the cluster function, numbers given in the f_{best} column should be multiplied by the number given after the names of data sets.

The error E of a given solution is computed using (10.10). We say that an algorithm finds the best known solutions to the clustering problem if its error $0 \leq E \leq 0.1$. If $0.1 < E \leq 1$, then an algorithm finds nearly the best known solution. For performance profiles, we select in (10.11) and (10.14) the constants $c = \bar{c} = 1$ and the number of iterations to solve a clustering problem $M = 100$.

12.3.1 Results for Extra Small Data Sets

We apply `GKM`, `MGKM`, `DG-Clust`, `NDC-Clust`, `IDCA-Clust`, `IS-Clust`, and `MS-KM` to extra small data sets. Other algorithms are not best suited for such data sets. Up to ten clusters are computed in all data sets. Results for accuracy are given in Table 12.2. Note that the best known solutions for all data sets, but Liver disorder data set, are also known to be the global solutions to the corresponding clustering problems.

Results presented in Table 12.2 demonstrate that `MS-KM` cannot be considered as an alternative to any other algorithm. It is able to find the best known solutions only when the number of clusters is small. Otherwise, `MS-KM` fails to find a good quality solution. Other algorithms show the good performance in finding accurate solutions. Nevertheless, most algorithms, except `IS-Clust`, failed to find solutions with high accuracy in Bavaria postal two data set.

Table 12.2 Accuracy results for extra small data sets

k	f_{best}	GKM	MGKM	DG-Clust	NDC-Clust	IDCA-Clust	IS-Clust	MS-KM
German towns ($\times 10^5$)								
2	1.21426	0.00	0.00	0.00	0.00	0.00	0.00	0.00
3	0.77009	1.45	0.00	0.00	0.00	0.00	0.00	0.00
4	0.49601	0.72	0.00	0.00	0.00	0.00	0.00	0.00
5	0.38716	0.00	0.00	0.00	0.00	0.00	0.00	0.00
6	0.30535	0.00	0.00	0.00	0.00	0.00	0.00	0.00
7	0.24433	0.09	0.09	0.00	0.09	0.09	0.00	0.00
8	0.21483	1.33	0.69	0.00	0.69	0.69	0.59	3.61
9	0.18669	1.48	1.48	0.00	1.48	1.48	1.37	8.66
10	0.16427	1.06	1.06	0.00	1.06	1.06	0.00	7.70
Bavaria postal 1 ($\times 10^{10}$)								
2	60.25472	7.75	0.00	0.00	0.00	0.00	0.00	7.75
3	29.45066	0.00	0.00	0.00	0.00	0.00	0.00	20.02
4	10.44747	0.00	0.00	0.00	0.00	0.00	0.00	0.08
5	5.97615	0.00	0.00	0.00	0.00	0.00	0.00	23.58
6	3.59085	0.00	28.02	27.79	27.65	28.02	27.65	28.02
7	2.19832	1.50	69.39	0.00	0.00	69.39	69.39	98.03
8	1.33854	0.00	141.13	0.00	0.00	141.13	0.00	225.23
9	0.84237	0.00	259.69	0.00	0.00	259.69	1.44	416.79
10	0.64465	0.00	350.66	0.00	0.00	350.66	0.00	452.52
Bavaria postal 2 ($\times 10^{10}$)								
2	1.99080	162.17	144.28	144.28	144.28	144.28	144.28	144.28
3	1.73988	0.00	0.00	0.00	0.00	0.00	0.00	106.79
4	0.75591	0.00	0.00	0.00	0.00	0.00	0.00	0.00
5	0.53429	1.86	1.14	1.14	1.14	1.14	0.00	1.14
6	0.31876	1.21	39.04	39.04	39.04	39.04	0.00	39.04
7	0.22159	0.50	77.07	76.94	76.94	77.07	0.00	95.00
8	0.17045	0.73	113.22	112.22	112.22	113.21	0.00	153.50
9	0.14011	0.14	142.69	142.48	142.48	142.69	0.00	208.41
10	0.11908	0.16	170.46	169.65	169.65	170.46	0.00	240.56
Iris plant($\times 10^2$)								
2	1.52348	0.00	0.00	0.00	0.00	0.01	0.00	0.00
3	0.78851	0.01	0.00	0.01	0.01	0.01	0.00	0.00
4	0.57228	0.05	0.05	0.00	0.00	0.02	0.00	0.00
5	0.46446	0.54	0.06	0.00	0.00	0.09	0.00	0.06
6	0.39040	1.44	0.07	0.00	0.00	0.10	0.00	0.07
7	0.34298	3.17	0.00	0.00	0.00	0.03	0.00	13.90
8	0.29989	1.71	0.09	0.00	0.00	0.29	0.00	30.27
9	0.27786	2.85	0.10	0.00	0.00	0.40	0.00	40.60
10	0.25834	3.56	0.51	0.00	0.50	0.34	0.06	42.70

(continued)

Table 12.2 (continued)

k	f_{best}	GKM	MGKM	DG-Clust	NDC-Clust	IDCA-Clust	IS-Clust	MS-KM
TSPLIB1060 ($\times 10^{10}$)								
2	0.98319	0.00	0.00	0.00	0.00	0.00	0.00	0.00
3	0.67058	0.00	0.00	0.00	0.00	0.00	0.00	0.00
4	0.47520	0.01	0.01	0.00	0.00	0.01	0.00	0.00
5	0.37910	0.01	0.01	0.01	0.01	0.06	0.00	0.00
6	0.31770	0.06	0.06	0.06	0.06	0.06	0.06	0.00
7	0.27042	0.02	0.02	0.02	0.00	0.02	0.03	0.01
8	0.22643	0.00	0.00	0.00	0.00	0.00	0.02	19.44
9	0.19910	0.30	0.30	0.14	0.00	0.30	0.02	35.81
10	0.17548	0.23	0.04	0.03	0.04	0.23	0.04	28.97
Liver disorders ($\times 10^{6}$)								
2	0.42398	0.00	0.00	0.00	0.00	0.00	0.00	0.00
3	0.32271	0.71	0.71	0.71	0.88	0.88	0.00	0.00
4	0.26066	0.49	0.49	0.11	0.22	0.48	0.22	0.00
5	0.21826	0.08	0.08	0.00	0.07	0.07	0.08	0.01
6	0.18709	0.97	0.05	0.00	0.00	0.14	0.00	0.28
7	0.16420	0.72	0.34	0.00	0.00	0.37	0.00	14.26
8	0.14778	0.41	0.41	0.00	0.00	0.29	0.01	26.95
9	0.13734	0.83	0.00	0.31	0.20	0.31	0.33	36.10
10	0.12742	0.21	0.01	0.00	0.15	0.29	0.00	16.76

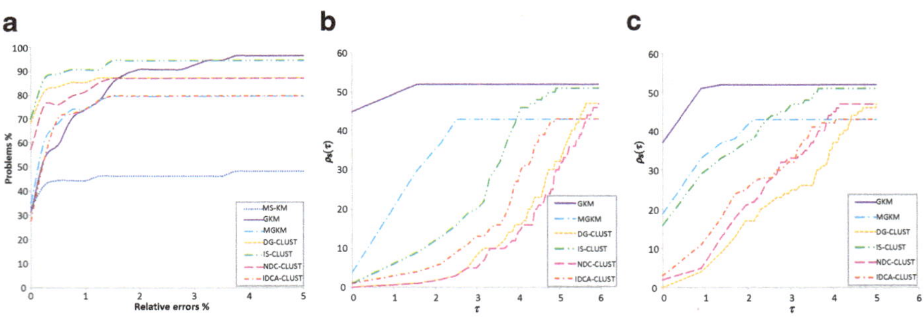

Fig. 12.1 Performance profiles for extra small data sets. (**a**) Relative errors. (**b**) Distance function evals. (**c**) CPU time

Performance profiles for extra small data sets are illustrated in Fig. 12.1. IS-Clust is the most successful in finding the best known solutions and GKM is the most successful in solving clustering problems with the error no more than 5%. MS-KM has the worst performance while GKM requires the least number of distance function evaluations and CPU time. On the other hand, NDC-Clust uses more distance function evaluations and DG-Clust requires more CPU time than any other algorithm.

In Fig. 12.2 graphs for three indices (DB, purity and NMI) and in Fig. 12.3 silhouette plots for $k = 2, 3, 5$ and ten clusters are illustrated using results obtained

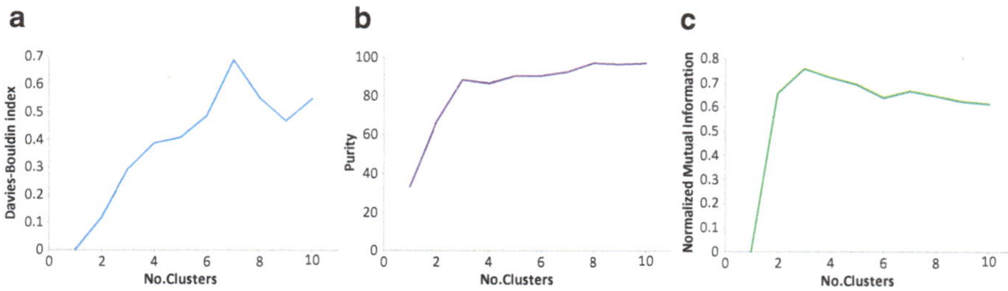

Fig. 12.2 Results for Iris Plant data set using different indices. (**a**) DB index. (**b**) Purity. (**c**) NMI index

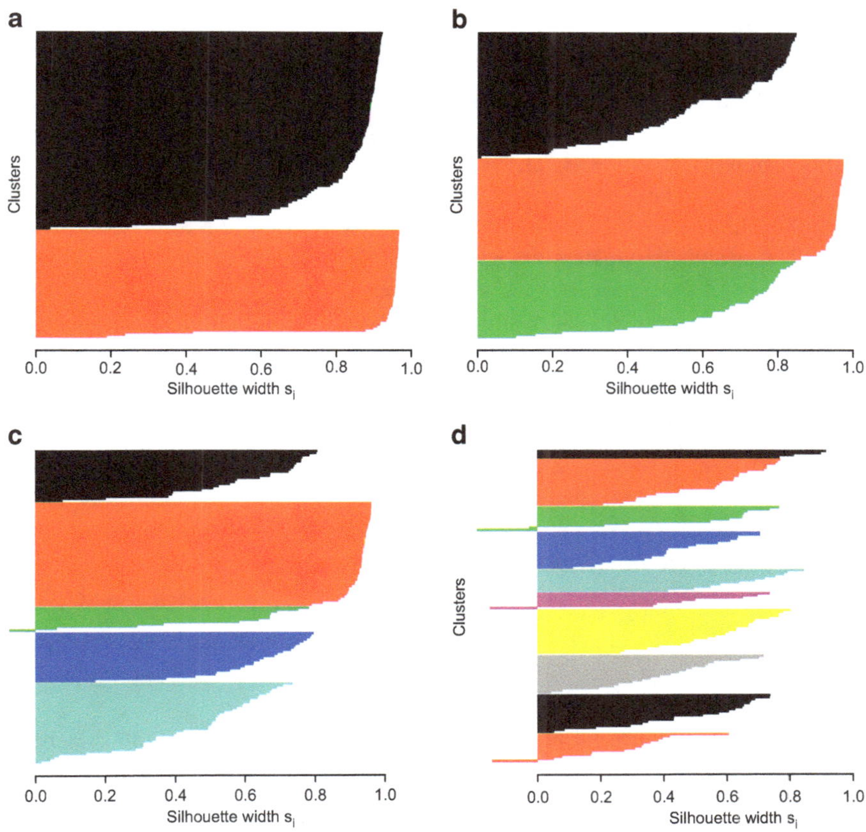

Fig. 12.3 Silhouette plots for Iris plant data set. (**a**) $k = 2$. (**b**) $k = 3$. (**c**) $k = 5$. (**d**) $k = 10$

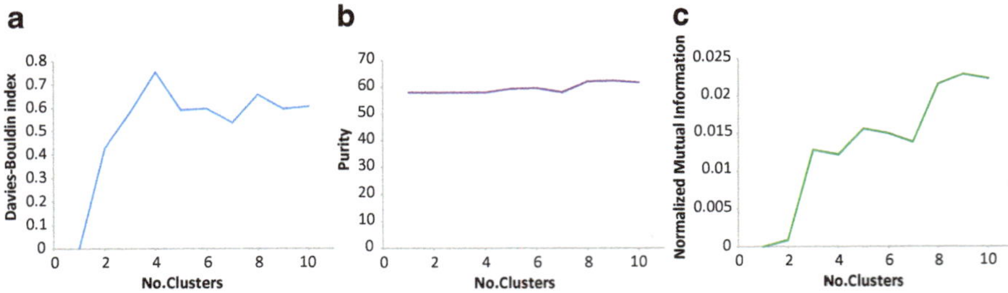

Fig. 12.4 Results for Liver disorder data set using different indices. (**a**) DB index. (**b**) Purity. (**c**) NMI index

by the IS-CLUST in Iris plant data set. The DB index has several knee points at $k = 3, 4, 5, 6$ and one local minimizer at $k = 9$. In general, the purity increases as the number of clusters increases; however, at $k = 3$ it has a local maximizer. The NMI index gets its highest value at $k = 3$. Finally, silhouette plots show that clusters are well-separated when $k = 3$. These results demonstrate a good consistency of the class and the cluster (with $k = 3$) distributions in Iris Plant data set.

In Fig. 12.4 graphs for the three indices and in Fig. 12.5 silhouette plots for $k = 2, 3, 5$ and ten clusters are given based on results obtained by `IS-Clust` in Liver disorder data set. Here, the DB index has three distinct local minimizers at $k = 5, 7$ and $k = 9$. The purity increases as the number of clusters increases; however, this increase is not significant. The NMI index is close to 0 for $k < 3$ and silhouette plots show that clusters are not compact and not well-separated for $k = 2, 3, 5, 10$. Summarizing these results we can conclude that the class and the cluster distributions in Liver disorder data set are incompatible.

12.3.2 Results for Small Data Sets

We apply `GKM`, `MGKM`, `DG-Clust`, `NDC-Clust`, `IDCA-Clust`, `IS-Clust`, and `MS-KM` to small data sets. Other algorithms are not well suited for these data sets. Up to 25 clusters are computed in these data sets.

Results for accuracy are given in Table 12.3. The results show that `MS-KM` can reach the best solutions only when the number of clusters is small. Otherwise, this algorithm fails to find high quality solutions. Other algorithms are, in general, successful in finding accurate solutions.

Performance profiles for small data sets are presented in Fig. 12.6. `IS-Clust` is the most successful in finding the best known solutions and in solving clustering problems with the error no more than 5%. As before, `MS-KM` has the worst performance while `GKM` requires the least number of distance function evaluations and CPU time. In addition, `NDC-Clust` uses more distance function evaluations and `DG-Clust` requires more CPU time than any other algorithm.

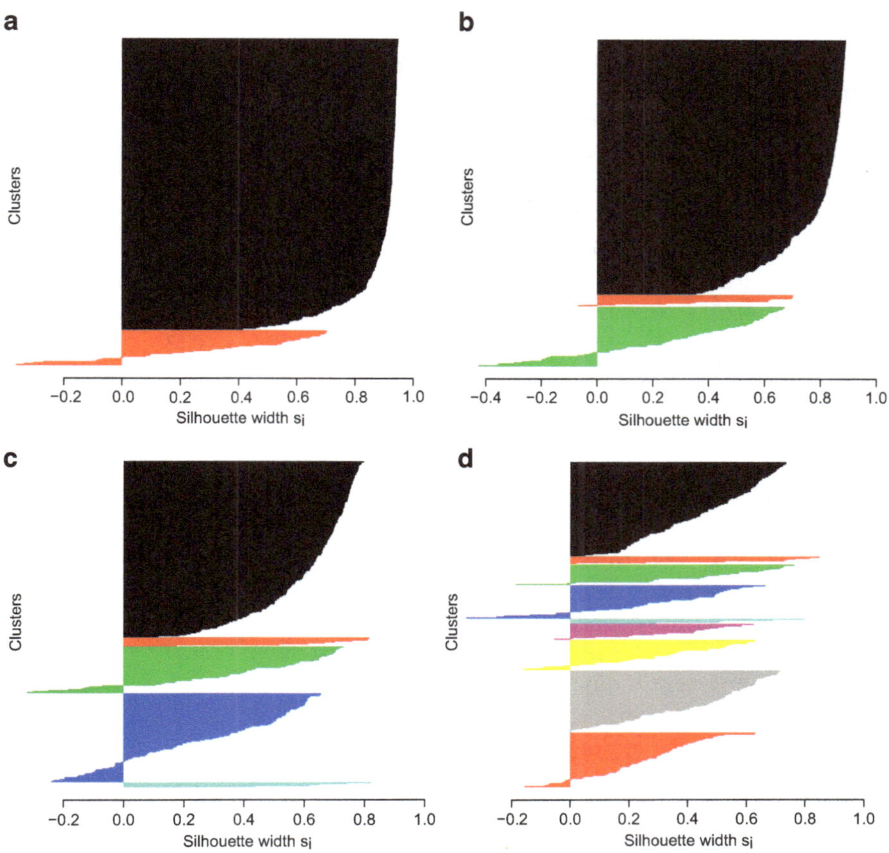

Fig. 12.5 Silhouette plots for Liver disorder data set. (**a**) $k = 2$. (**b**) $k = 3$. (**c**) $k = 5$. (**d**) $k = 10$

In Fig. 12.7, the graph of the DB index is given using the results obtained by `IS-Clust` in TSPLIB3038 data set. Note that the purity and the NMI index require the existence of the class labels and since this data set has no classes we only present the DB index in Fig. 12.7. The silhouette plots for TSPLIB3038 data set with $k = 2, 3, 5$ and ten clusters are given in Fig. 12.8. From Fig. 12.7, it can be observed that the DB index has local minimizers at $k = 5, 7, 11, 15, 20$. Silhouette plots show that in the 10-partition of the data set six clusters are compact and well-separated and other clusters contain some "misclassified" points. The similar observation is true for the k-partitions of the data set with $k = 2, 3$ and 5.

In Fig. 12.9 graphs for the three indices and in Fig. 12.10 silhouette plots for $k = 2, 3, 5, 10$, and 25 clusters are given using results obtained by `IS-Clust` in Vehicle silhouettes data set. The DB index has the distinct local minimizers at $k = 4, 7, 11, 14$ and $k = 19$. The purity increases as the number of clusters increases; however, even for 25 clusters it is only about 55%. The largest value for the NMI

Table 12.3 Accuracy results for small data sets

k	f_{best}	GKM	MGKM	DG-Clust	NDC-Clust	IDCA-Clust	IS-Clust	MS-KM
Heart disease ($\times 10^5$)								
2	5.98899	0.00	0.00	0.00	0.00	0.00	0.00	0.00
3	4.67508	5.02	3.96	5.14	5.14	5.14	4.67	0.00
5	3.27965	0.52	0.53	0.00	0.01	0.53	0.44	0.03
7	2.64942	2.59	0.00	0.32	0.02	0.00	0.32	4.44
10	2.00558	0.83	0.19	0.00	0.00	0.19	0.03	20.82
15	1.46895	0.55	0.18	0.13	0.18	0.43	0.00	25.37
20	1.16993	0.67	0.00	0.52	0.36	0.51	0.49	37.96
22	1.09199	2.45	0.08	0.40	0.00	0.94	0.23	47.81
25	0.99314	3.33	1.36	0.00	1.13	1.31	0.06	62.52
TSPLIB3038 ($\times 10^9$)								
2	3.16880	0.00	0.00	0.00	0.00	0.00	0.00	0.00
3	2.17634	3.43	3.43	3.43	3.43	3.43	3.43	0.00
5	1.19820	0.00	0.00	0.00	0.00	0.00	0.00	0.00
7	0.83967	1.87	1.73	1.85	1.73	1.73	1.73	0.00
10	0.56025	2.78	0.58	0.57	0.58	0.58	0.00	0.00
15	0.35604	0.07	0.05	0.00	0.00	0.07	0.06	0.00
20	0.26681	2.00	0.43	0.14	0.20	0.43	0.16	0.17
22	0.24295	1.64	0.54	0.02	0.00	0.55	0.03	1.19
25	0.21450	0.78	0.43	0.43	0.56	0.43	0.00	1.56
Pima Indians diabetes ($\times 10^6$)								
2	5.14238	0.00	0.00	0.00	0.00	0.00	0.00	0.00
3	2.91332	1.23	1.23	1.23	1.23	1.23	1.23	0.00
5	1.73687	0.15	0.15	0.00	0.15	0.15	0.01	0.01
7	1.30315	0.00	0.00	0.00	0.00	0.00	0.00	0.00
10	0.93066	1.84	0.06	0.00	1.66	0.06	1.12	12.66
15	0.69579	0.21	0.00	1.06	0.23	0.05	1.37	34.14
20	0.57278	0.28	0.00	0.18	0.10	0.35	0.27	47.39
22	0.53501	0.55	0.34	0.33	0.00	0.38	0.14	55.16
25	0.48874	0.38	0.38	0.35	0.00	0.43	0.17	67.21
Breast cancer Wisconsin ($\times 10^4$)								
2	1.93232	0.00	0.00	0.00	0.00	0.00	0.00	0.00
3	1.62555	0.01	0.01	0.00	0.00	0.01	0.00	0.00
5	1.37047	2.28	0.01	0.02	0.00	0.02	0.00	0.00
7	1.20497	1.44	0.12	0.00	0.02	0.10	0.00	6.27
10	1.01996	0.16	0.32	0.04	0.13	0.27	0.00	17.41
15	0.86928	1.02	0.58	0.55	0.68	0.97	0.00	23.39
20	0.76651	3.40	0.69	0.68	0.53	1.42	0.00	31.34
22	0.72906	5.37	2.12	0.48	1.35	2.42	0.00	32.59
25	0.69446	4.48	0.35	0.00	1.12	2.41	0.38	32.68

(continued)

Table 12.3 (continued)

k	f_{best}	GKM	MGKM	DG-Clust	NDC-Clust	IDCA-Clust	IS-Clust	MS-KM
Ionosphere ($\times 10^4$)								
2	0.24194	0.00	0.00	0.00	0.00	0.00	0.00	2.75
3	0.21933	0.96	0.03	0.89	0.02	0.03	0.00	2.45
5	0.18908	0.11	0.11	0.00	0.10	0.11	0.13	2.20
7	0.17382	0.46	0.53	0.00	0.38	0.53	0.03	3.28
10	0.15540	2.74	0.27	0.00	0.22	0.32	0.12	5.11
15	0.13729	6.48	2.10	0.92	1.61	1.43	0.00	6.47
20	0.12307	9.23	3.32	2.09	2.79	2.73	0.00	13.52
22	0.11839	9.70	2.80	1.62	2.00	2.92	0.00	15.27
25	0.11147	10.00	3.03	1.38	1.86	2.98	0.00	21.06
Vehicle silhouettes ($\times 10^6$)								
2	7.29088	0.00	0.00	0.00	0.00	0.00	0.00	0.00
3	4.87412	0.02	0.02	0.00	0.00	0.02	0.02	0.00
5	2.36484	0.00	0.00	0.00	0.04	0.00	0.00	0.00
7	1.71738	1.08	0.00	0.00	0.00	0.00	0.00	1.72
10	1.25217	0.68	0.04	0.52	0.01	0.22	0.00	21.48
15	0.89095	1.03	0.02	0.00	0.07	0.08	0.00	24.49
20	0.74221	1.17	0.26	0.09	0.00	0.28	0.12	15.44
22	0.69630	0.68	0.09	0.00	0.02	0.09	0.03	20.05
25	0.63106	1.10	0.07	0.00	0.03	0.08	0.01	31.18

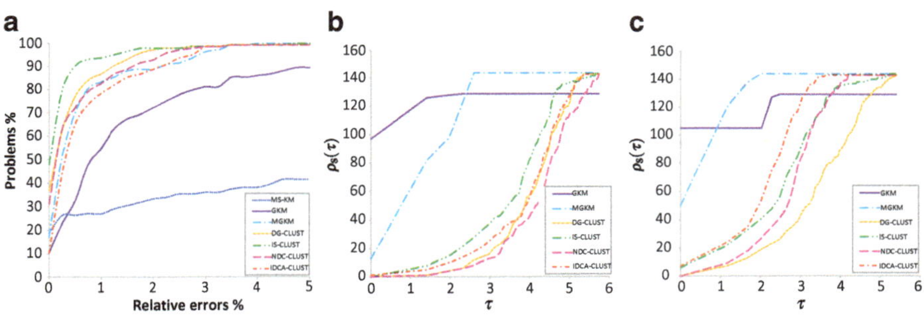

Fig. 12.6 Performance profiles for small data sets. (**a**) Relative errors. (**b**) Distance function evals. (**c**) CPU time

index is about 0.22 for $k = 5$ and $k = 7$. Silhouette plots show that not all clusters are well-separated in k-partitions with $k = 2, 3, 5, 10$ and 25. For instance, five clusters are not well-separated when $k = 25$. These results demonstrate that in vehicle silhouettes data set the class and the cluster distributions are not consistent.

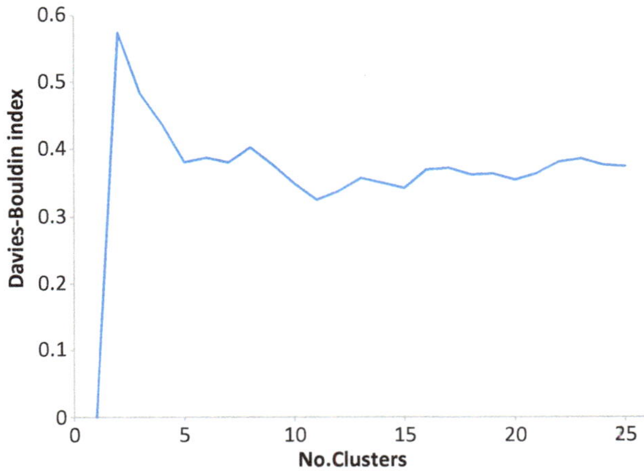

Fig. 12.7 *DB* index for TSPLIB3038 data set

12.3.3 Results for Medium Sized Data Sets

GKM, MGKM, DG-Clust, NDC-Clust, IDCA-Clust, IS-Clust, LMB-Clust and DCDB-Clust are applied to medium sized data sets. Results for accuracy are given in Table 12.4. These results demonstrate that, overall, all algorithms, except DCDB-Clust, are able to find the best known solutions.

Performance profiles for medium sized data sets are depicted in Fig. 12.11. They show that IS-Clust is the most successful algorithm in finding the best known solutions and GKM, MGKM, DG-Clust, NDC-Clust, IDCA-Clust and IS-Clust are all successful in solving clustering problems with the error no more than 5%. DCDB-Clust has the worst performance both in terms of errors and distance function calls. MGKM requires the least number of distance function evaluations whereas LMB-Clust requires the least CPU time among all algorithms. Finally, GKM uses more CPU time than other algorithms.

In Fig. 12.12 graphs of the three indices and in Fig. 12.13 silhouette plots for $k = 2, 3, 5, 10$ and 25 clusters are illustrated using results obtained by IS-Clust in Image segmentation data set. The DB index has local minimizers at $k = 3, 7, 16, 21$ and $k = 3$ is a global minimizer. Overall, the purity shows the steady increase as the number of clusters increases; however, it becomes almost a constant after $k = 16$. The NMI index has the large value at $k = 7$ and the largest value at $k = 14$. Note that the number of classes in this data set is 7. Results for silhouettes demonstrate that a large portion of clusters are not well-separated. Summarizing, we can say that in this data set there is some compatibility between the class and the cluster distributions but it is not very high.

The graph of the DB index for Pla85900 data set is depicted in Fig. 12.14. The DB index has local minimizers at $k = 4, 8, 11, 13$ and $k = 21$. Among them $k = 4$,

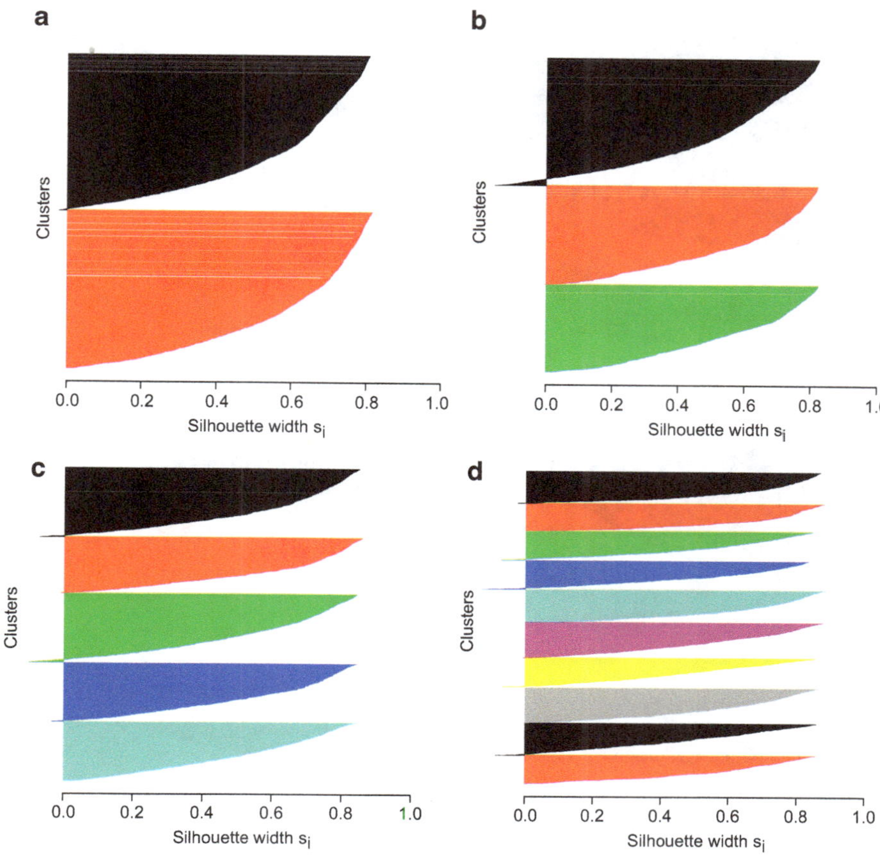

Fig. 12.8 Silhouette plots for TSPLIB3038 data set. (**a**) $k = 2$. (**b**) $k = 3$. (**c**) $k = 5$. (**d**) $k = 10$

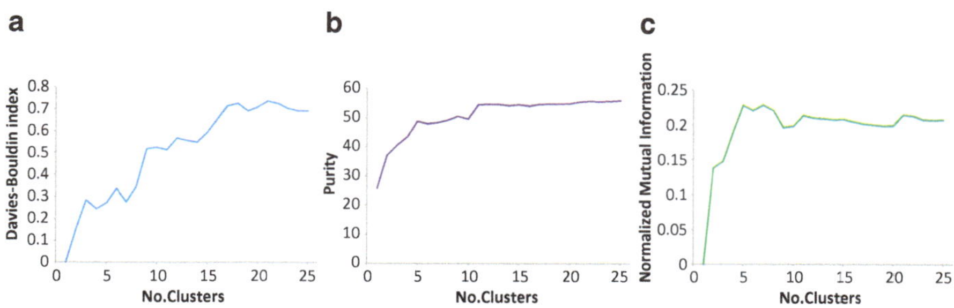

Fig. 12.9 Results for Vehicle silhouettes data set using different indices. (**a**) DB index. (**b**) Purity. (**c**) NMI index

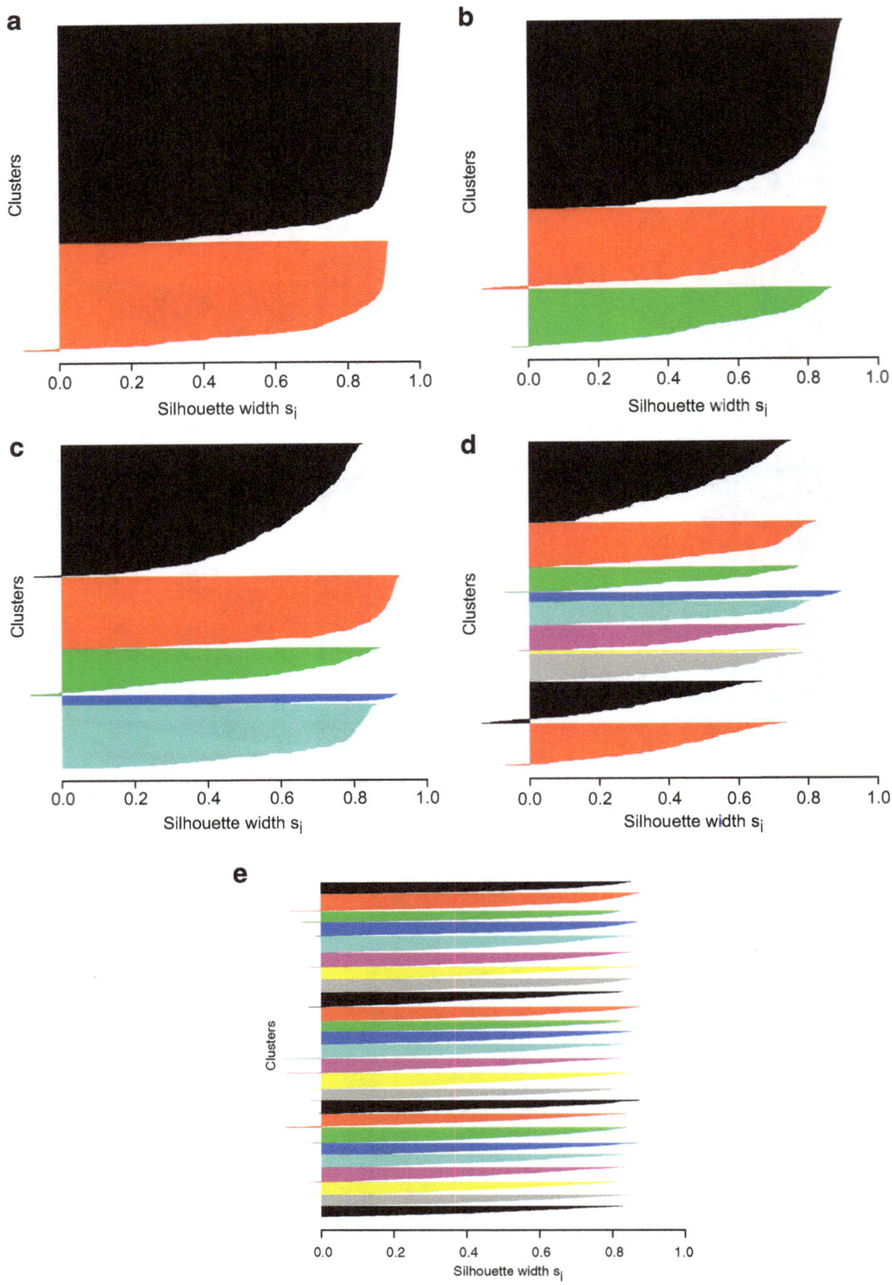

Fig. 12.10 Silhouette plots for Vehicle silhouettes data set. (**a**) $k = 2$. (**b**) $k = 3$. (**c**) $k = 5$. (**d**) $k = 10$. (**e**) $k = 25$

Table 12.4 Accuracy results for medium sized data sets

k	f_{best}	GKM	MGKM	DG-Clust	NDC-Clust	IDCA-Clust	IS-Clust	LMB-Clust	DCDB-Clust
DC15112 ($\times 10^{11}$)									
2	3.68403	0.00	0.00	0.00	0.00	0.00	0.00	0.00	0.00
3	2.53240	0.00	0.00	0.00	0.00	0.00	0.00	0.00	0.00
5	1.32707	0.00	0.00	0.00	0.00	0.00	0.00	0.00	0.00
7	0.93208	0.00	0.00	0.00	0.00	0.00	0.00	0.00	0.00
10	0.64491	1.41	1.41	1.41	1.41	1.41	1.41	1.41	1.41
15	0.43136	0.26	0.26	0.00	0.24	0.25	0.24	0.23	0.00
20	0.32178	0.24	0.24	0.24	0.24	0.24	0.24	0.24	0.00
22	0.28995	0.73	0.20	0.20	0.00	0.20	0.00	0.00	0.00
25	0.25428	0.01	0.01	0.01	0.00	0.01	0.18	0.00	0.01
Image segmentation ($\times 10^{7}$)									
2	3.56057	0.00	0.00	0.00	0.00	0.00	0.00	0.00	0.03
3	2.74163	0.00	0.00	0.00	0.00	0.00	0.00	0.00	0.76
5	1.71429	0.00	0.00	0.00	0.00	0.00	0.00	0.00	0.00
7	1.34043	2.30	2.30	2.30	2.30	2.31	1.79	2.62	5.26
10	0.97955	1.75	1.75	1.75	1.75	1.75	1.15	1.75	6.50
15	0.65554	0.10	1.91	1.89	1.89	1.90	1.72	1.90	11.98
20	0.51300	0.06	0.63	0.65	0.65	0.64	0.02	0.00	15.66
22	0.47112	0.06	0.67	0.71	0.67	0.70	0.00	0.39	23.14
25	0.41605	0.11	0.82	0.80	0.80	0.82	0.05	0.00	9.54
Page blocks ($\times 10^{10}$)									
2	5.79368	0.24	0.00	0.00	0.00	0.00	0.00	0.24	0.00
3	3.31337	0.00	0.00	0.01	0.00	0.00	0.00	0.00	0.00
5	1.32184	0.00	0.00	0.00	0.03	0.00	0.19	0.00	0.25
7	0.82934	0.18	0.18	0.18	0.16	0.18	0.00	0.03	0.19

(continued)

Table 12.4 (continued)

k	f_{best}	GKM	MGKM	DG-Clust	NDC-Clust	IDCA-Clust	IS-Clust	LMB-Clust	DCDB-Clust
10	0.45330	1.53	1.53	1.51	1.54	1.53	1.51	1.47	0.68
15	0.24936	1.02	1.87	1.65	1.72	1.65	1.84	4.57	4.84
20	0.17139	0.00	1.54	2.11	2.11	2.11	1.54	4.47	3.73
22	0.14988	0.75	0.79	0.93	0.93	0.79	0.80	3.58	0.00
25	0.12034	2.64	2.01	2.08	2.09	2.01	2.02	4.89	0.00
Pla85900 ($\times 10^{15}$)									
2	3.74908	0.00	0.00	0.00	0.00	0.00	0.00	1.44	0.00
3	2.28057	0.00	0.00	0.00	0.00	0.00	0.00	0.00	0.00
5	1.33972	0.00	0.00	0.00	0.00	0.00	0.00	2.77	0.00
7	0.97110	0.00	0.00	0.00	0.00	0.00	0.00	0.00	0.00
10	0.68294	0.00	1.03	0.00	0.00	0.00	0.00	0.40	0.00
15	0.46029	0.51	0.98	0.51	0.51	0.98	0.02	0.92	0.48
20	0.34986	0.29	0.29	0.52	0.52	0.52	0.52	0.95	0.52
22	0.31942	0.10	0.09	0.00	0.19	0.19	0.17	0.47	0.46
25	0.28223	1.09	0.14	0.13	0.30	0.30	0.00	0.30	0.30
Pen-based recognition of handwritten digits ($\times 10^{8}$)									
2	1.28119	0.39	0.00	0.00	0.00	0.00	0.00	0.00	0.39
3	1.01594	0.00	0.00	0.00	0.00	0.00	0.00	0.00	0.00
5	0.75304	0.00	0.00	0.00	0.00	0.00	0.00	0.00	1.17
7	0.59993	0.00	0.00	0.00	0.00	0.00	0.00	0.00	0.00
10	0.49302	0.00	0.00	0.00	0.00	0.00	0.00	2.63	0.00
15	0.39067	0.00	0.00	0.00	0.00	0.00	0.00	1.75	0.00
20	0.34123	0.00	0.17	0.16	0.17	0.17	0.17	0.94	0.17
22	0.32312	0.00	0.01	0.00	0.00	0.00	0.00	1.17	0.55
25	0.30109	0.00	0.00	0.00	0.00	0.00	0.00	0.87	0.00

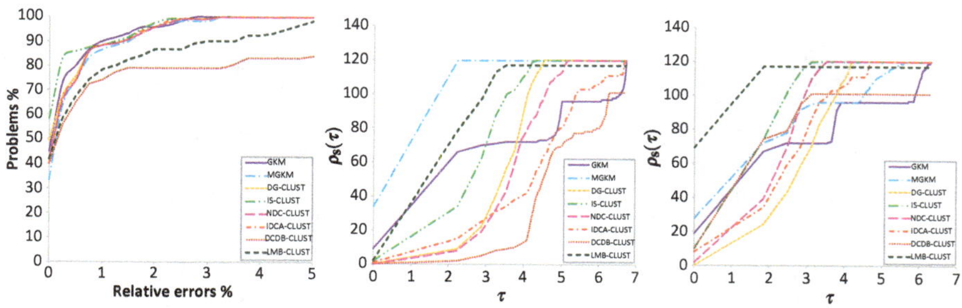

Fig. 12.11 Performance profiles for medium sized data sets. (**a**) Relative errors. (**b**) Distance function evals. (**c**) CPU time

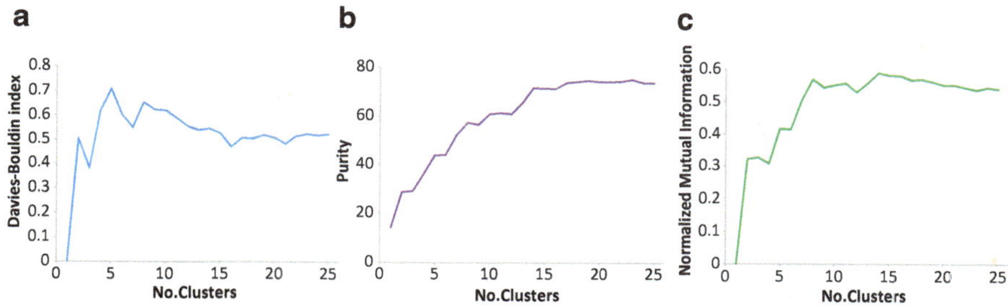

Fig. 12.12 Results for Image segmentation data set using different indices. (**a**) DB index. (**b**) Purity. (**c**) NMI index

$k = 8$ and $k = 21$ are global minimizers or nearly global minimizers. The deep global minimizer is located at $k = 4$.

12.3.4 Results for Large Data Sets

We apply GKM, MGKM, DG-Clust, NDC-Clust, IDCA-Clust, IS-Clust, LMB-Clust, and DCDB-Clust to large data sets. Results for accuracy are given in Tables 12.5 and 12.6. Note that in these tables the values of f_{best} are the best values obtained by algorithms used in our numerical experiments. Results demonstrate that all algorithms are successful in finding best known solutions.

Performance profiles for large data sets are presented in Fig. 12.15. As before, IS-Clust is the most successful in finding the best solutions, and DG-Clust, IS-Clust, NDC-Clust, IDCA-Clust are successful in solving clustering problems with the error no more than 5%. LMB-Clust requires the least and GKM the largest number of distance function evaluations. In addition, LMB-Clust requires the least CPU time among all algorithms. GKM uses more CPU time than other algorithms.

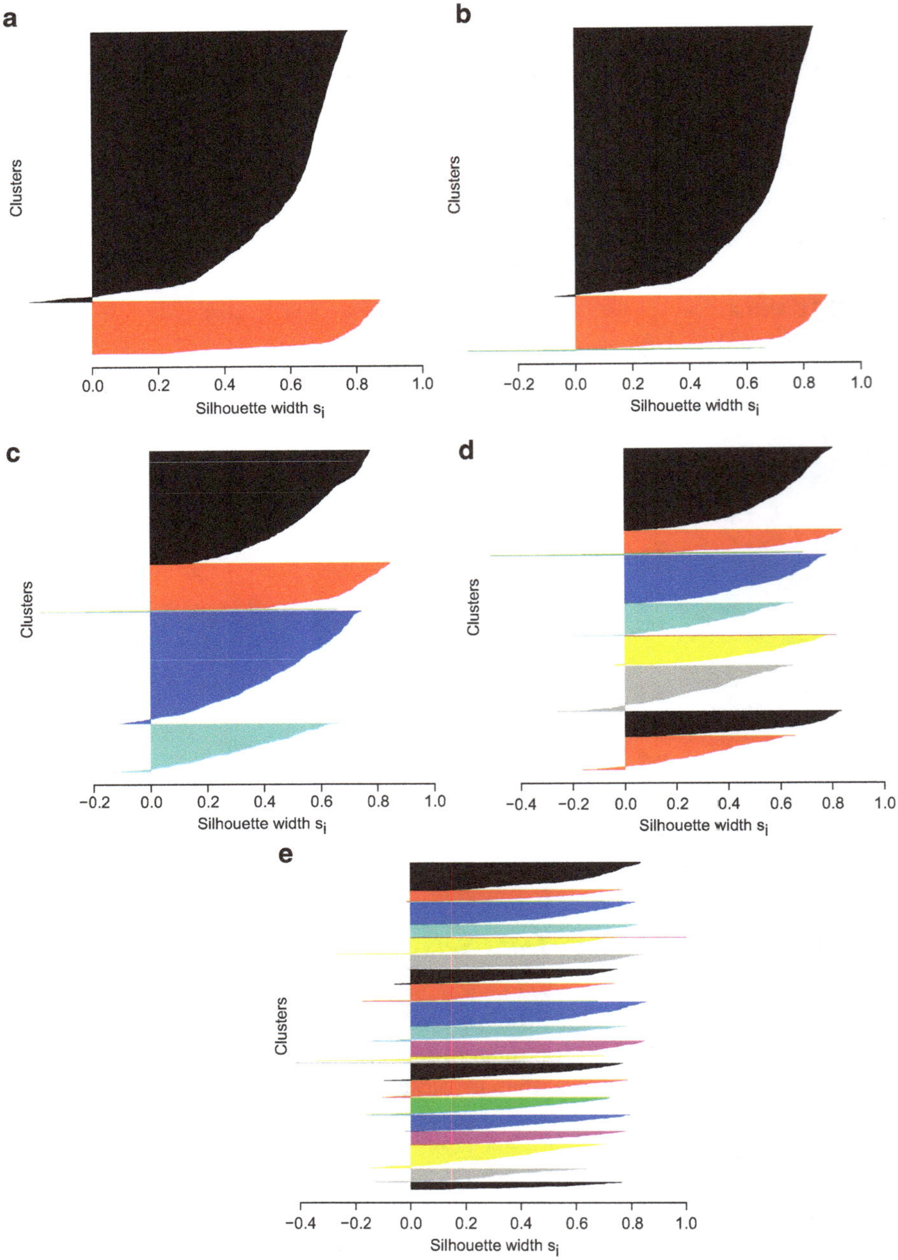

Fig. 12.13 Silhouette plots for Image segmentation data set. (**a**) $k = 2$. (**b**) $k = 3$. (**c**) $k = 5$. (**d**) $k = 10$. (**e**) $k = 25$

Table 12.5 Accuracy results for large data sets

k	f_{best}	`GKM`	`MGKM`	`DG-Clust`	`NDC-Clust`	`IDCA-Clust`	`IS-Clust`	`LMB-Clust`	`DCDB-Clust`
Landsat satellite ($\times 10^7$)									
2	5.12686	0.00	0.00	0.00	0.00	0.00	0.00	0.00	0.01
3	2.50603	0.00	0.00	0.00	0.00	0.00	0.00	0.00	0.01
5	1.82661	0.00	1.70	1.69	1.70	1.70	1.69	1.70	1.70
7	1.50121	0.00	0.00	0.00	0.00	0.00	0.00	0.00	0.02
10	1.23428	0.00	2.03	2.02	2.02	2.02	2.02	1.74	2.06
15	1.02265	0.06	0.10	0.07	0.00	0.08	0.88	0.88	0.00
20	0.91254	0.90	0.02	0.00	0.82	0.01	0.82	1.93	0.90
22	0.87952	0.11	0.27	0.24	0.22	0.26	0.00	0.26	1.25
25	0.83690	0.10	0.05	0.03	0.06	0.07	0.00	0.66	0.74
Optical recognition of handwritten digits ($\times 10^7$)									
2	0.60042	0.00	0.00	0.00	0.00	0.00	0.00	0.63	0.01
3	0.54582	0.00	0.00	0.00	0.00	0.00	0.00	0.75	0.01
5	0.47140	0.56	1.01	1.00	0.00	1.01	0.00	0.82	0.56
7	0.41601	0.00	0.00	0.00	0.88	0.00	0.89	0.86	0.08
10	0.36777	0.00	0.80	0.79	0.60	0.79	0.79	0.60	0.00
15	0.32522	0.92	0.90	0.90	0.90	0.90	0.00	1.86	0.90
20	0.30030	0.78	0.01	0.00	0.00	0.00	0.00	1.04	0.32
22	0.29376	0.00	0.18	0.16	0.16	0.17	0.16	0.19	0.17
25	0.28427	0.23	0.07	0.03	0.03	0.07	0.02	0.19	0.00

(continued)

Table 12.5 (continued)

k	f_{best}	GKM	MGKM	DG-Clust	NDC-Clust	IDCA-Clust	IS-Clust	LMB-Clust	DCDB-Clust
Letters recognition ($\times 10^6$)									
2	1.38189	0.00	0.00	0.00	0.00	0.00	0.00	0.00	0.04
3	1.25058	0.00	0.00	0.00	0.00	0.00	0.00	4.19	0.20
5	1.08652	1.06	0.00	0.00	0.00	0.00	0.00	0.00	1.06
7	0.97240	0.00	0.00	0.00	0.00	0.00	0.00	0.00	0.49
10	0.85750	0.00	0.19	0.18	0.19	0.19	0.00	1.10	0.21
15	0.74761	0.08	0.09	0.39	0.07	0.09	0.00	0.30	0.00
20	0.67601	0.22	0.03	0.00	0.03	0.04	0.02	1.13	0.31
22	0.65355	1.02	0.84	0.00	0.48	0.84	0.00	0.84	0.42
25	0.62338	1.39	0.41	0.00	0.36	0.41	0.41	0.65	0.89
EEG eye state ($\times 10^8$)									
2	8178.14	0.00	0.00	0.00	0.00	0.00	0.00	0.00	0.00
3	1833.88	0.00	0.00	0.00	0.00	0.00	0.00	0.00	0.00
5	1.33858	0.00	0.00	0.00	0.00	0.00	0.00	0.00	0.00
7	0.67714	0.00	0.00	0.00	0.00	0.00	0.01	0.00	0.01
10	0.45669	0.00	0.00	0.00	0.00	0.00	0.00	0.00	0.00
15	0.34653	0.00	1.09	0.69	0.69	1.09	0.93	0.05	0.28
20	0.28985	0.01	1.32	0.09	0.96	1.32	0.00	0.00	1.54
22	0.27622	0.28	0.78	1.02	0.69	0.75	0.00	0.29	0.81
25	0.25979	0.67	0.00	0.24	0.31	0.75	0.23	0.20	0.08

Shuttle control ($\times 10^8$)									
2	21.34329	0.00	0.00	0.00	0.00	0.00	0.00	0.00	0.00
3	10.85415	0.00	0.00	0.00	0.00	0.00	0.00	0.00	0.00
5	7.24479	0.00	0.00	0.00	0.07	0.01	0.00	0.09	5.39
7	4.33840	0.01	0.00	0.00	0.52	0.00	0.00	0.37	0.65
10	2.83166	0.02	0.00	0.00	0.59	0.01	0.00	0.57	0.85
15	1.53154	0.00	0.00	0.00	3.16	0.02	0.00	0.02	7.48
20	1.06012	0.00	0.00	0.15	1.32	0.02	0.00	0.03	4.33
22	0.94081	0.16	0.08	0.16	2.49	0.02	0.00	0.03	5.85
25	0.77978	2.48	2.48	2.47	2.05	2.52	2.48	0.00	10.12
Person activity ($\times 10^5$)									
2	1.03715	0.00	0.00	0.00	0.00	0.02	0.00	0.00	1.50
3	0.77228	0.00	0.00	0.00	0.00	0.03	0.00	0.00	4.29
5	0.56018	0.00	0.00	0.00	0.00	0.10	0.00	0.00	2.30
7	0.44514	0.00	0.00	0.00	0.00	0.15	0.00	1.20	1.10
10	0.33412	0.00	7.02	5.46	7.02	7.33	7.57	7.02	0.00
15	0.26151	0.54	1.09	0.00	0.00	2.05	0.00	0.00	0.00
20	0.21791	1.00	1.00	0.83	0.83	1.83	0.83	0.83	0.00
22	0.20573	0.01	1.22	1.32	1.32	1.38	0.00	1.15	0.70
25	0.18890	0.00	0.00	0.00	0.00	1.80	0.00	2.84	0.71

(continued)

Table 12.5 (continued)

k	f_{best}	GKM	MGKM	DG-Clust	NDC-Clust	IDCA-Clust	IS-Clust	LMB-Clust	DCDB-Clust
KEGG metabolic relation network ($\times 10^8$)									
2	11.38530	0.00	0.00	0.00	0.00	0.00	0.00	0.00	0.00
3	4.90060	0.01	0.00	0.00	0.00	0.00	0.00	0.00	0.00
5	1.88367	0.07	0.00	0.00	0.00	0.00	0.00	0.00	0.07
7	1.19630	0.19	0.03	0.08	0.82	0.06	0.00	2.20	1.69
10	0.63515	0.01	0.00	0.01	0.32	0.00	0.00	0.00	1.01
15	0.35122	3.41	4.34	1.04	1.04	1.03	0.00	1.48	2.21
20	0.24984	1.21	1.78	2.12	1.34	2.08	0.00	3.50	0.40
22	0.22365	1.00	0.00	1.00	0.29	1.00	2.03	4.95	3.53
25	0.19289	0.53	0.51	0.01	1.53	0.00	1.68	1.64	4.02
Skin segmentation ($\times 10^9$)									
2	1.32236	0.00	0.00	0.00	0.00	0.00	0.00	0.00	0.00
3	0.89362	0.00	0.00	0.00	0.00	0.00	0.00	0.00	0.00
5	0.50203	0.00	0.00	0.00	0.00	0.00	1.65	0.00	0.00
7	0.36308	0.00	0.00	0.00	0.00	0.00	0.00	11.43	0.00
10	0.25122	0.01	4.25	4.25	4.25	4.25	4.25	13.37	0.00
15	0.16963	0.02	0.19	0.00	0.00	0.00	0.05	3.84	0.00
20	0.12615	Fail	0.00	1.70	1.70	1.71	0.21	4.50	1.20
22	0.11654	Fail	0.57	0.73	0.73	0.75	0.00	6.36	0.75
25	0.10228	Fail	0.00	0.70	0.70	0.70	0.00	5.79	0.70

3D Road network ($\times 10^6$)									
2	49.13298	0.00	0.00	0.00	0.00	0.00	0.00	0.00	0.00
3	22.77818	0.00	0.00	0.00	0.00	0.00	0.00	0.00	0.03
5	8.82574	Fail	0.00	0.00	0.00	0.01	0.00	0.00	0.00
7	4.84744	Fail	0.00	0.00	0.01	0.04	0.00	0.00	0.94
10	2.56662	Fail	Fail	0.01	0.02	0.23	0.00	0.00	0.02
15	1.27069	Fail	Fail	0.00	0.01	0.62	0.00	0.00	0.98
20	0.80865	Fail	Fail	0.00	0.01	0.48	0.00	0.00	0.03
22	0.70328	Fail	Fail	0.00	0.00	0.42	0.00	0.00	4.14
25	0.59258	Fail	Fail	1.94	3.79	2.39	1.94	0.00	0.30
Gas sensor array drift ($\times 10^{13}$)									
2	7.91182	0.00	0.00	0.00	0.00	0.00	0.00	0.00	0.00
3	5.02409	0.00	0.00	0.00	0.00	0.00	0.00	0.00	0.00
5	3.22395	0.10	0.10	0.10	0.10	0.10	0.10	0.00	0.10
7	2.25010	0.00	0.05	0.00	0.00	0.00	0.00	0.00	0.00
10	1.65228	0.18	0.18	0.18	0.18	0.18	0.00	0.18	0.18
15	1.14212	0.00	0.00	0.00	0.00	0.00	0.15	0.00	1.25
20	0.87878	0.67	1.58	0.01	0.01	1.58	0.00	2.74	1.54
22	0.80882	1.12	3.22	0.00	0.00	3.21	0.27	3.49	1.37
25	0.72211	0.65	2.58	0.16	0.16	2.53	0.00	3.05	0.66

Table 12.6 Accuracy results for very large data sets

k	f_{best}	GKM	MGKM	DG-Clust	NDC-Clust	IDCA-Clust	IS-Clust	LMB-Clust	DCDB-Clust
Isolet ($\times 10^6$)									
2	0.72194	0.00	0.00	0.00	0.00	0.00	0.00	0.00	0.00
3	0.67878	0.00	0.00	0.00	0.00	0.00	0.00	0.00	0.69
5	0.61365	2.44	1.04	0.39	1.04	1.04	1.04	0.00	0.00
7	0.57029	0.00	0.00	0.00	0.00	0.00	0.00	0.00	0.22
10	0.53478	1.64	0.12	0.12	0.12	0.12	0.14	1.64	0.00
15	0.48739	0.10	0.00	0.00	0.00	0.00	0.55	0.54	0.10
20	0.46045	0.04	0.03	0.00	0.00	0.03	0.17	0.12	0.16
22	0.45459	0.00	0.60	0.04	0.04	0.57	0.74	0.09	0.11
25	0.44389	0.23	0.18	0.15	0.14	0.15	3.16	0.40	0.00
Online news popularity ($\times 10^{14}$)									
2	9.53913	0.00	0.00	0.00	0.00	0.00	0.00	0.00	0.00
3	5.91077	0.00	0.00	0.00	0.00	0.00	0.00	0.00	0.00
5	3.09885	0.00	0.00	0.00	0.00	0.00	0.00	0.00	0.00
7	1.79526	0.00	0.00	0.00	0.00	0.00	0.00	0.00	0.00
10	1.17247	0.00	0.00	0.02	0.00	0.00	0.00	2.57	0.00
15	0.77637	0.00	0.00	0.00	0.00	0.00	0.00	14.77	0.00
20	0.59812	0.02	0.11	0.12	0.10	0.11	0.00	7.40	0.11
22	0.55266	0.61	2.24	2.24	0.11	0.11	0.00	5.26	0.69
25	0.49615	0.13	0.26	0.25	0.13	0.13	0.00	7.01	0.99

Sensorless drive diagnosis ($\times 10^7$)									
2	3.88116	0.00	0.00	0.00	0.00	0.00	0.00	0.00	0.00
3	2.91313	0.00	0.00	0.00	0.02	0.17	0.00	0.00	8.31
5	1.93652	0.01	0.01	0.00	0.03	0.23	0.00	5.32	10.67
7	1.53488	0.01	0.01	0.00	0.03	0.27	0.00	0.00	4.18
10	0.96091	0.00	0.00	0.14	3.82	0.44	0.00	4.41	12.62
15	0.62816	0.00	0.00	0.11	2.83	0.46	0.00	0.00	3.69
20	0.49989	0.00	3.92	6.83	5.63	4.25	3.89	0.80	5.29
22	0.46915	3.36	3.53	13.83	4.53	3.57	3.30	0.00	4.93
25	0.42232	14.82	6.22	26.45	7.16	6.42	6.18	0.00	2.85
Covertype ($\times 10^{11}$)									
2	1.34188	Fail	0.00	0.00	0.00	0.00	0.00	0.00	0.00
3	0.95287	Fail	0.00	0.00	0.00	0.00	0.00	0.00	0.00
5	0.58977	Fail	Fail	0.00	0.00	0.00	0.00	0.00	0.00
7	0.44828	Fail	Fail	0.00	0.00	0.00	0.00	0.00	0.00
10	0.33878	Fail	Fail	0.00	0.00	0.00	0.00	0.00	0.00
15	0.24669	Fail	Fail	0.00	0.00	0.87	0.87	0.00	0.87
20	0.20395	Fail	Fail	0.00	0.00	0.00	0.00	0.05	Fail
22	0.18951	Fail	Fail	0.00	0.00	3.58	0.00	1.55	Fail
25	0.17362	Fail	Fail	0.12	0.12	13.06	0.12	0.00	Fail

(continued)

Table 12.6 (continued)

k	f_{best}	GKM	MGKM	DG-Clust	NDC-Clust	IDCA-Clust	IS-Clust	LMB-Clust	DCDB-Clust
MiniBooNE particle identification ($\times 10^{10}$)									
2	8.92236	0.00	0.00	0.00	0.00	0.00	0.00	0.00	0.00
3	5.22601	0.00	0.00	0.00	0.00	0.00	0.00	21.68	0.00
5	1.82252	0.01	0.00	0.00	0.00	0.00	0.00	0.00	0.00
7	1.29369	Fail	0.00	0.00	0.00	0.00	0.00	0.00	0.00
10	0.92406	Fail	0.01	0.01	2.85	2.85	5.50	0.00	0.03
15	0.63507	Fail	0.00	0.00	0.01	0.00	0.00	0.00	0.01
20	0.50871	Fail	0.35	0.38	0.38	Fail	0.35	1.15	0.00
22	0.47966	Fail	1.48	Fail	0.23	Fail	Fail	1.49	0.00
25	0.44425	Fail	0.01	Fail	0.04	Fail	Fail	0.00	0.02
Gisette ($\times 10^{13}$)									
2	0.41994	Fail	0.00	Fail	0.00	Fail	Fail	0.00	0.00
3	0.41160	Fail	0.00	Fail	0.00	Fail	Fail	0.00	0.69
5	0.40232	Fail	0.00	Fail	0.00	Fail	Fail	0.00	1.34
7	0.39535	Fail	1.76	Fail	0.01	Fail	Fail	0.00	2.45
10	0.38843	Fail	Fail	Fail	0.00	Fail	Fail	0.00	2.95
15	0.38177	Fail	Fail	Fail	0.34	Fail	Fail	0.00	Fail
20	0.38144	Fail	Fail	Fail	0.43	Fail	Fail	0.00	Fail
22	0.37843	Fail	Fail	Fail	1.23	Fail	Fail	0.00	Fail
25	0.37344	Fail	Fail	Fail	Fail	Fail	Fail	0.40	Fail

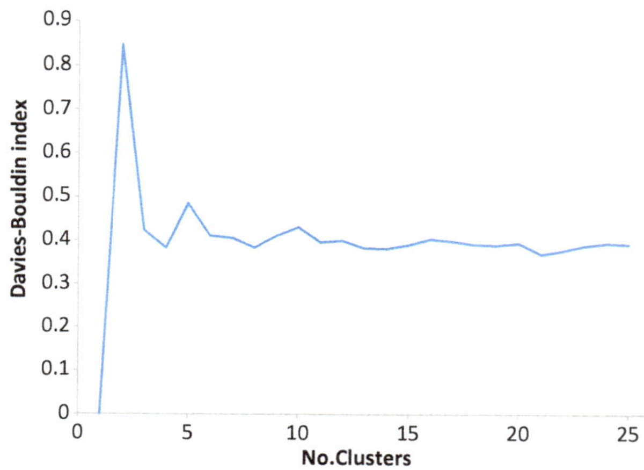

Fig. 12.14 DB index for Pla85900 data set

In Fig. 12.16 graphs of the three indices are depicted using results obtained by `IS-Clust` in Gas sensor array drift data set. It can be seen that the DB index has local minimizers at $k = 4, 6, 11, 17, 24$ and $k = 4$ is the global minimizer. The purity increases up to about 55% as the number of clusters increases. The NMI index has the largest values at $k = 17$ and $k = 22, 23, 24, 25$ with the value 0.34. Note that the number of classes in this data set is 6. The results show that in this data set the level of the compatibility between the class and the cluster distributions is not high.

The graph of the DB index for KEGG metabolic relation network data set is presented in Fig. 12.17. Here, the DB index has many local minimizers. Two of them are global minimizers ($k = 7$ and $k = 10$). This means that the most compact and well-separated clusters for this data set obtained for the 7- and 10-partitions.

12.3.5 Results for Very Large Data Sets

`GKM`, `MGKM`, `DG-Clust`, `NDC-Clust`, `IDCA-Clust`, `IS-Clust`, `LMB-Clust` and `DCDB-Clust` are applied to very large data sets. Results for accuracy are given in Table 12.6. Note that in these tables the values of f_{best} are the best values obtained by all algorithms used in the numerical experiments.

We can see that not all algorithms are able to solve clustering problems within the given 5 h time limit. `GKM` fails in three largest data sets, `MGKM` fails in two of them, `IDCA-Clust` and `IS-Clust` fail in one of them. This means that these algorithms are not always applicable to solve clustering problems in very large data sets. However, `LMB-Clust` succeeds to solve all problems within the given 5 h time limit.

Performance profiles for very large data sets are presented in Fig. 12.18. It can be observed that `LMB-Clust` is the most successful in finding the best known

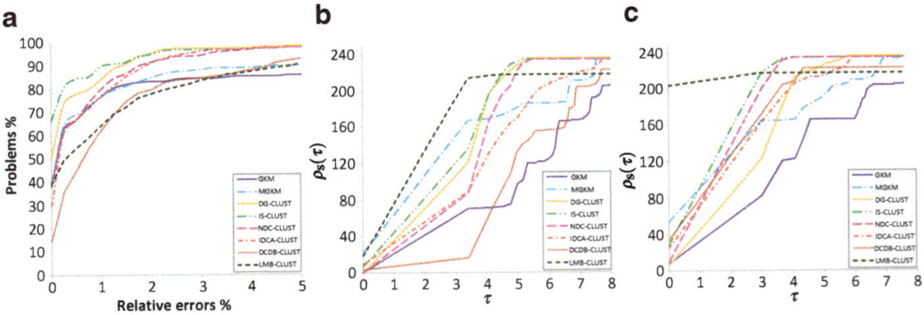

Fig. 12.15 Performance profiles for large data sets. (**a**) Relative errors. (**b**) Distance function evals. (**c**) CPU time

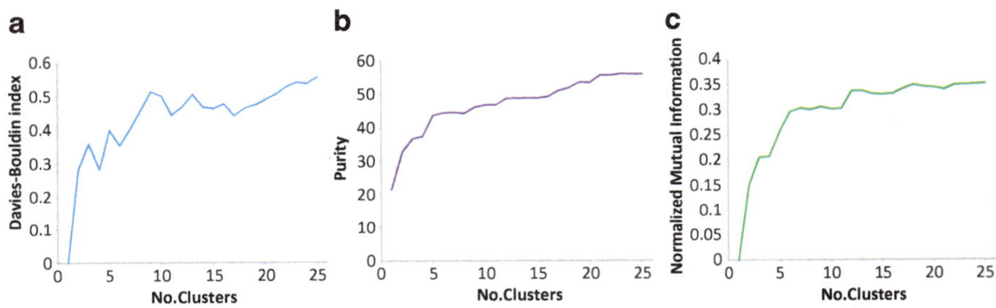

Fig. 12.16 Results for Gas sensor array drift data set using different indices. (**a**) DB index. (**b**) Purity. (**c**) NMI index

Fig. 12.17 DB index for KEGG metabolic relation network data set

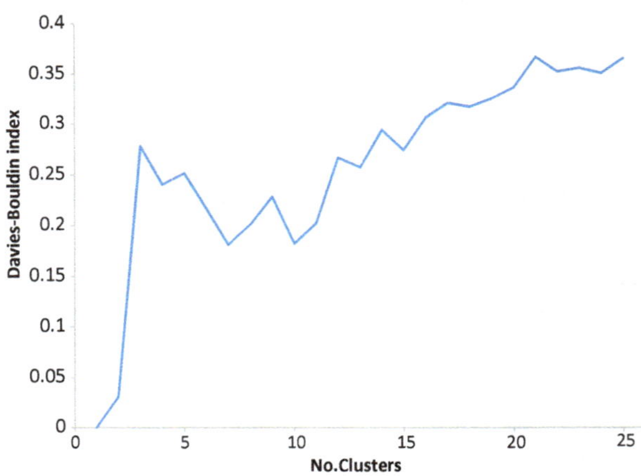

solutions and `NDC-Clust` is the most successful in solving clustering problems with the error no more than 5%. In addition, `LMB-Clust` has the least number of distance function evaluations and CPU time. The results also show that `GKM` is not applicable to very large data sets, it requires largest number of distance function evaluations and CPU time among all algorithms.

In Fig. 12.19 graphs of the DB index and purity are presented based on results obtained by `IS-Clust` in Covertype data set. It can be seen from the figure that the DB index has local minimizers at $k = 3, 8, 21$ and $k = 8$ is a global minimizer. The DB index tends to increase as the number of clusters increases. This can be considered as an indication that according to the DB index the 8-partition of Covertype data set has the best separated clusters among all k-partitions ($k = 2, \ldots, 25$).

The purity tends to increase starting from 48% up to about 52% as the number of clusters increases from 2 to 25. This means that we should calculate a large number of clusters to get a significant increase of the purity in Covertype data set.

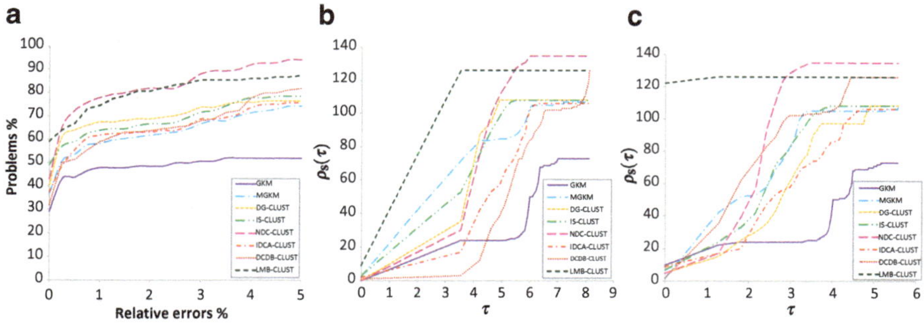

Fig. 12.18 Performance profiles for very large data sets. (**a**) Relative errors. (**b**) Distance function evals. (**c**) CPU time

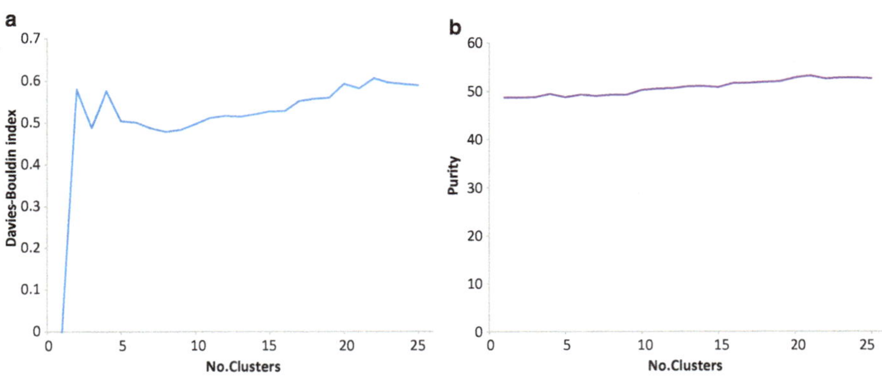

Fig. 12.19 DB index and purity for Covertype data set. (**a**) DB index. (**b**) Purity

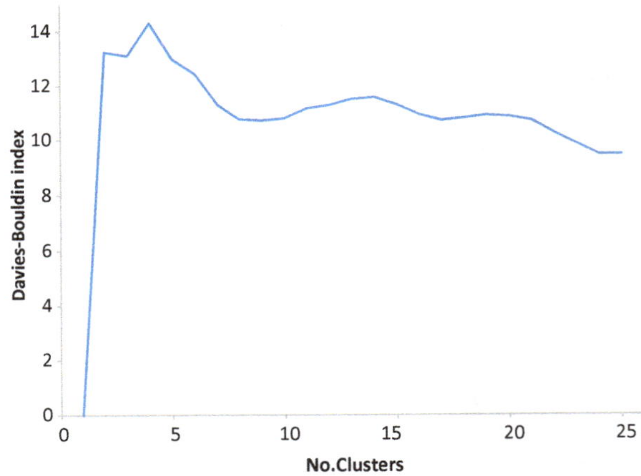

Fig. 12.20 DB index for Gisette data set

The graph of the DB index for Gisette data set is presented in Fig. 12.20. This index has three distinct local minimizers at $k = 8$, 16 and $k = 24$. However, it tends to decrease as the number of clusters increases. This means that we need to compute a large number of clusters to find a cluster distribution with well-separated clusters. This is not unexpected for Gisette data set as it has 5000 attributes and is sparse.

12.4 Comparative Results with Different Similarity Measures

In this section, we discuss the performance of the incremental clustering algorithms when different similarity measures—d_1, d_2, and d_∞—are used in the clustering functions. For this aim, we apply `DG-Clust` on different sizes of data sets: Bavaria postal 1, Bavaria postal 2, Iris plant, TSPLIB1060, Breast cancer Wisconsin, TSPLIB3038, D15112, Image segmentation, Page blocks, Pla85900, EEG eye state, and KEGG metabolic relation network. We use the cluster function values, the CPU time and Voronoi diagrams to compare results. The maximum number of clusters in extra small data sets is 10, in small size data sets 15 and it is 20 in all other data sets.

12.4.1 *Optimal Values for Cluster Functions*

Table 12.7 presents optimal values of the cluster function f_k obtained using similarity measures $d_1, d_2,\ d_\infty$ and different number k of clusters. Note that these values are multiplied by m—the number of points in a data set—and also by numbers shown under names of data sets. The results show that in all cases, except

Iris plant data set, the values of the cluster function with d_∞ are the smallest among all three similarity measures.

12.4.2 Computational Time

The dependence of the CPU time used by `DG-Clust` for similarity measures d_1, d_2, and d_∞ are depicted in Fig. 12.21. The following conclusions can be made based on these results:

Table 12.7 Optimal values for cluster functions with different similarity measures

k	d_1	d_2	d_∞	d_1	d_2	d_∞	d_1	d_2	d_∞
	Bavaria postal 1			Bavaria postal 2			Iris plant		
	$\times 10^6$	$\times 10^{10}$	$\times 10^6$	$\times 10^6$	$\times 10^{10}$	$\times 10^6$	$\times 10^2$	$\times 10^2$	$\times 10^2$
2	4.0249	60.2547	3.9940	1.8600	5.2192	0.9456	2.1670	1.5235	0.9715
3	2.8284	29.4507	2.7892	1.2607	1.7399	0.6594	1.5920	0.7885	0.7420
5	1.7208	5.9762	1.6948	0.7872	0.5442	0.4221	1.2460	0.4645	0.5860
7	1.0704	2.1983	1.0368	0.5659	0.2215	0.2946	1.0620	0.3430	0.4915
10	0.6037	0.6447	0.5828	0.4340	0.1181	0.2173	0.9070	0.2583	0.4245
	TSPLIB1060			TSPLIB3038			Breast cancer		
	$\times 10^7$	$\times 10^9$	$\times 10^6$	$\times 10^6$	$\times 10^9$	$\times 10^6$	$\times 10^4$	$\times 10^4$	$\times 10^4$
2	0.3864	9.8319	2.6809	3.7308	3.1688	2.5651	0.6401	1.9323	0.1831
3	0.3139	6.7058	2.1508	3.0056	2.1763	2.1221	0.5702	1.6256	0.1607
5	0.2310	3.7915	1.6546	2.2551	1.1982	1.5576	0.5165	1.3707	0.1460
10	0.1563	1.7553	1.1048	1.5508	0.5634	1.0738	0.4270	1.0212	0.1278
15	0.1198	1.1219	0.8827	1.2295	0.3560	0.8592	0.3872	0.8711	0.1172
	D15112			Image segmentation			Page blocks		
	$\times 10^8$	$\times 10^{11}$	$\times 10^8$	$\times 10^6$	$\times 10^7$	$\times 10^5$	$\times 10^7$	$\times 10^{10}$	$\times 10^6$
2	0.8860	3.6840	0.6109	0.5192	3.5606	1.4929	0.8414	5.7937	4.1746
3	0.6908	2.5324	0.4896	0.4160	2.7416	1.3284	0.6747	3.3134	3.4309
5	0.4998	1.3271	0.3619	0.3400	1.7143	1.1081	0.4882	1.3218	2.4671
10	0.3618	0.6489	0.2524	0.2575	0.9967	0.8170	0.3152	0.4533	1.4446
15	0.2930	0.4324	0.2065	0.2188	0.6556	0.6966	0.2555	0.2495	1.1784
20	0.2501	0.3218	0.1768	0.1942	0.5137	0.6200	0.2200	0.1672	1.0160
	Pla85900			EEG eye state			KEGG metabolic		
	$\times 10^{10}$	$\times 10^{15}$	$\times 10^{10}$	$\times 10^7$	$\times 10^8$	$\times 10^6$	$\times 10^7$	$\times 10^8$	$\times 10^6$
2	2.0656	3.7491	1.4533	0.5289	8178.1381	1.5433	0.3586	11.3853	1.9821
3	1.6262	2.2806	1.1434	0.4197	1833.8806	0.9049	0.2800	4.9006	1.5112
5	1.2587	1.3397	0.8712	0.2944	1.3386	0.5183	0.2095	1.8837	1.0549
10	0.8950	0.6829	0.6218	0.2191	0.4567	0.3947	0.1459	0.6352	0.6667
15	0.7335	0.4625	0.5082	0.1965	0.3500	0.3562	0.1231	0.3512	0.5114
20	0.6374	0.3517	0.4443	0.1827	0.2899	0.3292	0.1108	0.2654	0.4440

- DG-Clust requires the largest CPU time with d_∞ in all data sets except Bavaria postal 1 and TSPLIB1060, and the least CPU time with d_2 in all data sets. The clustering problem with d_∞ is the most complex one and DG-Clust requires a large number of approximate subgradient evaluations to find search directions in this problem. On the other hand, the clustering problem with d_2 is the easiest

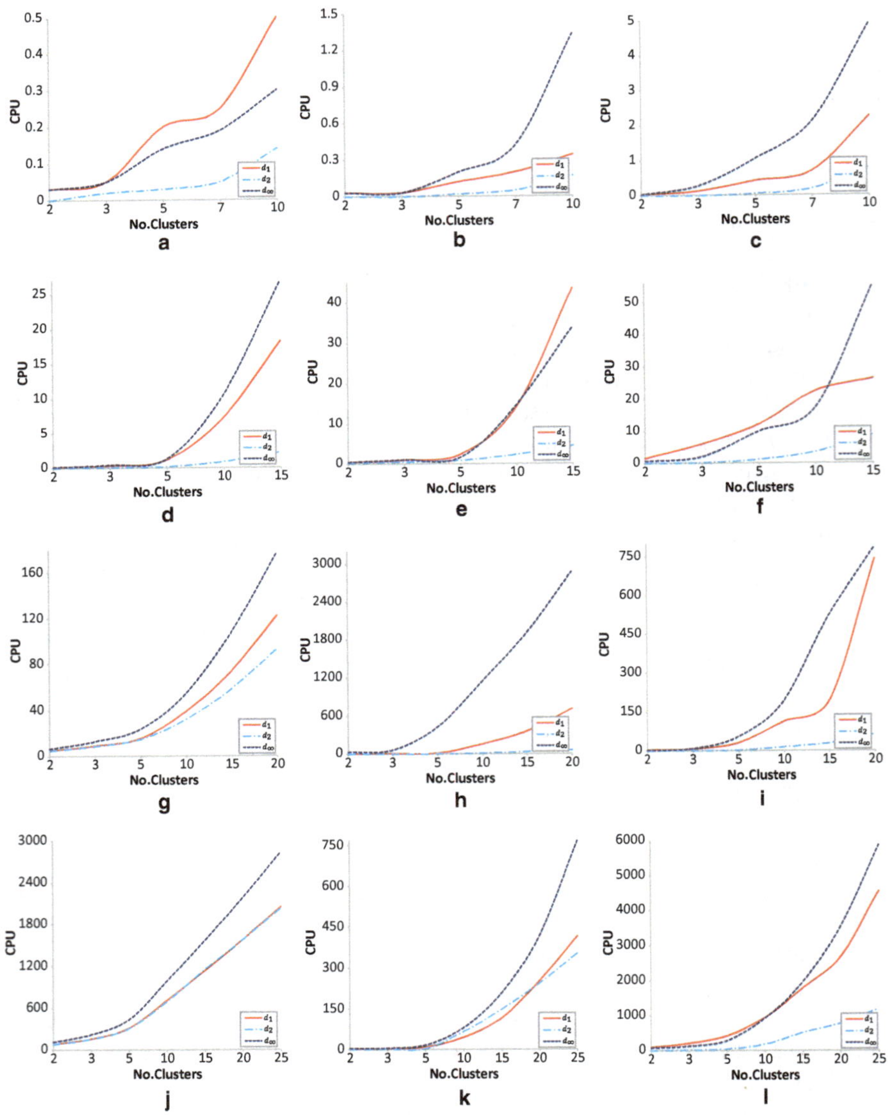

Fig. 12.21 CPU time with different similarity measures. (**a**) Bavaria postal 1. (**b**) Bavaria postal 2. (**c**) Iris plant. (**d**) TSPLIB1060. (**e**) TSPLIB3038. (**f**) Breast cancer. (**g**) D15112. (**h**) Image segmentation. (**i**) Page blocks. (**j**) Pla85900. (**k**) EEG eye state. (**l**) KEGG metabolic

as d_2 is smooth. In this case, the optimization method does not require a large number of approximate subgradient evaluations to find search directions;

- the CPU time required by `DG-Clust` depends more strongly on the number of attributes than on the number of data points. This claim is confirmed by comparing results for data sets with the similar number of data points and significantly different number of attributes: Image segmentation, TSPLIB3038, D15112, EGE eye state, Pla85900, and KEGG metabolic relation network. The comparison shows that `DG-Clust` becomes time-consuming in large data sets with the large number of attributes. In such data sets the size of the optimization problem increases rapidly as the number of clusters increase; and
- for all similarity measures the CPU time required at each iteration of the incremental algorithm, in general, is more than that of required at the previous iterations. This is due to the fact that the size of the optimization problem for finding all cluster centers increases at each iteration of the incremental algorithm.

12.4.3 *Visualization of Results*

Voronoi diagrams are used to visualize results obtained by `DG-Clust` in three data sets: German towns, TSPLIB1060 and TSPLIB3038. We utilize the software from [259] for this purpose. Figures 12.22, 12.23 and 12.24 present Voronoi diagrams for these data sets with five clusters. We can see that cluster structures for similarity measures d_1, d_2, and d_∞ are different in all data sets, although the distributions of cluster centers for d_1 and d_2 functions in TSPLIB1060 data set are similar.

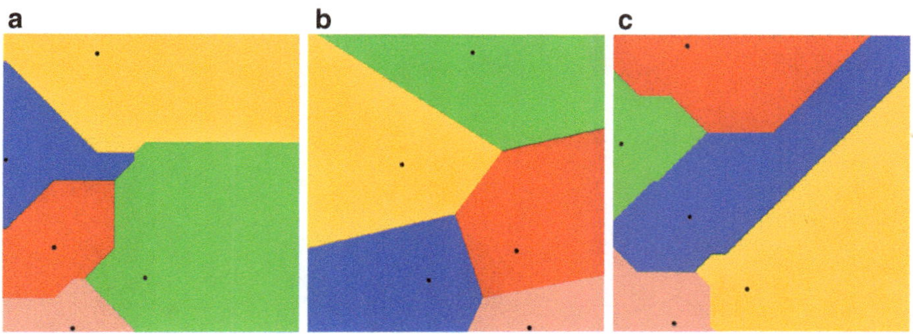

Fig. 12.22 Visualization of clusters in German towns data set. (**a**) L_1-norm. (**b**) L_2-norm. (**c**) L_∞-norm

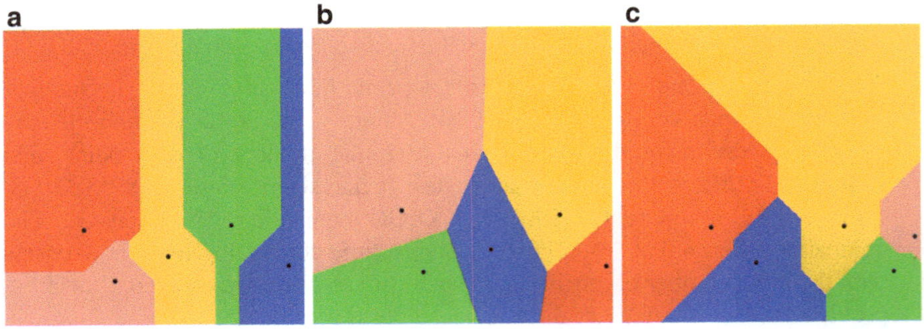

Fig. 12.23 Visualization of clusters in TSPLIB1060 data set. (**a**) L_1-norm. (**b**) L_2-norm. (**c**) L_∞-norm

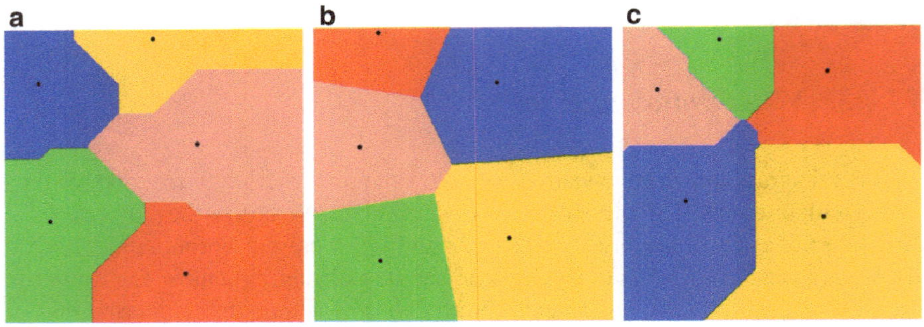

Fig. 12.24 Visualization of clusters in TSPLIB3038 data set. (**a**) L_1-norm. (**b**) L_2-norm. (**c**) L_∞-norm

Chapter 13
Concluding Remarks

Clustering is an important technique in exploratory data analysis. There are different types of clustering and the focus in this book is on partitional hard clustering. We considered various models of the clustering problem and demonstrated that the nonconvex nonsmooth optimization approach provides a model with significantly less number of decision variables than other approaches. In addition, this approach allows us to easily examine the clustering problems with different similarity measures, in particular, those based on the L_1-, L_2- and L_∞-norms. Furthermore, the use of the nonconvex nonsmooth optimization model in clustering problems enables us to formulate the optimality conditions in a compact form.

Partitional hard clustering problems are global optimization problems. They have many local minimizers and global or nearly global minimizers are of interest. These minimizers provide cluster distributions where clusters are better separated and more compact than clusters represented by any other local minimizers. However, conventional global optimization methods are prohibitively time-consuming for solving clustering problems in large and very large data sets with relatively large number of clusters.

Since 1950s different algorithms have been proposed to solve partitional hard clustering problems. Most of them aim to solve clustering problems with the similarity measure defined using the squared Euclidean norm. Such problems are also known as the minimum sum-of-squares clustering problems. These algorithms are, predominantly, local search algorithms and they can obtain only local solutions. Clustering solutions obtained by these algorithms might be significantly different from global solutions, especially in large data sets.

To improve the quality of solutions obtained by clustering algorithms, we need to apply a special procedure to generate the diverse set of starting cluster centers. Most of the successful local search clustering algorithms involve such a procedure. Usually, these procedures try to find starting clusters centers among data points

A. M. Bagirov et al., *Partitional Clustering via Nonsmooth Optimization*, Unsupervised and Semi-Supervised Learning,
https://doi.org/10.1007/978-3-030-37826-4_13

which is reasonable as clusters centers are usually located in the dense areas of a data set. Starting cluster centers are generated using either randomization or deterministic schemes.

The incremental approach allows us to design efficient procedures for finding starting cluster centers. Clustering algorithms based on this approach construct clusters starting from one cluster, which is the whole data set, and by adding one cluster center at each iteration of the incremental algorithm. Using this approach we introduced the auxiliary clustering problem that is applied to generate promising starting cluster centers from the whole search space.

We designed incremental clustering algorithms by combining the incremental approach with the heuristic clustering algorithms such as k-means, k-medians and also with nonsmooth optimization methods. These algorithms and methods are applied at each iteration of the incremental algorithm to solve both the clustering and the auxiliary clustering problems. Nonsmooth optimization methods that are applied to solve these problems include semi-derivative-free discrete gradient method, methods based on smoothing techniques, and bundle-type methods such as the diagonal bundle method and the limited memory bundle method. Using the difference of convex decomposition of the objective functions in both the clustering and the auxiliary clustering problems two clustering algorithms: the nonsmooth difference of convex clustering and the DCA (difference of convex algorithm) clustering algorithms were developed.

The auxiliary clustering problem is nonsmooth and nonconvex. Its number of decision variables is the same at all iterations of the incremental clustering algorithm, whereas the number of variables in the clustering problem increases as the number of clusters increases.

Results of numerical experiments show that the use of the auxiliary clustering problem allows us to improve the quality of the clustering solutions. Solutions to this problem are in some proximity of solutions of the clustering problem, and therefore, the application of the auxiliary clustering problem leads to the significant reduction of computational effort. This means that the auxiliary clustering problem is an important step of the incremental clustering algorithms.

Results of numerical experiments also demonstrate that incremental clustering algorithms, based on the heuristics like k-means and k-medians, are only efficient for extra small to large data sets (containing hundreds of thousands of instances). However, they become very time-consuming in very large data sets. At the same time, the incremental clustering algorithms based on nonsmooth optimization techniques such as the limited memory bundle method and those based on smoothing techniques are accurate and efficient in very large data sets. These results show that clustering algorithms utilizing special structures of the clustering problems, such as their difference of convex decompositions and piecewise separability, are real-time clustering algorithms in very large data sets.

In summary, we can conclude that nonsmooth optimization approaches in clustering provide better models and also efficient and accurate algorithms. The results presented in this book demonstrate that clustering algorithms based on these approaches constitute a solid basis to develop such model algorithms for solving clustering problems in large and very large data sets when the whole data cannot be stored in the memory of a computer.

References

1. Aarts, E., Korst, J.: Simulated Annealing and Boltzmann Machines: A Stochastic Approach to Combinatorial Optimization and Neural Computing. Wiley Interscience Series in Discrete Mathematics and Optimization. Wiley, New York, NY (1989)
2. Aggarwal, C.C., Hinneburg, A., Keim, D.: On the surprising behavior of distance metrics in high dimensional space. In: ICDT '01 Proceedings of the 8th International Conference on Database Theory, pp. 420–434 (2001)
3. Aggarwal, C.C., Reddy, C.K.: Data Clustering: Algorithms and Applications. CRC Press, Boca Raton (2014)
4. Al-Daoud, M.B., Roberts, S.A.: New methods for the initialisation of clusters. Pattern Recogn. Lett. **17**(5), 451–455 (1996)
5. Aliguliyev, R.M.: Performance evaluation of density-based clustering methods. Inf. Sci. **179**(20), 3583–3602 (2009)
6. Aliguliyev, R.M.: Clustering of document collection: a weighting approach. Expert Syst. Appl. **36**(4), 7904–7916 (2009)
7. Aloise, D., Deshpande, A., Hansen, P., Popat, P.: NP-hardness of Euclidean sum-of-squares clustering. Mach. Learn. **75**(2), 245–248 (2009)
8. Alok, A.K., Saha, S., Ekbal, A.: Multi-objective semi-supervised clustering for automatic pixel classification from remote sensing imagery. Soft Comput. **20**(12), 4733–4751 (2016)
9. Al-Sultan, K.S.: A tabu search approach to the clustering problem. Pattern Recogn. **28**(9), 1443–1451 (1995)
10. Al-Sultan, K.S., Fedjki, C.A.: A tabu search-based algorithm for the fuzzy clustering problem. Pattern Recogn. **30**(12), 2023–2030 (1997)
11. Al-Sultan, K.S., Khan, M.M.: Computational experience on four algorithms for the hard clustering problem. Pattern Recogn. Lett. **17**, 295–308 (1996)
12. Anderberg, M.R.: Cluster Analysis for Applications. Academic, New York, NY (1973)
13. Andritsos, P., Tsaparas, P., Miller, R.J., Servcik, K.C.: LIMBO: a linear algorithm to cluster categorical data. Technical Report CSRG-467, Department of Computer Science, UofT (2003)
14. Arthur, D., Vassilvitskii, S.: k-means++: the advantages of careful seeding. In: Bansal, N., Pruhs, K., Stein, C. (eds.) SODA '07 Proceedings of the Eighteenth Annual ACM-SIAM Symposium on Discrete Algorithms, pp. 1027–1035 (2007)
15. Audet, C., Hare, W.: Derivative-Free and Blackbox Optimization. Springer Series in Operations Research and Financial Engineering. Springer, Berlin (2017)

A. M. Bagirov et al., *Partitional Clustering via Nonsmooth Optimization*, Unsupervised and Semi-Supervised Learning,
https://doi.org/10.1007/978-3-030-37826-4

16. Babu, G.P., Murty, M.N.: A near optimal initial seed value selection in the k-meanws algorithm using a genetic algorithm. Pattern Recogn. Lett. **14**(10), 763–769 (1993)
17. Baeza-Yates, R.A.: Introduction to data structures and algorithms related to information retrieval. In: Frakes, W.B., Baeza Yates, R. (eds.) Information Retrieval: Data Structures and Algorithms. Prentice Hall, Upper Saddle River, NJ, pp. 13–27 (1992)
18. Bagirov, A.M.: Continuous subdifferential approximations and their applications. J. Math. Sci. **115**(5), 2567–2609 (2003)
19. Bagirov, A.M.: Modified global k-means algorithm for sum-of-squares clustering problem. Pattern Recogn. **41**, 3192–3199 (2008)
20. Bagirov, A.M.: An incremental DC algorithm for the minimum sum-of-squares clustering. Iran. J. Oper. Res. **5**(1), 1–14 (2014)
21. Bagirov, A.M., Mardaneh, K.: Modified global k-means algorithm for clustering in gene expression data sets. In: Boden, M., Bailey, T. (eds.) Proceedings of the AI 2006 Workshop on Intelligent Systems of Bioinformatics, pp. 23–28 (2006)
22. Bagirov, A.M., Mohebi, E.: Nonsmooth optimization based algorithms in cluster analysis. In: Celebi, E. (ed.) Partitional Clustering Algorithms. Springer, New York, pp. 99–146 (2015)
23. Bagirov, A.M., Taheri, S.: A DC optimization algorithm for clustering problems with L_1-norm. Iran. J. Oper. Res. **8**(2), 2–24 (2017)
24. Bagirov, A.M., Ugon, J.: An algorithm for minimizing clustering functions. Optimization **54**(4–5), 351–368 (2005)
25. Bagirov, A.M., Ugon, J.: Piecewise partially separable functions and a derivative-free algorithm for large scale nonsmooth optimization. J. Glob. Optim. **35**(2), 163–195 (2006)
26. Bagirov, A.M., Yearwood, J.: A new nonsmooth optimization algorithm for minimum sum-of-squares clustering problems. Eur. J. Oper. Res. **170**(2), 578–596 (2006)
27. Bagirov, A.M., Rubinov, A., Soukhoroukova, N., Yearwood, J.: Unsupervised and supervised data classification via nonsmooth and global optimization, Top **11**, 1–93 (2003)
28. Bagirov, A.M., Karasözen, B., Sezer, M.: Discrete gradient method: derivative-free method for nonsmooth optimization. J. Optim. Theory Appl. **137**, 317–334 (2008)
29. Bagirov, A.M., Ugon, J., Webb, D.: Fast modified global k-means algorithm for sum-of-squares clustering problems. Pattern Recogn. **44**, 866–876 (2011)
30. Bagirov, A.M., Al Nuaimat, A., Sultanova, N.: Hyberpolic smoothing function methodfor minimax problems. Optimization **62**(6), 759–784 (2013)
31. Bagirov, A.M., Ugon, J., Mirzayeva, H.: Nonsmooth nonconvex optimization approach to clusterwise linear regression problems. Eur. J. Oper. Res. **229**(1), 132–142 (2013)
32. Bagirov, A.M., Karmitsa, N., Mäkelä, M.M.: Introduction to Nonsmooth Optimization: Theory, Practice and Software. Springer, New York (2014)
33. Bagirov, A.M., Ordin, B., Ozturk, G., Xavier, A.: An incremental clustering algorithm based on hyperbolic smoothing. Comput. Optim. Appl. **61**(1), 219–241 (2015)
34. Bagirov, A.M., Ugon, J., Mirzayeva, H.: Nonsmooth optimization algorithm for solving clusterwise linear regression problems. J. Optim. Theory Appl. **164**(3), 755–780 (2015)
35. Bagirov, A.M., Ugon, J., Mirzayeva, H.: An algorithm for clusterwise linear regression based on smoothing techniques. Optim. Lett. **9**(2), 375–390 (2015)
36. Bagirov, A.M., Taheri, S., Ugon, J.: Nonsmooth DC programming approach to the minimum sum-of-squares clustering problems. Pattern Recogn. **53**, 12–24 (2016)
37. Bagirov, A.M., Mahmood, A., Barton, A.: Prediction of monthly rainfall in Victoria, Australia: clusterwise linear regression approach. Atmos. Res. **188**, 20–29 (2017)
38. Bagirov, A.M., Al Nuaimat, A., Sultanova, N., Taheri, S.: Solving minimax problems: local smoothing versus global smoothing. In: Al-Baali, M., Grandinetti, L., Purnama, A. (eds.) Numerical Analysis and Optimization. Springer Proceedings in Mathematics and Statistics, vol. 235, pp. 23–43. Springer, Cham (2018)
39. Bai, Q.: Analysis of particle swarm optimization algorithm. Comput. Inf. Sci. **3**(1), 180–184 (2010)
40. Ball, G.H., Hall, D.J.: ISODATA, a novel method of data analysis and pattern classification. Technical Report NTIS AD 699616, Stanford Research Institute, Menlo Park, CA (1965)

41. Ball, G.H., Hall, D.J.: A clustering technique for summarizing multivariate data. Behav. Sci. **12**(2), 153–155 (1967)
42. Bandyopadhyay, S., Saha, S.: A point symmetry-based clustering technique for automatic evolution of clusters. IEEE Trans. Knowl. Data Eng. **20**(11), 1441–1457 (2008)
43. Banerjee, A., Dhillon, I.S., Ghosh, J., Sra, S.: Clustering on the unit hypesphere using von Mises-Fisher distributions. J. Mach. Learn. Res. **6**, 1345–1382 (2005)
44. Basturk, B., Karaboga, D.: An artificial bee colony (ABC) algorithm for numeric function optimization. In: IEEE Swarm Intelligence Symposium, pp. 12–14 (2006)
45. Baudry, J.P., Raftery, A., Celeux, G., Lo, K., Gottardo, R.G.: Combining mixture components for clustering. J. Comput. Graph. Stat. **19**(2), 332–353 (2010)
46. Bertsekas., D.P.: Convex Optimization Algorithms, 2nd edn. Athena Scientific, Belmont, MA (2015)
47. Bezdek, J.C.: Pattern Recognition with Fuzzy Objective Function Algoritms. Plenum Press, New York (1981)
48. Bhuyan, N.J., Raghavan, V.V. Venkatesh, K.E.: Genetic algorithms for clustering with an ordered representation. In: Proceedings of the Fourth International Conference on Genetic Algorithms, pp. 408–415 (1991)
49. Biernacki, C., Celeux, G., Gold, E.M.: Assessing a mixture model for clustering with the integrated completed likelihood. IEEE Trans. Pattern Anal. Mach. Intell. **22**, 719–725 (2000)
50. Bobrowski, L., Bezdek, J.C.: c-means clustering with the L_1 and L_∞ norms. IEEE Trans. Syst. Man Cybern. **21**, 545–554 (1991)
51. Bock, H.H.: Probabilistic models in cluster analysis. Comput. Stat. Data Anal. **23**, 5–28 (1996)
52. Bock, H.H.: Clustering and neural networks. In: Rizzi, A., Vichi, M., Bock, H.H. (eds.) Advances in Data Science and Classification, pp. 265–277. Springer, Berlin (1998)
53. Brauksa, I.: Use of cluster analysis in exploring economic indicator differences among regions: the case of latvia. J. Econ. Bus. Manag. **1**(1), 42–45 (2013)
54. Brown, D.E., Entail, C.L.: A practical application of simulated annealing to the clustering problem. Pattern Recogn. **25**, 401–412 (1992)
55. Brown, M., Grundy, W., Lin, D., Christianini, N., Sugnet, C., Furey, T., Ares, M., Haussler, D.: Knowledg-based analysis of microarray gene expression data using support vector machines. Proc. Natl. Acad. Sci. **97**, 262–267 (2000)
56. Byrd, R.H., Nocedal, J., Schnabel, R.B.: Representations of quasi-Newton matrices and their use in limited memory methods. Math. Program. **63**, 129–156 (1994)
57. Calinski, T., Harabasz, J.: A dendrite method for cluster analysis. Commun. Stat. **3**, 1–27 (1974)
58. Cariou, C., Chehdi, K.: Unsupervised nearest neighbors clustering with application to hyperspectral images. IEEE J. Sel. Top. Sign. Process. **9**(6), 1105–1116 (2015)
59. Carmichael, J., Sneath, P.: Taxometric maps. Syst. Zool. **18**, 402–415 (1969)
60. Carpenter, G.A., Grossberg, S.: A massively parallel architecture for a self-organizing neural pattern recognition machine. Comput. Vis. Graph. Image Process. **37**, 54–115 (1987)
61. Carpenter, G.A., Grossberg, S.: Art3: hierarchical search using chemical transmitters in self organising pattern recognition architectures. Neural Netw. **3**, 129–152 (1990)
62. Carpenter, G.A., Grossberg, S., Reynolds, J.H.: ARTMAP: supervised real-time learning and classification of nonstationary data by a self-organizing neural network. Neural Netw. **4**, 565–588 (1991)
63. Celebi, M.E.: Improving the performance of k-means for color quantization. Image Vis. Comput. **29**(4), 260–271 (2011)
64. Celebi, M.E., Kingravi, H.A., Vela, P.A.: A comparative study of efficient initialization methods for the k-means clustering algorithm. Expert Syst. Appl. **40**(1), 200–210 (2013)
65. Cerny, V.: Thermodynamical approach to the travelling salesman problem: an efficient simulation algorithm. J. Optim. Theory Appl. **45**, 41–51 (1985)
66. Chaudhuri, B.B., Garai, G.: Grid clustering with genetic algorithm and tabu search process. J. Pattern Recogn. Res. **4**(1), 152–168 (2009)

67. Cheng, Y., Church, G.M.: Biclustering of expression data. In: Proceedings of the Eighth International Conference on Intelligent Systems for Molecular Biology, vol. 8, pp. 93–103 (2000)
68. Chipman, H., Tibshirani, R.: Hybrid hierarchical clustering with applications to microarray data. Biostatistics **7**(2), 286–301 (2006)
69. Christofor, D., Simovici, D.A.: An information theoretic approach to clustering categorical databases using genetic algorithms. In: Second SIAM ICDM Workshop on Clustering High Dimensional Data, pp. 37–46 (2002)
70. Clarke., F.H.: Optimization and Nonsmooth Analysis. Wiley, New York (1983)
71. Courvisanos, J., Jain, A., Mardaneh, K.: Economic resilience of regions under crises: a study of the Australian Economy. Reg. Stud. **50**(4), 629–643 (2016)
72. Cowgill, M.C., Harvey, R.J., Watson, L.T.: A genetic algorithm approach to cluster analysis. Comput. Math. Appl. **37**, 99–108 (1999)
73. Cura, T.: A particle swarm optimization approach to clustering. Expert Syst. Appl. **39**(1), 1582–1588 (2012)
74. Das, S., Abraham, A., Konar, A.: Clustering using multi-objective differential evolution algorithms. In: Metaheuristic Clustering. Studies in Computational Intelligence, vol. 178, pp. 213–238. Springer, Berlin/Heidelberg (2009)
75. Davies, D.L., Bouldin, D.W.: A cluster separation measure. IEEE Trans. Pattern Anal. Mach. Intell. **1**(4), 224–227 (1979)
76. Dempster, A.P., Laird, N.M., Rubin, D.B.: Maximum likelihood for incomplete data via the EM algorithm (with discussion). J. R. Stat. Soc. Ser. B **39**(1), 1–38 (1977)
77. Demyanov., V.F.: On codifferentiable functionals. Vestn. Leningr. Univ. **2**(8), 22–26 (1988)
78. Demyanov, V.F., Rubinov., A.M.: On quasidifferentiable functionals. Proc. USSR Acad. Sci. **250**(1), 21–25 (1980)
79. Demyanov, V.F., Rubinov., A.M.: Constructive Nonsmooth Analysis. Verlag Peter Lang, Frankfurt am Main (1995)
80. Demyanov, V.F., Bagirov, A.M., Rubinov, A.M.: A method of truncated codifferential with application to some problems of cluster analysis. J. Glob. Optim. **23**(1), 63–80 (2002)
81. DeSarbo, W.S., William, L.C.: A maximum likelihood methodology for clusterwise linear regression. J. Classif. **5**(2), 249–282 (1988)
82. De Souza, R.M.C.R., de Carvalho, F.A.T.: Clustering of interval data based on city-block distances. Pattern Recogn. Lett. **25**, 353–365 (2004)
83. Dhillon, I.S., Fan, J., Guan, Y.: Efficient clustering of very large document collections. In: Kamath, C., Kumar, V., Grossman, R., Namburu, R. (eds.) Data Mining for Scientific and Engineering Applications, Massive Computing, vol. 2, pp. 357–381. Springer, Boston, MA (2001)
84. Diehr, G.: Evaluation of a branch and bound algorithm for clustering. SIAM J. Sci. Stat. Comput. **6**, 268–284 (1985)
85. Doherty, K.A.J., Adams, R.G., Davey, N.: Non-Euclidean norms and data normalisation. In: Proceedings of ESANN, pp. 181–186 (2004)
86. Dolan, E., Moré, J.: Benchmarking optimization software with performance profiles. Math. Program. **91**, 201–213 (2002)
87. Dolnicar, S.: Using cluster analysis for market segmentation - typical misconceptions, established methodological weaknesses and some recommendations for improvement. Australasian J. Mark. Res. **11**(2), 5–12 (2003)
88. Dorigo, M., Blum, Ch.: Ant Colony optimization theory: a survey. Theor. Comput. Sci. **344**, 243–278 (2005)
89. Dorigo, M., Maniezzo, V., Colorni, A.: Ant system: optimization by a colony of cooperating agents. IEEE Trans. Syst. Man Cybern. B **26**(1), 29–41 (1996)
90. Dorigo, M., Caro, G.D., Gambardella, L.M.: Ant algorithms for discrete optimization. Artif. Life **5**(2), 137–172 (1999)
91. Dua, D., Graff, C.: UCI Machine Learning Repository. School of Information and Computer Sciences, University of California, Irvine. http://archive.ics.uci.edu/ml (2017)

92. Duda, R.O., Hart, P.E.: Pattern Classification and Scene Analysis. Wiley, New York (1973)
93. Dunn, J.C.: A fuzzy relative of the ISODATA process and its use in detecting compact well-separated clusters. J. Cybern. **3**, 32–57 (1973)
94. Dunn, J.C.: Well-separated clusters and optimal fuzzy partitions. J. Cybern. **4**(1), 95–104 (1974)
95. Eberhart, R., Shi, Y., Kennedy, J.: Swarm Intelligence. Morgan Kaufmann, Burlington (2001)
96. Eisen, M.B., Spellman, P.T., Brown, P.O., Botstein, D.: Cluster analysis and display of genome-wide expression patterns. Proc. Natl. Acad. Sci. **95**, 14863–14868 (1998)
97. Eren, K., Deveci, M., Kücüktunc, O., Catalyürek, U.V.: A comparative analysis of biclustering algorithms for gene expression data. Brief. Bioinform. **14**(3), 279–292 (2013)
98. Ermoliev, Y.M., Norkin, V.I., Wets, R.J-B.: The minimization of semicontinuous functions: mollifier subgradients. SIAM J. Control Optim. **33**, 149–167 (1995)
99. Ester, M., Kriegel, K.P., Sander, J., Xu, X.: A density-based algorithm for discovering clusters in large spatial databases with noise. In: Proceedings of the Second International Conference on Knowledge Discovery and Data Mining, Portland, OR, pp. 226–231 (1996)
100. Ester, M., Kriegel, H.P., Sander, J., Xu, X.: A density-based algorithm for discovering clusters in large spatial databases with noise. In: Simoudis, E., Han, J., Fayyad, U.M. (eds.) Proceedings of the Second International Conference on Knowledge Discovery and Data Mining, Portland, OR, pp. 226–231 (1996)
101. Evans, L.C., Gariepy, V.: Measure Theory and Fine Properties of Functions. CRC Press, Boca Raton, FL (1992)
102. Falkenauer, E.: Genetic Algorithms and Grouping Problems. Wiley, New York (1998)
103. Fayyad, U.M., Piatetsky-Shapiro, G., Smyth, P.: Advances in knowledge discovery and data mining. In: American Association for Artificial Intelligence, pp. 1–34 (1996)
104. Figueiredo, M.A.T., Jain, A.K.: Unsupervised learning of finite mixture models. IEEE Trans. Pattern Anal. Mach. Intell. **24**(3), 381–396 (2002)
105. Finnie, G., Sun, Z.: r^5 model for case-based reasoning. Knowl. Based Syst. **16**, 59–65 (2003)
106. Fisher, D.: Knowledge acquisition via incremental conceptual clustering. Mach. Learn. **2**, 139–172 (1987)
107. Fogel, D.B.: An introduction to simulated evolutionary optimization. IEEE Trans. Neural Netw. **5**(1), 3–14 (1994)
108. Forgy, E.W.: Cluster analysis of multivariate data: efficiency versus interpretability of classifications. Biometrics **21**, 768–769 (1965)
109. Fowlkes, E., Mallows, C.: A method for comparing two hierarchical clusterings. J. Am. Stat. Assoc. **78**, 553–569 (1983)
110. Frismantas, V., et al.: Ex vivo drug response profiling detects recurrent sensitivity patterns in drug-resistant acute lymphoblastic leukemia. Blood **129**(11), e26–e37 (2017)
111. Fu, H.: A novel clustering algorithm with ant colony optimization. In: IEEE Pacific-Asia Workshop on Computational Intelligence and Industrial Application, pp. 66–69 (2008)
112. Ganti, V., Gehrke, J., Ramakrishnan, R.: CACTUS: clustering categorical data using summaries. In: Knowledge Discovery and Data Mining, pp. 73–83 (1999)
113. Gaudioso, M., Gorgone, E.: Gradient set splitting in nonconvex nonsmooth numerical optimization. Optim. Methods Softw. **25**, 59–74 (2010)
114. Gaudioso, M., Giallombardo, G., Miglionico, G., Bagirov, A: Minimizing nonsmooth DC functions via successive DC piecewise-affine approximations. J. Glob. Optim. **71**(1), 37–55 (2018)
115. Gen, M., Cheng, R.: Genetic Algorithms and Engineering Design. Wiley, New York (1997)
116. Ghorbani, M.: Maximum entropy-based fuzzy clustering by using L_1-norm space. Turk. J. Math. **29**, 431–438 (2005)
117. Gibson, D., Kleinberg, J., Raghavan, P.: Clustering categorical data: an approach based on dynamical systems. In: Proceedings of the 24th International Conference on Very Large Databases (VLDB), pp. 103–114 (1998)
118. Glover, F.: Future paths for integer programming and links to artificial intelligence. Comput. Oper. Res. **13**(5), 533–549 (1986)

119. Glover, F.: Tabu search - part 1. ORSA J. Comput. **1**(2), 190–206 (1989)
120. Glover, F.: Tabu search - part 2. ORSA J. Comput. **2**(1), 4–32 (1990)
121. Glover, F.: Artificial intelligence, heuristic frameworks and tabu search. Manag. Decis. Econ. **11**, 365–375 (1990)
122. Glover, F., Laguna, M.: Tabu Search. Kluwer Academic, Dordrecht (1997)
123. Goldberg, D.E.: Genetic Algorithms in Search, Optimization & Machine Learning. Addison-Wesley, Boston, MA (1989)
124. Goldberg, D.E., Deb, K.: A comparative analysis of selection schemes used in genetic algorithms. In: Rawlins, G.J.E. (ed.) Foundations of Genetic Algorithms, pp. 69–93. Morgan Kaufmann, San Mateo, CA (1991)
125. Gong, M. Zhang, L., Jiao, L., Gou, S.: Solving multi-objective clustering using animmune-inspired algorithm. In: Proceedings of IEEE Conference on Evolutionary Computation, pp. 15–22 (2007)
126. Gonzalez, T.: Clustering to minimize the maximum intercluster distance. Theor. Comput. Sci. **38**(2–3), 293–306 (1985)
127. Grefenstette, J.: Optimization of control parameters for genetic algorithms. IEEE Trans. Syst. Man Cybern. **1**, 122–128 (1986)
128. Guha, S., Rastogi, R., Shim, K.: CURE: an efficient clustering algorithm for large databases. In: Proceedings of ACM SIGMOD International Conference on Management of Data, pp. 73–84. ACM Press, New York (1998)
129. Guha, S., Rastogi, R., Shim, K.: ROCK: a robust clustering algorithm for categorical attributes. Inf. Syst. **25**(5), 345–366 (2000)
130. Guha, S., Meyerson, A., Mishra, N., Motwani, R., O'Callaghan, L.: Clustering data streams: theory and practice. IEEE Trans. Knowl. Data Eng. **15**(3), 515–528 (2003)
131. Haarala, M.: Large-Scale Nonsmooth Optimization: Variable Metric Bundle Method with Limited Memory. PhD thesis, Department of Mathematical Information Technology, University of Jyväskylä (2004)
132. Haarala, M., Miettinen, K., Mäkelä, M.M.: New limited memory bundle method for large-scale nonsmooth optimization. Optim. Methods Softw. **19**(6), 673–692 (2004)
133. Haarala, N., Miettinen, K., Mäkelä, M.M.: Globally convergent limited memory bundle method for large-scale nonsmooth optimization. Math. Program. **109**(1), 181–205 (2007)
134. Halkidi, M., Batistakis, Y., Vazirgiannis, M.: Cluster validity methods: part I. ACM SIGMOD Rec. **31**(2), 40–45 (2002)
135. Halkidi, M., Batistakis, Y., Vazirgiannis, M.: Cluster validity methods: part II. ACM SIGMOD Rec. **31**(3), 19–27 (2002)
136. Han, J., Kamber, M., Pei, J.: Data Mining: Concepts and Techniques. The Morgan Kaufmann Series in Data Management Systems, 3rd edn. Morgan Kaufmann, San Francisco, CA (2011)
137. Handl, J., Knowles, J.: Evolutionary multi-objective clustering. In: Proceedings of 8th International Conference on Parallel Problem Solving from Nature, pp. 1081–1091 (2004)
138. Handl, J., Knowles, J.: An evolutionary approach to multi-objective clustering. IEEE Trans. Evol. Comput. **11**(1), 56–76 (2007)
139. Hanilci, C., Ertas, F.: Comparison of the impact of some Minkowski metrics on VQ/GMM based speaker recognition. Comput. Electr. Eng. **37**, 41–56 (2011)
140. Hansen, P., Jaumard, B.: Cluster analysis and mathematical programming. Math. Program. **79**(1–3), 191–215 (1997)
141. Hansen, P., Mladenovic, N.: J-means: a new local search heuristic for minimum sum of squares clustering. Pattern Recogn. **34**(2), 405–413 (2001)
142. Hansen, P., Ngai, E., Cheung, B., Mladenovic, N.: Analysis of global k-means, an incremental heuristic for minimum sum of squares clustering. J. Classif. **22**, 287–310 (2005)
143. Hartigan, J.A.: Clustering Algorithms. Wiley, New York, NY (1975)
144. Hartigan, J.A., Wong, M.A.: Algorithm AS 136: a k-means clustering algorithm. J. R. Stat. Soc. Ser. C (Appl. Stat.) **28**, 100–108 (1979)
145. Hiriart-Urruty, J.-B., Lemarechal, C.: Convex Analysis and Minimization Algorithms I and II. Springer, Heidelberg (1993)

146. Holland, J.H.: Adaptation in Natural and Artificial Systems. University of Michigan Press, Ann Arbor, MI (1975)
147. Hruschka, H., Natter, M.: Comparing performance of feedforward neural nets and k-means for cluster-based market segmentation. Eur. J. Oper. Res. **114**(2), 346–353 (1999)
148. Hruschka, E.R., Campello, R.J.G.B., Freitas, A.A., de Carvalho, A.C.P.L.F.: A survey of evolutionary algorithms for clustering. IEEE Trans. Syst. Man Cybern. C (Appl. Rev.) **39**, 133–155 (2009)
149. Huang, J.J., Tzeng, G.H., Ong, C.Sh.: Marketing segmentation using support vector clustering. Expert Syst. Appl. **32**(2), 313–317 (2007)
150. Hubert, L., Arabie, P.: Comparing partitions. J. Classif. **2**(1), 193–218 (1985)
151. Inkaya, T., Kayaligil, S., Özdemirel, N.E.: An adaptive neighbourhood construction algorithm based on density and connectivity. Pattern Recogn. Lett. **52**, 17–24 (2015)
152. Inkaya, T., Kayaligil, S., Özdemirel, N.E.: Ant colony optimization based clustering methodology. Appl. Soft Comput. **28**, 301–311 (2015)
153. Ismkhan, H.: $i - k$-means-+: an iterative clustering algorithm based on an enhanced version of the k-means. Pattern Recogn. **79**, 402–413 (2018)
154. Jain, A.K.: Data clustering: 50 years beyond k-means. Pattern Recogn. Lett. **31**(8), 651–666 (2010)
155. Jain, A.K., Dubes, R.: Algorithms for Clustering Data. Prentice Hall, Upper Saddle River, NJ (1988)
156. Jain, A.K., Murty, M.N., Flynn, P.J.: Data clustering: a review. ACM Comput. Surv. **31**(3), 264–323 (1999)
157. Jajuga, K.: A clustering method based on the L_1-norm. Comput. Stat. Data Anal. **5**, 357–371 (1987)
158. Jardine, N., Sibson, R.: Mathematical Taxonomy. Wiley, London/New York (1971)
159. Jensen, R.E.: A dynamic programming algorithm for cluster analysis. Oper. Res. **17**, 1034–1057 (1969)
160. Ji, J., Pang, W., Zheng, Y., Wang, Z., Ma, Z.: A novel artificial bee colony based clustering algorithm for categorical data. PLoS ONE **10**(5), 1–17 (2015)
161. Joki, K., Bagirov, A.M., Karmitsa, N., Mäkelä, M.M., Taheri, S.: Double bundle method for finding Clarke stationary points in nonsmooth DC programming. SIAM J. Optim. **28**, 1892–1919 (2018)
162. Jones, D., Beltramo, M.A.: Solving partitioning problems with genetic algorithms. In: Proceedings of the Fourth International Conference on Genetic Algorithms, pp. 442–449 (1991)
163. Kao, Y.T., Zahara, E., Kao, I.W.: A hybridized approach to data clustering. Expert Syst. Appl. **34**, 1754–1762 (2008)
164. Karaboga, D.: An idea based on honey bee swarm for numerical optimization. Technical Report-TR06, Erciyes University, Engineering Faculty, Computer Engineering Department (2005)
165. Karaboga, D., Basturk, B.: A powerful and efficient algorithm for numerical function optimization: artificial bee colony algorithm. J. Glob. Optim. **39**(3), 459–471 (2007)
166. Karaboga, D., Ozturk, C.: A novel clustering approach: artificial bee colony algorithm. Appl. Soft Comput. **11**, 652–657 (2011)
167. Karmitsa, N.: Diagonal bundle method for nonsmooth sparse optimization. J. Optim. Theory Appl. **166**(3), 889–905 (2015)
168. Karmitsa, N., Mäkelä, M.M, Ali, M.M.: Limited memory interior point bundle method for large inequality constrained nonsmooth minimization. Appl. Math. Comput. **198**(1), 382–400 (2008)
169. Karmitsa, N., Bagirov, A.M., Mäkelä, M. M.: Comparing different nonsmooth optimization methods and software. Optim. Methods Softw. **27**(1), 131–153 (2012)
170. Karmitsa, N., Bagirov, A.M., Taheri, S.: New diagonal bundle method for clustering problems in large data sets. Eur. J. Oper. Res. **263**(2), 367–379 (2017)

171. Karmitsa, N., Bagirov, A.M., Taheri, S.: Clustering in large data sets with the limited memory bundle method. Pattern Recogn. **83**, 245–259 (2018)
172. Katsavounidis, I., Kuo, C.-C.J., Zhang, Z.: A new initialization technique for generalized Lloyd iteration. IEEE Signal Process. Lett. **1**(10), 144–146 (1994)
173. Kaufman, L., Rousseeuw, P.J.: Clustering by means of medoids. In: Dodge, Y. (ed.) Statistical Data Analysis Based on the L_1-Norm and Related Methods, pp. 405–416. North-Holland, Amsterdam (1987)
174. Kaufman, L., Rousseeuw, P.J.: Finding Groups in Data: An Introduction to Cluster Analysis. Wiley Series in Probability and Statistics. Wiley, New York (1990)
175. Kennedy, J., Eberhart, R.C.: Particle swarm optimization. In: Proceedings of IEEE International Conference on Neural Networks, vol. 1, pp. 1942–1948 (1995)
176. Ketchen, D.J., Shook, C.L.: The application of cluster analysis in strategic management research: an analysis and critique. Strateg. Manag. J. **17**(6), 441–458 (1996)
177. King, B.: Step-wise clustering procedures. J. Am. Stat. Assoc. **69**, 86–101 (1967)
178. Kirkpatrick, S., Gelatt, C.D., Vecchi, M.P.: Optimization by simulated annealing. Science **220**(4598), 671–680 (1983)
179. Kiwiel, K.C.: Methods of Descent for Nondifferentiable Optimization. Lecture Notes in Mathematics, vol. 1133. Springer, Berlin (1985)
180. Kiwiel, K.C.: Proximity control in bundle methods for convex nondifferentiable minimization. Math. Program. **46**(1–3), 105–122 (1990)
181. Kiwiel, K.C.: Improved convergence result for the discrete gradient and secant methods for nonsmooth optimization. J. Optim. Theory Appl. **144**(1), 69–75 (2010)
182. Klein, R.W., Dubes, R.C.: Experiments in projection and clustering by simulated annealing. Pattern Recogn. **22**, 213–220 (1989)
183. Kogan, J.: Introduction to Clustering Large and High-Dimensional Data. Cambridge University Press, Cambridge (2007)
184. Kohonen, T.: Self-organization formation of topologically correct feature maps. Biol. Cybern. **43**(1), 59–69 (1982)
185. Kohonen, T.: Self Organization and Associative Memory. Springer Information Sciences Series, 3rd edn. Springer, Heidelberg (1989)
186. Koontz, W.L.G., Narendra, P.M., Fukunaga, K.: A branch and bound clustering algorithm. IEEE Trans. Comput. **24**(9), 908–915 (1975)
187. Krzanowski, W., Lai, Y.: A criterion for determining the number of groups in a data set using sum-of-squares clustering. Biometrics **44**(1), 23–34 (1988)
188. Kuo, R.J., Ho, L.M., Hu, C.M.: Integration of self-organizing feature map and k-means algorithm for market segmentation. Comput. Oper. Res. **29**(11), 1475–1493 (2002)
189. Kvalseth, T.O.: Entropy and correlation: some comments. IEEE Trans. Syst. Man Cybern. **17**(3), 517–519 (1987)
190. Lai, J.Z.C., Huang, T.J.: Fast global k-means clustering using cluster membership and inequality. Pattern Recogn. **43**(5), 1954–1963 (2010)
191. Larsen, B., Aone, Ch.: Fast and effective text mining using linear-time document clustering. In: Proceedings of the fifth ACM SIGKDD International Conference on Knowledge Discovery and Data Mining, pp. 16–22 (1999)
192. Le-Khac, N., Cai, F., Kechadi, M.: Clustering approaches for financial data analysis: a survey. In: Abou-Nasr, M. Arabnia, H. (eds.) Proceedings of the International Conference on Data Mining, Las Vegas, Nevada (2012)
193. Lemaréchal, C., Strodiot, J.-J., Bihain, A.: On a bundle algorithm for nonsmooth optimization. In: Mangasarian, O.L., Mayer, R.R., Robinson, S.M. (eds.) Nonlinear Programming, pp. 245–281. Academic Press, New York (1981)
194. Le Thi, H.A., Pham Dinh, T.: Solving a class of linearly constrained indefinite quadratic problems by D.C. algorithms. J. Glob. Optim. **11**(3), 253–285 (1997)
195. Le Thi, H.A., Pham Dinh, T.: The DC (differnece of convex functions) programming and DCA revised with DC models of real world nonconvex optimization problems. Ann. Oper. Res. **133**(1–4), 23–46 (2005)

196. Li, J.: Clustering based on multi-layer mixture model. J. Comput. Graph. Stat. **14**(3), 547–568 (2005)
197. Likas, A., Vlassis, M., Verbeek, J.: The global k-means clustering algorithm. Pattern Recogn. **36**(2), 451–461 (2003)
198. Liu, X., Hu, F.: An effective clustering algorithm with ant colony. J. Comput. **5**(4), 598–605 (2010)
199. Liu, Y., Wu, X., Shen, Y.: Automatic clustering using genetic algorithms. Appl. Math. Comput. **218**(4), 1267–1279 (2011)
200. Lloyd, S.: Least squares quantization in PCM. IEEE Trans. Inf. Theory **28**(2), 129–137 (1982)
201. Locatelli, M.: Simulated annealing algorithms for continuous global optimization: convergence conditions. J. Optim. Theory Appl. **104**(1), 121–133 (2000)
202. Lu, S.Y., Fu, K.S.: A sentence to sentence clustering procedure for pattern analysis. IEEE Trans. Syst. Man Cybern. **8**(5), 381–389 (1978)
203. Lukšan, L., Vlček, J.: Globally convergent variable metric method for convex nonsmooth unconstrained minimization. J. Optim. Theory Appl. **102**(3), 593–613 (1999)
204. MacQueen, J.: Some methods for classification and analysis of multivariate observations. In: Cam, L.M.L., Neyman, J. (eds.) Proceedings of the fifth Berkeley Symposium on Mathematical Statistics and Probability, vol. 1, pp. 281–297. University of California Press, Berkeley, CA (1967)
205. Mahajan, M., Nimbhorkar, P., Varadarajan, K.: The planar k-means problem is NP-hard. Theor. Comput. Sci. **442**, 13–21 (2012)
206. Mäkelä., M.M.: Survey of bundle methods for nonsmooth optimization. Optim. Methods Softw. **17**(1), 1–29 (2002)
207. Mäkelä, M.M., Neittaanmäki, P.: Nonsmooth Optimization: Analysis and Algorithms with Applications to Optimal Control. World Scientific, Singapore (1992)
208. Maulik, U., Bandyopadhyay, S.: Performance evaluation of some clustering algorithms and validity indices. IEEE Trans. Pattern Anal. Mach. Intell. **24**(12), 1650–1654 (2002)
209. Maulik, U., Bandyopadhyay, S.: Genetic algorithm-based clustering technique. Pattern Recogn. **33**(9), 1455–1465 (2000)
210. McLachlan, G., Krishnan, T.: The EM Algorithm and Extensions. Wiley, New York (1997)
211. McLachlan, G.J., Basford, K.E.: Mixture Models: Inference and Applications to Clustering. Marcel Dekker, New York (1988)
212. Melnykov, V., Maitra, R.: Finite mixture models and model-based clustering. Stat. Surv. **4**, 80–116 (2010). Digital Repository, Statistics Publications, Iowa State University
213. Mifflin, R.: A modification and an extension of Lemaréchal's algorithm for nonsmooth minimization. Math. Program. Stud. **17**, 77–90 (1982)
214. Milligan, G.W., Cooper, M.C.: An examination of procedures for determining the number of clusters in a data set. Psychometrika **50**(2), 159–179 (1985)
215. Milligan, G.W., Cooper, M.C.: A study of the comparability of external criteria for hierarchical cluster analysis. Multivar. Behav. Res. **21**, 441–458 (1986)
216. Mirkin, B.: Mathematical Classification and Clustering. Springer, Berlin/Heidelberg (1996)
217. Mohebi, E., Bagirov, A.M.: A convolutional recursive modified self organizing map for handwritten digits recognition. Neural Netw. **60**, 104–118 (2014)
218. Mohebi, E., Bagirov, A.M.: Modified self organising maps with a new topology and initialisation algorithm. J. Exp. Theor. Artif. Intell. **27**(3), 351–372 (2015)
219. Mohebi, E., Bagirov, A.M.: Constrained self organizing maps for data clusters visualization. Neural Process. Lett. **43**(3), 849–869 (2016)
220. Mordukhovich, B.: Variational Analysis and Generalized Differentiation I and II. Springer, Heidelberg (2006)
221. Murtagh, F.: A survey of recent advances in hierarchical clustering algorithms which use cluster centres. Comput. J. **26**(4), 354–359 (1984)
222. Mustjoki, S., et al.: Discovery of novel drug sensitivities in T-PLL by high-throughput ex vivo drug testing and mutation profiling. Leukemia **32**, 774–787 (2017)

223. Nagy, G.: State of the art in pattern recognition. Proc. IEEE **56**(5), 836–862 (1968)
224. Naldi, M.C., de Carvalho, A.C.P.L.F., Campello, R.J.G.B., Hruschka, E.R.: Genetic clustering for data mining. In: Maimon, O., Rokach, L. (eds.) Soft Computing for Knowledge Discovery and Data Mining, pp. 113–132. Springer, Berlin (2007)
225. Nappa, S.D., Wang, X., Nair, S.: A comparison of machine learning techniques for phishing detection. In: Proceedings of the Anti-Phishing Working Groups 2nd Annual eCrime Researchers Summit (eCrime 07), New York, pp. 60–69 (2007)
226. Newcomb, S.: A generalized theory of the combination of observations so as to obtain the best result. Am. J. Math. **8**(4), 343–366 (1886)
227. Nesterov, Y.: Smooth minimization of nonsmooth functions. Math. Program. **103**(1), 127–152 (2005)
228. Ordin, B., Bagirov, A.M.: A heuristic algorithm for solving the minimum sum-of-squares clustering problems. J. Glob. Optim. **61**(2), 341–361 (2015)
229. Ordin, B., Bagirov, A.M., Mohebi, E.: An incremental nonsmooth optimization algorithm for clustering using L_1- and L_∞- norms. J. Ind. Manag. Optim. Accepted for publication. http://dx.doi.org/10.3934/jimo.2019079
230. Oyelade, J., Isewon, I., Oladipupo, F., Aromolaran, O., Uwoghiren, E. Ameh, F., Achas, M., Adebiyi, E.: Clustering algorithms: their application to gene expression data. Bioinf. Biol. Insights **10**, 237–253 (2016)
231. Park, H.S., Jun, C.H.: A simple and fast algorithm for k-medoids clustering. Expert Syst. Appl. **36**(2), 3336–3341 (2009)
232. Parsons, L., Haque, E., Liu, H.: Subspace clustering for high dimensional data: a review. ACM SIGKDD Explorations Newsletter - Special issue on learning from imbalanced datasets **6**(1), 90–105 (2004)
233. Pearson, K.: Contribution to the mathematical theory of evolution. Philos. Trans. R. Soc. **185**, 71–110 (1894)
234. Pelleg, D., Moore, A.W.: X-means: extending k-means with efficient estimation of the number of clusters. In: Langley, P. (ed.) Proceedings of the Seventeenth International Conference on Machine Learning, pp. 727–734. Morgan Kaufmann, San Francisco, CA (2000)
235. Pemovska, T., et al.: Individualized systems medicine strategy to tailor treatments for patients with chemorefractory acute myeloid leukemia. Cancer Discov. **3**(12), 1416–1429 (2013)
236. Penot, J.: Variations on the theme of nonsmooth analysis: another subdifferential. In: Demyanov, V.F, Pallaschke, D. (eds.) Nondifferentiable Optimization: Motivations and Applications, pp. 41–54. Springer, Berlin (1985)
237. Pizzuti, C., Talia, D., Vonella, G.: A divisive initialisation method for clustering algorithms. In: Proceedings of the 3rd European Conference on Principles and Practice of Knowledge Discovery in Databases, pp. 484–491 (1999)
238. Poggi, J.M., Portier, B.: PM10 forecasting using clusterwise regression. Atmos. Environ. **45**(38), 7005–7014 (2011)
239. Punj, G., Stewart, D.W.: Cluster analysis in marketing research: review and suggestions for application. J. Mark. Res. **20**(2), 134–148 (1983)
240. Quandt, R.E.: A new approach to estimating switching regressions. J. Am. Stat. Assoc. **67**(338), 306–310 (1972)
241. Raghavan, V.V., Birchand, K.: A comparison of the stability characteristics of some graph theoretic clustering methods. In: Proceedings of the Second international Conference on Information Storage and Retreival, pp. 10–22 (1979)
242. Rahman, M.A., Islam, M.Z.: A hybrid clustering technique combining a novel genetic algorithm with k-means. Knowl. Based Syst. **71**, 345–365 (2014)
243. Rand, W.M.: Objective criteria for the evaluation of clustering methods. J. Am. Stat. Assoc. **66**(336), 846–850 (1971)
244. Rao, M.R.: Cluster analysis and mathematical programming. J. Am. Stat. Assoc. **66**(335), 622–626 (1971)

245. Redmond, S.J., Heneghan, C.: A method for initialising the k-means clustering algorithm using kd-trees. Pattern Recogn. Lett. **28**(8), 965–973 (2007)
246. Reinelt, G.: TSP-LIB-A travelling salesman library. ORSA J. Comput. **3**, 319–350 (1991)
247. Rezanková, H.: Cluster analysis of economic data. Statistica **94**(1), 73–86 (2014)
248. Robinson, S.M.: Linear convergence of epsilon-subgradient descent methods for a class of convex functions. Math. Program. **86**(1), 41–50 (1999)
249. Rockafellar, R.T.: Convex Analysis. Princeton University Press, Princeton, NJ (1970)
250. Rosch, E.: Principles of Categorization. MIT Press, Cambridge (1999)
251. Rousseeuw, P.J.: Silhouettes: a graphical aid to the interpretation and validation of cluster analysis. Comput. Appl. Math. **20**, 53–65 (1987)
252. Rui, X., Wunsch, D.: Survey of clustering algorithms. IEEE Trans. Neural Netw. **16**(3), 645–678 (2005)
253. Runkler, T.A.: Ant colony optimization of clustering models. Int. J. Intell. Syst. **20**(12), 1233–1251 (2005)
254. Sabo, K., Scitovski, R., Vazler, I.: One-dimensional center-based L_1-clustering method. Optim. Lett. **7**(1), 5–22 (2013)
255. Saha, S., Bandyopadhyay, S.: Some connectivity-based cluster validity indices. Appl. Soft Comput. **12**(5), 1555–1565 (2012)
256. Salvador, S., Chan, P.: Determining the number of clusters/segments in hierarchical clustering/segmentation algorithms. In: Proceedings of the 16th IEEE International Conference on Tools with Artificial Intelligence, pp. 576–584 (2004)
257. Santos, J.M., Embrechts, M.: On the use of the adjusted Rand index as a metric for evaluating supervised classification. In: Alippi, C., Polycarpou, M., Panayiotou, C., Ellinas, G. (eds) Artificial Neural Networks – ICANN 2009. Lecture Notes in Computer Science, vol. 5769, pp. 175–184. Springer, Berlin/Heidelberg (2009)
258. Schramm, H., Zowe, J.: A version of the bundle idea for minimizing a nonsmooth function: conceptual idea, convergence analysis, numerical results. SIAM J. Optim. **2**(1), 121–152 (1992)
259. Sedgewick, R., Wayne, K.: Introduction to Programming in Java. Addison-Wesley, New York (2007)
260. Seifollahi, S., Bagirov, A.M. Layton, R., Gondal, I.: Optimization based clustering algorithms for authorship analysis of phishing emails. Neural Process. Lett. **46**(2), 411–425 (2017)
261. Selim, S.Z.: A global algorithm for the clustering problem. In: Presentation at the ORSA/TIMS Joint Meeting, San Diego, CA (1982)
262. Selim, S.Z., Alsultan, K.: A simulated annealing algorithm for the clustering problem. Pattern Recogn. Lett. **24**(10), 1003–1008 (1991)
263. Selim, S.Z., Ismail, M.A.: k-means-type algorithms: a generalized convergence theorem and characterization of local optimality. IEEE Trans. Pattern Anal. Mach. Intell. **6**(1), 81–87 (1984)
264. Sethi, I., Jain, A.K. (eds.): Artificial Neural Networks and Pattern Recognition: Old and new Connections. Elsevier, New York (1991)
265. Shang, Y., Wah, B.W.: Global optimization for neural network training. IEEE Comput. **29**(3), 31–44 (1996)
266. Shelokar, P.S., Jayaraman, V.K., Kulkarni, B.D.: An ant colony approach for clustering. Anal. Chim. Acta **509**(2), 187–195 (2004)
267. Shor, N.Z.: Minimization Methods for Non-differentiable Functions. Springer, Berlin (1985)
268. Slonm, N., Tishby, N.: Document clustering using word clusters via the information bottleneck method. In: Proceedings of the ACM SIGIR Conference on Research and Development in Information Retrieval, pp. 208–215 (2000)
269. Sneath, P.H.A., Sokal, R.R.: Numerical Taxonomy. Freeman, London (1973)
270. Späth, H.: Algorithm 30: L_1 cluster analysis. Computing **16**(4), 379–387 (1976)
271. Späth, H.: Algorithm 39: clusterwise linear regression. Computing **22**(4), 367–373 (1979)
272. Späth, H.: Cluster Analysis Algorithms for Data Reduction and Classification of Objects. Computers and Their applications. Ellis Horwood Limited, Chichester (1980)

273. Späth, H.: The Cluster Dissection and Analysis Theory FORTRAN Programs Examples. Prentice-Hall, Upper Saddle River, NJ (1985)
274. Sriperumbudur, B.K., Lanckriet, R.G.: On the convergence of the concave-convex procedure. In: Bengio, Y., Schuurmans, D., Lafferty, J.D., Williams, C.K.I., Culotta, A. (eds.) Proceedings of the 22nd International Conference on Neural Information Processing Systems, pp. 1759–1767. Curran Associates Inc., Red Hook (2009)
275. Sun, L.X., Xie, Y.L., Song, X.H., Wang, J.H., Yu, R.Q.: Cluster analysis by simulated annealing. Comput. Chem. **18**(2), 103–108 (1994)
276. Suresh, K., Kundu, D., Ghosh, S., Das, S., Abraham, A.: Data clustering using multi-objective differential evolution algorithms. Fund. Inform. **97**(4), 381–403 (2009)
277. Steinhaus, H.: Sur la division des corp materiels en parties. Bull. Acad. Polon. Sci. **4**(12), 801–804 (1956)
278. Tao, P.D.: Duality in d.c. (difference of convex functions) optimization. Subgradient methods. In: Hoffmann, K.H., et al. (ed.) Trends in Mathematical Optimization. International Series of Numer Math., vol. 84. Birkhauser, Basel (1988)
279. Teboulle, M.: A unified continuous optimization framework for center-based clustering methods. J. Mach. Learn. Res. **8**, 65–102 (2007)
280. Thalamuthu, A., Mukhopadhyay, I., Zheng, X., Tseng, G.C.: Evaluation and comparison of gene clustering methods in microarray analysis. Bioinformatics **22**(19), 2405–2412 (2006)
281. Torzcon, V.: On the convergence of pattern search algorithms. SIAM J. Optim. **7**(1), 1–25 (1997)
282. Tran, T.N., Wehrens, R., Buydens, L.M.C.: KNN-kernel density-based clustering for high-dimensional multivariate data. Comput. Stat. Data Anal. **51**(2), 513–525 (2006)
283. Tsai, C.Y., Chiu, C.C.: A purchase-based market segmentation methodology. Expert Syst. Appl. **27**(2), 265–276 (2004)
284. Tuy, H.: Convex Analysis and Global Optimization, 1st edn. Kluwert Academic, Dordrecht (1998)
285. Uryasév, S.P.: New variable metric algorithms for nondifferentiable optimization problems. J. Optim. Theory Appl. **71**(2), 359–388 (1991)
286. Van der Laan, M., Pollard, K., Bryan, J.: A new partitioning around medoids algorithm. J. Stat. Comput. Simul. **73**(8), 575–584 (2003)
287. Van der Merwe, D.W., Engelbrecht, A.P.: Data clustering using particle swarm optimization. In: The 2003 Congress on Evolutionary Computation, Canberra, ACT (2003)
288. Venkateswarlu, N., Raju, P.: Fast ISODATA clustering algorithms. Pattern Recogn. **25**(3), 335–342 (1992)
289. Vlček, J., Lukšan, L.: Globally convergent variable metric method for nonconvex nondifferentiable unconstrained minimization. J. Optim. Theory Appl. **111**(2), 407–430 (2001)
290. Wang, X., Qiu, W., Zamar, R.H.: CLUES: a non-parametric clustering method based on local shrinking. Comput. Stat. Data Anal. **52**(1), 286–298 (2007)
291. Ward, J.H.: Hierarchical grouping to optimize and objective function. J. Am. Stat. Assoc. **58**(301), 236–244 (1963)
292. Wedel, M., Kistemaker, C.: Consumer benefit segmentation using clusterwise linear regression. Int. J. Res. Mark. **6**(1), 45–59 (1989)
293. Weiszfeld, E.: Sur le point pour lequel la somme des distances de n points donnes est minimum. Tohoku Math. J. **43**, 355–386 (1937)
294. Weiszfeld, E., Plastria, F.: On the point for which the sum of the distances to n given points is minimum. Ann. Oper. Res. **167**(1), 7–41 (2009)
295. Wierzchon, S.T., Klopotek, M.A.: Modern Algorithms of Cluster Analysis. Springer, Cham (2018)
296. Wolfe, J.H.: Pattern clustering by multivariate mixture analysis. Multivar. Behav. Res. **5**(3), 329–350 (1970)
297. Wolfe, P.: A method of conjugate subgradients for minimizing nondifferentiable functions. In: Balinski, M.L., Wolfe, P. (eds.) Nondifferentiable Optimization, pp. 145–173. Springer, Heidelberg (1975)

298. Wu, X., Kumar, V., Quinlan, J.R., Ghosh, J., Yang, Q., Motoda, H., McLachlan, G.J., Ng, A., Liu, B., Yu, P.S., Zhou, Z.-H., Steinbach, M., Hand, D.J., Steinberg, D.: Top 10 algorithms in data mining. Knowl. Inf. Syst. **14**(1), 1–37 (2007)
299. Xavier, A.E.: Penalizaćao hiperbólica. In: I Congresso Latino-Americano de Pesquisa Operacional e Engenharia de Sistemas, 8 a 11 de Novembro, Rio de Janeiro, pp. 468–482 (1982)
300. Xavier, A.E.: The hyperbolic smoothing clustering method. Pattern Recogn. **43**(3), 731–737 (2010)
301. Xavier, A.E., Oliveira, A.A.F.D.: Optimal covering of plane domains by circles via hyperbolic smoothing. J. Glob. Optim. **31**(3), 493–504 (2005)
302. Xiao, Y., Yu, B.: A truncated aggregate smoothing Newton method for minimax problems. Appl. Math. Comput. **216**(6), 1868–1879 (2010)
303. Xie, X.L., Beni, G.: A validity measure for fuzzy clustering. IEEE Trans. Pattern Anal. Mach. Intell. **13**(8), 841–847 (1991)
304. Xu, L.: Bayesian ying-yang machine, clustering and number of clusters. Pattern Recogn. Lett. **18**(11–13), 1167–1178 (1997)
305. Xu, S.: Smoothing method for minimax problems. Comput. Optim. Appl. **20**(3), 267–279 (2001)
306. Yang, M.-Sh., Hung, W.-L., Chung, T.-I.: Alternative fuzzy clustering algorithms with L_1-norm and covariance matrix. In: Blanc-Talon J., Philips W., Popescu D., Scheunders P. (eds) Advanced Concepts for Intelligent Vision Systems, ACIVS 2006. Lecture Notes in Computer Science, vol. 4179, pp. 654–665. Springer, Berlin/Heidelberg, (2006)
307. Ye, F., Liu, H., Zhou, Sh., Liu, S.: A smoothing trust-region Newton-CG method for minimax problem. Appl. Math. Comput. **199**(2), 581–589 (2008)
308. Yeung, K.Y., Haynor, D.R., Ruzzo, W.L.: Validating clustering for gene expression data. Bioinformatics **17**(4), 309–318 (2001)
309. Yuille, A.L., Rangarajan, A.: The concave-convex procedure. Neural Comput. **15**(4), 915–936 (2003)
310. Zang, I.: A smoothing-out technique for min-max optimization. Math. Program. **19**(1), 61–77 (1980)
311. Zhang, C., Ouyang, D., Ning, J.: An artificial bee colony approach for clustering. Expert Syst. Appl. **37**(7), 4761–4767 (2010)
312. Zhang, J., Peng, L., Zhao, X., Kuruoglu, E.E.: Robust data clustering by learning multi-metric L_q-norm distances. Expert Syst. Appl. **39**(1), 335–349 (2012)
313. Zhang, T., Ramakrishnan, R., Livny, M.: BIRCH: an efficient data clustering method for very large databases. In: Proceedings of the ACM SIGMOD Conference on Management of Data, pp. 103–114 (1996)
314. Zhao, Q., Fränti, P.: WB-index: a sum-of-squares based index for cluster validity. Data Knowl. Eng. **92**, 77–89 (2014)
315. Zhao, Q., Xu, M., Fränti, P.: Knee point detection on Bayesian information criterion. In: 20th IEEE International Conference on Tools with Artificial Intelligence, pp. 431–438 (2008)
316. Zhao, Q., Xu, M., Fränti, P.: Sum-of-squares based cluster validity index and significance analysis. In: Proceedings of the 17th International Conference on Adaptive and Natural Computing Algorithms, pp. 313–322 (2009)

Index

A. M. Bagirov et al., *Partitional Clustering via Nonsmooth Optimization*,
Unsupervised and Semi-Supervised Learning,
https://doi.org/10.1007/978-3-030-37826-4

D

E

F

G

H

I

J

K

L

M

N

O

P